Dardo Oscar Guaraglia, Jorge Lorenzo Pousa

Introduction to Modern Instrumentation for Hydraulics and Environmental Sciences

Dardo Oscar Guaraglia,
Jorge Lorenzo Pousa

Introduction to Modern Instrumentation

for Hydraulics and Environmental Sciences

Managing Editor: Alessandro Pezzoli

Language Editor: Robert G. Watts

DE GRUYTER
OPEN

Published by De Gruyter Open Ltd, Warsaw/Berlin
Part of Walter de Gruyter GmbH, Berlin/Munich/Boston

ISBN: 978-3-11-040171-4
e-ISBN: 978-3-11-040172-1

Bibliographic information published by the Deutsche Nationalbibliothek The Deutsche Nationalbibliothek lists this publication in the Deutsche Nationalbibliografie; detailed bibliographic data are available in the Internet at http://dnb.dnb.de.

Managing Editor: Alessandro Pezzoli
Language Editor: Robert G. Watts

www.degruyteropen.com

Cover illustration: © Sergio Schmidt and Horacio Ezcurra from Ezcurra & Schmidt Engineering Firm

To my wife, daughters and son
To my parents
(Dardo Oscar Guaraglia)
To my mother María, in memoriam
To Nelly, in memoriam
(Jorge Lorenzo Pousa)

I often say that when you can measure what you are speaking about, and express it in numbers, you know something about it; but when you cannot express it in numbers, your knowledge is of a meagre and unsatisfactory kind; it may be the beginning of knowledge, but you have scarcely, in your thoughts, advanced to the stage of science, whatever the matter may be.

William Thomson, 1st Baron Kelvin (1824 - 1907) (Lord Kelvin)

Although this may seem a paradox, all exact science is dominated by the idea of approximation. When a man tells you that he knows the exact truth about anything, you are safe in inferring that he is an inexact man. Every careful measurement in science is always given with the probable error ... every observer admits that he is likely wrong, and knows about how much wrong he is likely to be.

Bertrand Russell (1872-1970,) in The Scientific Outlook (1931, 2009), 42.

Preface

Natural hazards and anthropic activities threaten the quality of the environment surrounding the human being, risking life and health. Among the different actions that must be taken to control the quality of the environment, the gathering of field data is a basic one.

In order to obtain the needed data for environmental research, a great variety of new instruments based on electronics is used by professionals and researchers. Sometimes, the potentials and limitations of this new instrumentation remain somewhat unknown to the possible users. In order to better utilize modern instruments it is very important to understand how they work, avoiding misinterpretation of results. All instrument operators must gain proper insight into the working principles of their tools, because this internal view permits them to judge whether the instrument is appropriately selected and adequately functioning.

Frequently, manufacturers have a tendency to show the great performances of their products without advising their customers that some characteristics are mutually exclusive. Car manufacturers usually show the maximum velocity that a model can reach and also the minimum fuel consumption. It is obvious for the buyer that both performances are mutually exclusive, but it is not so clear for buyers of measuring instruments. This book attempts to make clear some performances that are not easy to understand to those uninitiated in the utilization of electronic instruments.

Technological changes that have occurred in the last few decades are not yet reflected in academic literature and courses; this material is the result of a course prepared with the purpose of reducing this shortage. The content of this book is intended for students of hydrology, hydraulics, oceanography, meteorology and environmental sciences.

Most of the new instruments presented in the book are based on electronics, special physics principles and signal processing; therefore, basic concepts on these subjects are introduced in the first chapters (Chapters 1 to 3) with the hope that they serve as a complete, yet easy-to-digest beginning. Because of this review of concepts it is not necessary that the reader have previous information on electronics, electricity or particular physical principles to understand the topics developed later. Those readers with a solid understanding of these subjects could skip these chapters; however they are included because some students could find them as a useful synthesis.

Chapter 4 is completely dedicated to the description of transducers and sensors frequently used in environmental sciences. It is described how electrical devices are modified by external parameters in order to become sensors. Also an introduction to oscillators is presented because they are used in most instruments. In the next chapters all the information presented here is recurrently referred to as needed to explain operating principles of instruments.

Chapters 1 to 4 are bitter pills that could discourage readers interested in the description of specific instruments. Perhaps, those readers trying this book from the beginning could abandon it before arriving at the most interesting chapters. Therefore, they could read directly Chapters 5 to 11, going back as they feel that they need the knowledge of the previous chapters. We intended to make clear all the references to the previous subjects needed to understand each one of the issues developed in the later chapters.

Chapter 5 contributes to the understanding of modern instrumentation to measure flow in industrial and field conditions. Traditional mechanical meters are avoided to focus the attention on electronic ones, such as vortex, electromagnetic, acoustic, thermal, and Coriolis flowmeters. Special attention is dedicated to acoustic Doppler current profilers and acoustic Doppler velocimeters.

Chapter 6 deals with two great subjects; the first is devoted to instruments for measuring dynamic and quasi static levels in liquids, mainly water. Methods to measure waves at sea and in the laboratory are explained, as well as instruments to measure slow changes such as tides or piezometric heads for hydrologic applications. The second subject includes groundwater measurement methods with emphasis on very low velocity flowmeters which measure velocity from inside a single borehole. Most of them are relatively new methods and some are based on operating principles described in the previous chapter. Seepage meters used to measure submarine groundwater discharge are also presented.

Chapter 7 presents methods and instruments for measuring rain, wind and solar radiation. Even though the attention is centered on new methods, some traditional methods are described not only because they are still in use, and it is not yet clear if the new technologies will definitely replace them, but also because describing them permits their limitations and drawbacks to be better understood. Methods to measure solar radiation are described from radiation detectors to complete instruments for total radiation and radiation spectrum measurements.

Chapter 8 is a long chapter where we have tried to include most remote measuring systems useful for environmental studies. It begins with a technique called DTS (Distributed Temperature Sensing) that has the particularity of being remote, but where the electromagnetic wave propagates inside a fibre optic. The chapter follows with atmosphere wind profilers using acoustic and electromagnetic waves. Radio acoustic sounding systems used to get atmospheric temperature profiles are explained in detail as well as weather radar. Methods for ocean surface currents monitoring are also introduced. The chapter ends with ground penetrating radars.

Chapter 9 is an introduction to digital transmission and storage of information. This subject has been reduced to applications where information collected by field instruments has to be conveyed to a central station where it is processed and stored. Some insight into networks of instruments is developed; we think this information will help readers to select which method to use to transport information from field to office, by means of such diverse communication media as fibre optic, digital telephony,

GSM (Global System for Mobile communications), satellite communications and private radio frequency links.

Chapter 10 is devoted to satellite-based remote sensing. Introductory concepts such as image resolution and instrument's scanning geometry are developed before describing how passive instruments estimate some meteorological parameters. Active instruments are presented in general, but the on-board data processing is emphasized due to its importance in the quality of the measurements. Hence, concepts like Synthetic Aperture Radar (SAR) and Chirp Radar are developed in detail. Scatterometers, altimeters and Lidar are described as applications of the on-board instruments to environmental sciences.

Chapter 11 attempts to transfer some experiences in field measuring to the readers. A pair of case studies is included to encourage students to perform tests on the instruments before using them. In this chapter we try to condense our ideas, most of them already expressed throughout the book, about the attitude a researcher should have with modern instruments before and after a measuring field work.

As can be inferred from the foregoing description the book aims to provide students with the necessary tools to adequately select and use instruments for environmental monitoring. Several examples are introduced to advise future professionals and researchers on how to measure properly, so as to make sure that the data recorded by the instruments actually represents the parameters they intend to know. With this purpose, instruments are explained in detail so that their measuring limitations are recognized. Within the entire work it is underlined how spatial and temporal scales, inherent to the instruments, condition the collection of data. Informal language and qualitative explanations are used, but enough mathematical fundamentals are given to allow the reader to reach a good quantitative knowledge.

It is clear from the title of the book that it is a basic tool to introduce students to modern instrumentation; it is not intended for formed researchers with specific interests. However, general ideas on some measuring methods and on data acquisition concepts could be useful to them before buying an instrument or selecting a measuring method. Those readers interested in applying some particular method or instrument described in this book should consider these explanations just as an introduction to the subject; they will need to dig deeper in the specific bibliography before putting hands on.

Acknowledgements

The authors are very grateful to Enrique Spinelli and Pablo Landea for having reviewed Chapters 2 and 9, respectively, and for their useful comments. They would also like to express their gratitude to Sergio Schmidt and Horacio Ezcurra from Ezcurra & Schmidt Engineering Firm for having provided the cover photo of the book.

Contents

1 Introduction to Measuring Techniques

1.1 Introduction

A desirable condition of any measuring method is to disturb as little as possible the phenomenon being measured. If this condition is not satisfied, the measuring method would be altering the event being measured, and the measurement would not be representative of the phenomenon. For example, if it is desired to measure flow in a small channel with a propeller or a rotor whose cross-sectional area is similar to that of the channel, the flow will be surely altered by the presence of the rotor. So this instrument would not allow measurement in the original flow conditions.

Assuming that care is taken to ensure that the phenomenon is not disturbed by the measuring method, the next step is to analyze to what extent measurements are representative of the researched phenomenon. Each phenomenon has its particular spatial and temporal characteristics that must be taken into account by the measuring method. To do so, users must analyze these characteristics and verify that the selected method fulfills them. Below, some examples are examined to convey these ideas.

Measurement methods require using instruments, equipment, devices, etc. based on certain operational principles. The user should select the principle that is most suitable for the phenomenon being studied. Each principle defines the sensor to be used, which is the first component of a chain of elements through which the signal containing the desired information will have to pass. The passage by these elements modifies the information to some extent. Even if the sensor were perfect, the data obtained by the whole instrument might not be representative of the phenomenon if an inappropriate signal processing inside the instrument is carried out. In order to clarify these ideas the constituent parts of an instruments and the task performed by each part will be described below.

1.2 The Spatial and Temporal Characteristics of the Phenomena to be Measured

The spatial and temporal characteristics of the physical phenomena to be measured determine the technical specifications of the instruments required to perform the measurements. Before selecting a measuring method it is necessary to know some key features about the phenomenon under study to avoid the risk of acquiring unsuitable data. Among these features, the spatial and temporal characteristics are fundamental. Some measurement examples will be presented to help understanding these concepts.

1.2.1 Sea Surface Measurements

Sea level measurements may be needed when studying tides and ordinary waves. The temporal and spatial characteristics of these two phenomena are largely different and will lead to measuring instruments with very different characteristics.

The tide is a phenomenon with an average period of 12.4 h (24.8 h in some places) and, in general, it is desirable that ordinary waves do not appear superimposed on the recorded level. Since the tide changes in a relatively slow way, it would be adequately characterized by samples taken every minute. Therefore, because tide requires a few samples per unit time, it needs a low **temporal resolution**.

From the spatial point of view, it is known that sea level is almost the same over an area of several thousand square meters (e.g. 500 m × 500 m), then the sea level average calculated over such an area would be representative of the phenomenon being measured. Taking a "big picture" of the sea surface is enough. Therefore, the **spatial resolution** necessary to measure tides is also low. The spatial resolution of a sensor is related to the area or volume from which the sensor takes the information, in this example, a sea surface area.

If it is desired to study ordinary waves to evaluate their energy, it could be of interest to measure the dynamics of the sea surface and not merely its average value. If waves of periods equal to or greater than 1 s are to be considered in this study, successive measurements of sea surface have to be taken several times per second, so the needed **temporal resolution increases** markedly with respect to the tide case. Furthermore, a wave of such a period may have a short wavelength (about 2 m), depending on the water depth. Then, the area of the sea from which the sensor takes the information should be reduced, perhaps to a few square centimeters; then the required **spatial resolution increases.** For example, if an acoustic wave meter were used and the acoustic wave intercept the sea surface forming a circumference of 1 m in diameter, the reflected energy will carry information from all this surface, thus the instrument would spatially average the sea surface information. This spatial integration would preclude the user from acquiring the details of the small period waves with wavelengths of a few meters.

As expected, the study of ordinary waves requires instruments with a greater spatial and temporal resolution than in the tide case, but the important, non-trivial point to solve is how much sea surface the instrument should "sense", and how many samples per unit time it should take. The answer depends on the particular user's needs. One early conclusion from this example is that before deciding which instrument will be used to measure a phenomenon, it is necessary to establish the **temporal and spatial characteristics** of interest for the study under consideration. Also, the user should know from beforehand how the instruments will measure in order to select the most adequate one for the desired research.

1.2.2 Water Table Measurements

Measurements of the water table at various boreholes to determine the velocity of groundwater flow in a non exploited unconfined aquifer located in a plain could require just one measurement a day at each well, as the change in level and flow velocity are very slow. Moreover, if the soil is sufficiently homogeneous it would be enough to measure at points several hundred meters apart.

Soils in mountain areas generally have greater heterogeneities due to changes in grain size. Even the presence of paleochannels could be identified. The hydraulic conductivity can thus change in a few meters, and because the slope is greater than in plain areas the groundwater flow speed is higher. Therefore, if it were required to know the direction of a contaminant spill in such an area, measurements should be performed at smaller spatial intervals, for example, distances of a few tens of meters. Also, if it were desired to study the level of the unconfined aquifer when exploited or recharged, it might be necessary to take multiple samples of the level per hour. Thus, the goal of the measurements defines the spatial and temporal resolution.

Once again, the conclusion is that before planning a field measuring campaign it is necessary to know very well the purpose of the study and the expected changes in time and space of the information that is desired to be collected. The time and spatial resolution will determine the method to be used, and eventually the technical characteristics of the instruments.

1.2.3 Wind and Temperature Examples

Assume that it is desired to study how rapid wind speed fluctuations modify the surface temperature of a sand beach. Because wind gusts and the beach surface temperature have fast temporal changes, it would be necessary to measure both several times per second. In contrast, if we wish to study the temperature at a depth of 10 cm, it would be enough to take temperature measurements once a minute or less, because the top layer of sand prevents a rapid temperature variation at that depth.

1.2.4 Summary

In short, before selecting a measurement method or instrument it is necessary to establish the changes in time and space of the phenomenon to be studied. Users should select instruments keeping in mind the spatial and time limitations imposed by the measuring method (e.g. time and spatial averaging). This requires knowledge of how instruments work and how manufacturers present instrument specifications. These minimum but essential precautions would help to prevent collecting unsuitable data.

If no previous knowledge of the temporal and spatial variation of the phenomena is available, when possible, some tests should be conducted before choosing the instruments for a measuring campaign.

1.3 A Generic Instrument

Figure 1.1 shows a generic instrument on which it was intended to display the constituent parts of most instruments. As it happens in any generalization attempt, there might be some parts of this scheme that certain instruments will not have. This scheme is presented with the intention of early identifying the stages that constitute an instrument in order to anticipate the issues to be dealt with in the following chapters.

The sensor is the element that perceives the phenomenon (**P**) to be measured, which is frequently called **the measurand**. Generally, the sensor produces an electrical signal containing the information of the measurand.

The signal conditioner (**SC**) input takes the sensor output and provides a signal easy to handle and process at its output. In many cases it is an analog electronic stage in which signals are amplified and filtered. Sometimes this stage is incorporated within the sensor, i.e. the manufacturer of the sensor provides suitable conditioned outputs for the next stages. In some essentially digital sensors this stage is reduced drastically.

If the signal that contains the information of interest is analog, it is usually converted into digital in order to facilitate its processing and storage. This function is performed by the analog-to-digital converter (**ADC**).

The signal processor (**SP**) may have different functions. In some cases it is a mathematical processing aimed at rescuing the signal from the background noise. By noise it is meant any unwanted signal that inevitably affects the quality of any wanted signal. Processing aims therefore at eliminating any unwanted information, saving the desired one. It may happen that this processing defines the main characteristics of the instrument. In sea level measurements the signal processor (**SP**) separates waves from tide. If both phenomena are of interest, once separated each is stored in a different block of data memory.

Processing can also be used to statistically treat the signal, for example to average several samples and then to save only the result of the calculation. This process reduces the information to be stored in the memory. In some instruments the **SP** calculates the signal-to-noise ratio (**S/N**) to evaluate the quality of the data.

The data storage (**DS**) is the place where the information is temporarily or permanently stored. It is also known as the memory. There are basically two types of memories: volatile and non-volatile. In volatile memories, the information is stored temporarily until processed and moved to permanent storage. While the information is in a temporary memory, it is lost if power supply is turned off. When in non-volatile or permanent memory, information is not missed even if the instrument is disconnected from the power supply.

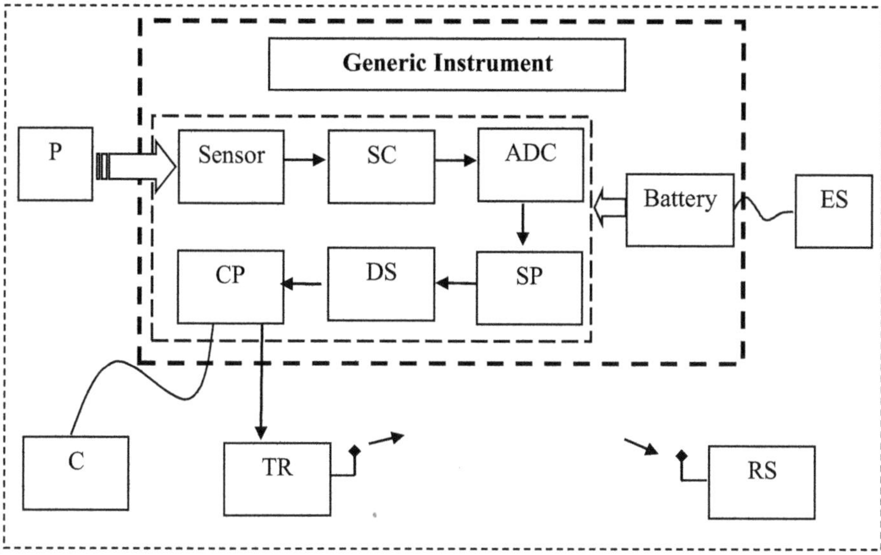

Fig. 1.1: Schematic of a generic instrument.

The permanent memory of the instrument is accessed regularly by users; usually through a communication port (**CP**), to unload the data stored onboard the instrument to a computer (**C**) (as it happens with a digital camera). This transference of data requires that the instrument have standard communications ports. The **CP** could also be connected to a transmitter / receiver (**TR**) that sends the data to a remote station (**RS**).

Any instrument requires an energy source (**ES**) to work. The energy may come from a main AC power supply or from a battery. In the second case, the battery can be recharged from a variety of **ES** devices, e.g. AC/DC converters, solar panels, wind generators, etc. Batteries may or may not be included in the instrument; the same happens with transmitters.

Each step of the signal processing inside the instrument, from the sensor to the signal processor, modifies to some extent the information being measured. Later on we will study how these steps restrict the signals that pass through them and how manufacturers specify these limitations that are usually not easy to understand.

1.4 Precision and Accuracy

There is some tendency to confuse precision and accuracy. **Precision** is a measure of the closeness of independent determinations of a value; it is an evaluation of the repeatability of a measuring process. An instrument may be very precise but inaccurate. An example of a precise but inaccurate fire arm is shown in Figure 1.2a.

Shots are close to each other, but far from the center. **Accuracy** refers to how close the average of the determinations is to the real value. A result both accurate and precise is shown in Figure 1.2b, where all shots are close to the center. The case shown in Figure 1.2c is neither accurate nor precise.

All measurements in science and industry must be reproducible if they are to be valid determinations. Precise and accurate instruments are thus needed. To determine the error and precision of an instrument it is required to perform several measurements of a known quantity.

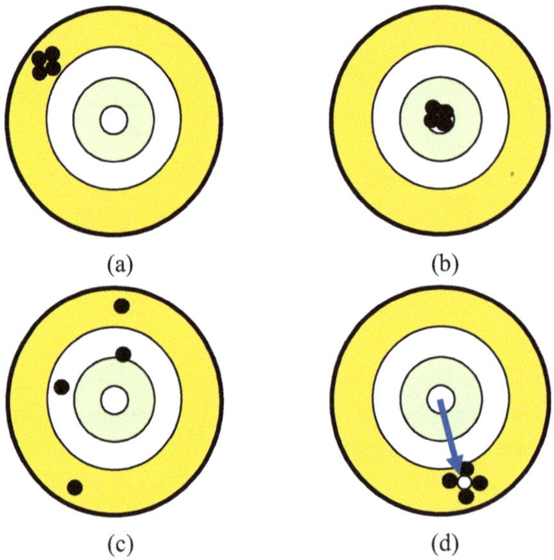

Fig. 1.2: (a), (b), (c) Distinction between precision and accuracy, and (d) between systematic and random errors.

1.5 Types of Errors

Three types of measuring errors may be mentioned: personal errors, systematic errors and random errors. **Personal errors** are mistakes on the part of the user responsible for the measurement. They may be errors in data recording or calculations. They are in general gross errors and are avoided by paying attention during the measuring process.

Systematic errors tend to bias all measurements in a similar way. For instance a lack of calibration is a systematic error. They can be known and corrected through a calibration process. When a meter stick made of metal is calibrated at 25 °C and used to measure at 50 °C a systematic error will be introduced due to the metal linear thermal expansion with temperature. Systematic errors are often hard to be identified.

Random errors are known as statistical uncertainties. They are a series of small unknowns, generally with negative and positive signs, whose average is close to zero. Figure 1.2d shows four (black dots) shots and their average (white dot). The distances between the white dot and the shots are random errors, whereas the arrow from the center to the white dot represents the systematic error.

As another example consider four measures of a length made with the aforementioned meter: 2.52; 2.56; 2.55 and 2.53 m. The average value is 2.54 m and the real length is 2.65 m. In this case, the measurements are precise because there is not much spread in the determinations, but the average value obtained is not accurate because there is a large deviation from the real value. The small spread implies small random errors, but the large deviation involves some sort of systematic error. In this case the systematic error could be attributable to an enlargement of the meter stick due to the thermal expansion caused by the increase in temperature. It is always important to analyze the errors and their possible sources. Systematic errors could be compensated if their causes are known. In the previous example the solution could be to calibrate the meter at the same temperature at which it will be used.

In general, if an instrument lacks accuracy but has good precision this implies the existence of systematic errors. A precise but inaccurate instrument might be converted into an accurate one by calibration. On the other hand, if an instrument lacks precision (has large random errors), the accuracy of the estimated value may be increased by averaging many samples. Samples have to be taken during a short time in which the measurand and the measuring conditions can be considered constant.

Unfortunately, many phenomena are time variable and it is not possible to take many samples in the same conditions. Then, the precision imposes a limit on the accuracy when a few measures can be made in similar conditions. It is quite difficult to get accurate measurements with an imprecise instrument.

1.6 Noise

As stated above, noise is any unwanted perturbation on a wanted signal. The name comes from the audible noise heard when listening to a weak radio signal. In a measuring process, noise is an undesired disturbance that affects signals or data and that may distort the information carried by them. It is always wished to minimize the noise and maximize the signal. Noise sources affecting modern instruments are generally of electromagnetic origin and are divided into internal and external to the instruments.

The internal sources are out of the control of the user. They are mostly electronic noise which exists in all circuits and devices as a result of thermal noise.

The random movement of electrons in a conductor material induces small temperature-dependent currents. The magnitude of thermal noise is generally in the μV range (10^{-6} V). As a signal passes by the successive stages of a system, it may be corrupted by different sources of noise. Generally, the first stages are the most vulnerable. Cables, sensors and input amplifiers are susceptible to noise. The procedure of converting an analog signal into a digital one also adds some noise due to sampling and quantification processes. Once in the digital domain, signals are less prone to noise.

The internal noise places a limit on the measuring capabilities of instruments, and this limit should be informed by the manufacturer as part of the instrument's specifications. External noise sources may be coupled to the desired signal and added to it during the measuring process. Two important sources of electrical noise that introduce errors in measuring processes are power lines and radio frequency interferences.

Instrument users should understand the importance of noise sources and take preventive actions to minimize their effects. There are two kinds of actions that could reduce noise effects on measurements. One is to prevent noise from corrupting the signal; the other is to remove noise from the corrupted signal. The second one could be ineffective if some characteristics of the signal and noise, such as frequency content, are similar.

1.6.1 Preventing Electromagnetic Noise

Devices producing electromagnetic fields can interfere with the electronic circuits of instruments. Usually, noise is introduced to the measuring device through sensors and cables. The measuring circuit acts as an antenna that receives the electromagnetic energy of the noise source. Nowadays, radio frequency from transmitters is a very common source of noise. Transmitters that often interfere with instruments include mobile phones, radio stations, TV stations and radars.

High voltage power lines produce strong electric fields that could jeopardize an instrument's measurements; also, power lines supplying electric trains are sources of noise. Varying magnetic fields such as those produced by an electric motor or generator will induce currents in a measuring circuit. The induced currents are proportional to the magnetic field strength, the rate at which it changes, the loop area of the cables exposed to the magnetic field and the distance between noise sources and instruments.

To reduce the induced noise from external sources, users should increase the distance between the noise source and the measuring circuit by locating instruments away from transmitters, microwave systems, radar and high voltage power lines. Also they should reduce cable lengths and use cables less prone to electromagnetic noise such as twisted pairs and shielded cables.

1.7 Why Sensors With Electrical Outputs Are Preferred?

Instruments to measure and record data have some kind of interface that takes the information from the phenomena of interest and converts it into **electrical signals**. For the moment we will call these interfaces **sensors**, but a deeper discussion of these devices will be developed in the next chapter. Not only sensors with electrical outputs exist, but below is an example that explains the advantages of having the information converted into electrical signals.

In the past decades several ingenious mechanical devices were invented to control manufacturing processes; among them is the bimetallic thermal switch. This is a device used to convert a temperature change into a mechanical displacement. It uses two metallic strips with different thermal expansion coefficients (usually steel and copper) (Fig. 1.3). At some "reference temperature", the two strips are welded together along their lengths, and the strips are flat. If the temperature is changed, the different expansions of metals will force the welded strips to bend, converting a change in temperature into a displacement of the bar. This device is still used to make or break an electrical contact according to the difference between the "reference" and the actual temperatures.

If it were desired to change the reference temperature, it would be necessary to change the shape of the strips or the metal from which they are made of. This means that there is not a simple way to adjust the switch, which is a severe limitation of these mechanical sensors.

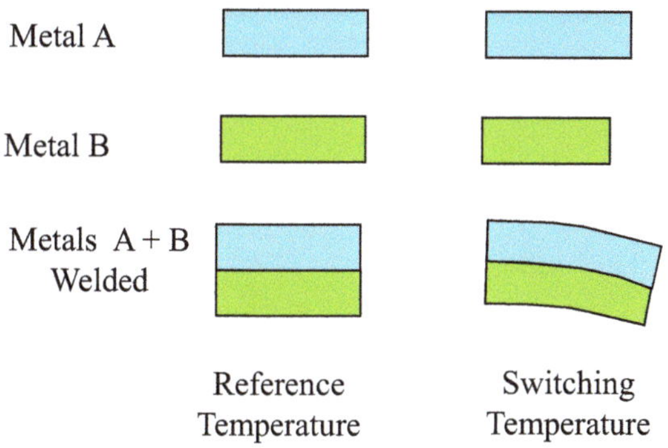

Fig. 1.3: (Left) Both metal bars are at the reference temperature, showing the same length. When they are welded at the reference temperature they are flat. (Right) The temperature is different from reference and because the bars tend to have different lengths the ensemble bends.

The bimetallic bar can also be used to read temperature over a scale by using a needle attached to it and arranged to work with a pin and a lever arm (Fig. 1.4). By calibrating the scale, a thermometer is obtained.

With the purpose of automatically recording the temperature as a function of time, a pen could be placed at the needle's tip and make a paper tape advance at a constant speed below it, drawing the temperature graph. In order to use the recorded temperature in an equation or in a mathematical model, the graph should be manually read by a person, and data introduced to a computer by typing. This is a tedious work very prone to errors that old researchers know very well.

As seen, mechanical solutions are somewhat rudimentary and the resulting data is difficult to store and process. If temperature information could be "impressed on" or "transferred to" an electrical signal, it could be treated as explained in the previous generic instrument. It would thus be easy to change the switching point of a temperature switch, to change a reading scale and to store the temperature data in solid state memories. Also, if it were desired to know the temperature at places located several kilometers away, the electrical signal containing the temperature information could be transmitted by modulating a high frequency electromagnetic wave.

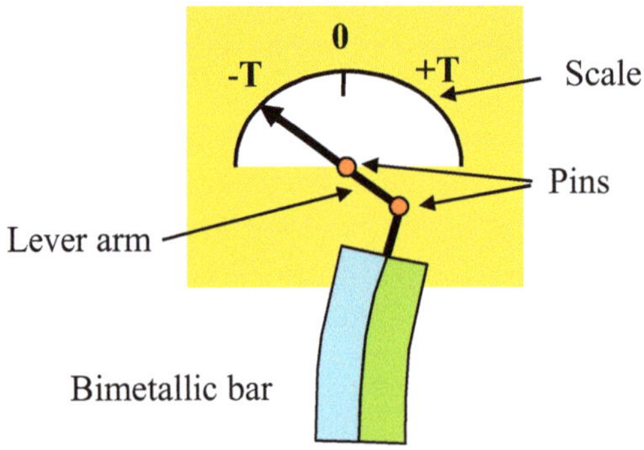

Fig. 1.4: The bimetallic bar associated with a needle, a lever arm and a scale becomes a thermometer when calibrated.

In summary, if we had the measurand converted into electrical signals we could take advantage of all the resources available nowadays to store, process and transmit the information. Sensors to convert physical or chemical information into electrical signals are then very useful and will be studied in the following chapters.

2 Introduction to Transducers and Sensors

2.1 Definitions of Transducer, Sensor, Actuator and Detector

In recent decades, advances in physics and electronics have enabled the development of devices that take information from a physical or chemical phenomenon and create or modify an electrical signal upon which this information is "copied". These devices are known by different names in instrumentation literature making unavoidable an introduction to some terminology frequently used.

In the electronic instrumentation literature some authors use the words transducer, sensor, actuator and detector in a way that could confuse the reader. This practice is often correct, because a device could meet more than one definition, but this is not always the case. Formal definitions from dictionaries provide some guidelines about how to use these words.

The Webster's on-line dictionary (http://www.merriam-webster.com/dictionary) defines **transducer** as:

> *"A device that is actuated by power from one system and supplies power usually in another form to a second system (a loudspeaker is a transducer that transforms electrical signals into sound energy)"*

The same on-line dictionary defines **sensor** as:

> *"A device that responds to a physical stimulus (as heat, light, sound, pressure, magnetism, or a particular motion) and transmits a resulting impulse (as for measurement or operating a control)"*

Dictionary definitions associate **transducer** with a device that converts one form of energy (or power) into another, and **sensor** with a device that perceives a physical stimulus giving a signal as a result. For example, microphones and hydrophones convert vibration into electricity, and thermocouples transform temperature into electricity, therefore they are examples of transducers (Fig. 2.1a).

Pressure sensors, resistance temperature sensors, thermistors, strain gauges and photoresistors are examples of sensors, because they are supplied with electrical energy and give an electrical signal when subjected to a stimulus (Fig. 2.1b).

Other authors (http://digital.ni.com) prefer to call **active transducers** those that generate an electric current or voltage in response to environmental stimulation, and **passive transducers** those that produce a change in some passive electrical quantity, such as capacitance, resistance, or inductance as a result of stimulation.

Generally, the word **actuator** refers to a device that converts an electrical signal into a mechanical motion. For example, an electric motor fits this definition because when powered by voltage produces a mechanical rotation of an axis. Also, the automatic locking of the car's doors is made by an actuator.

In the case of a car's horn, the electric circuit produces a mechanical vibration (motion) which is transferred to the environment (the sound), and so the horn could be said to be an **actuator**. Also, because the horn converts electrical energy into sound (mechanical energy) it might also be called a **transducer**. But it is not a **sensor** because it does not produce an electrical signal as a response to a physical stimulus.

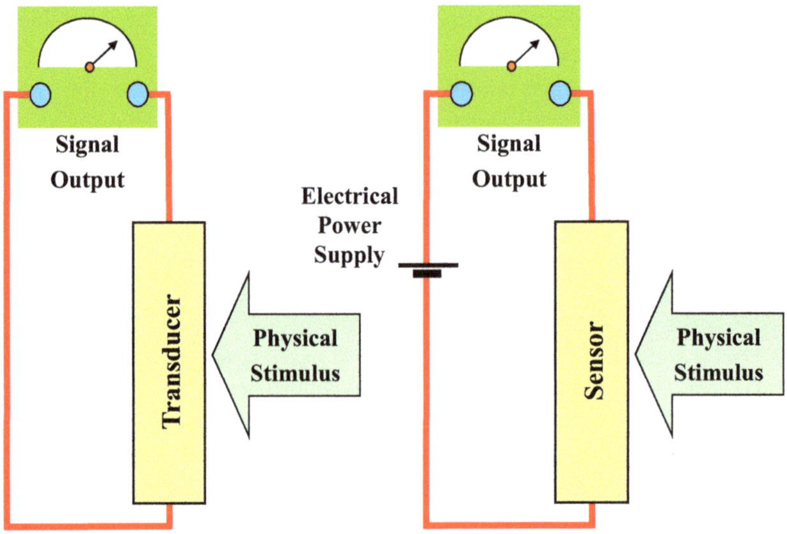

Fig. 2.1: (a) A transducer generates electrical energy from a physical stimulus. (b) A sensor "modifies" the electrical energy provided by the power supply.

To add more confusion, some articles use the term **detector**, defined as a device that recovers information of interest contained in a physical phenomenon and "impresses" it on an electrical signal. Therefore, it is similar to, and sometimes indistinguishable from, the concept of **sensor**.

These definitions are, however, disregarded or used interchangeably in some instrumentation literature, so the reader should be warned about these misuses of the words **sensor** and **transducer** just to avoid being confused. For example, photoelectric cells modify their electrical output signal when illuminated, so they are referred to as optical **detectors** or optical **sensors.** By means of the photovoltaic effect these devices transform the photon energy into electrical energy, and then they are also called **transducers.** As it happens with this example, there are many devices which are referred to in several ways, but it is useless to discuss each case. It is enough that the reader knows that one device can be referred to in different ways. In order to avoid any semantic discussions, the words sensor and transducer will be preferentially used in this work. Therefore, do not worry about names. More important than adopting one or another name for a concept, is to understand it. Henceforth, our efforts will be focused in explaining the concepts.

2.2 Transfer Functions

The intrinsic function of a sensor is to produce an output in response to a stimulus. Therefore, in order that the sensor could have a practical use, that is to say, that the output could be used to measure the stimulus which produced it, the relation between input and output should be time invariant and well known. The stimulus (or the parameter which is being measured) will also be referred to as sensor input or measurand.

The transfer function (or transference) is a curve representing the relationship between the input and the output. In general, the input is the physical quantity being measured (physical variable, **Inp**) and the output is the magnitude of the electrical output (voltage or current, **Out**). Examples of these relationships are shown in Figure 2.2. The transfer functions of sensors are thus mathematical representations of the relation between their inputs and outputs.

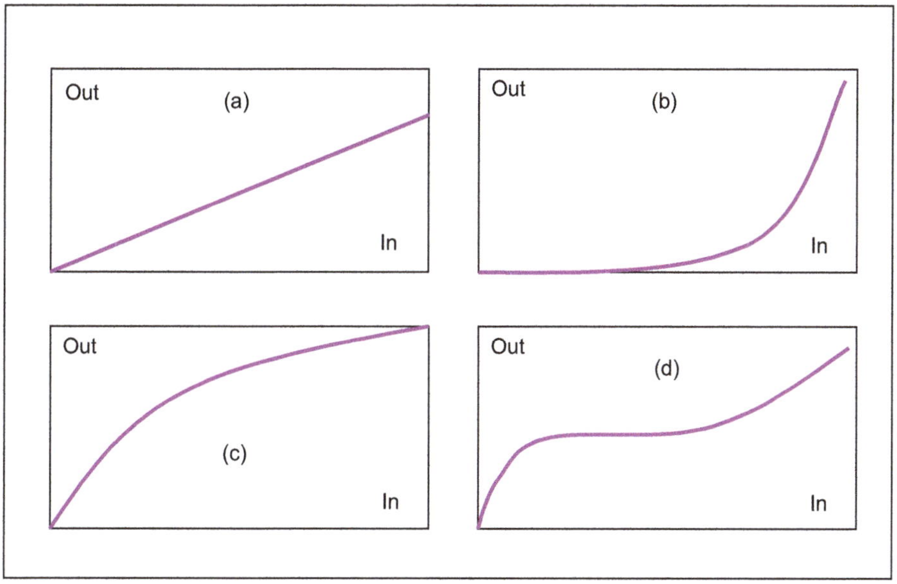

Fig. 2.2: Transfer functions: (a) linear transfer; (b) low sensitivity for small inputs; (c) low sensitivity for large inputs, (d) low sensitivity for middle inputs.

A transfer function may be regarded from a static or dynamic point of view. The static transference is the relation between the input and the output when the input is constant (does not vary with time). The dynamic transference describes the same relationship when the input varies with time. In other words, the output of a sensor can vary in a predictable way for different frequencies of the input; this idea is clarified below. We will first analyze the characteristics of static transfer functions.

Some real systems have linear input/output characteristics, so the representation is simply a straight line (Fig. 2.2a). The slope of a static transfer function curve at any point is called the **sensitivity** at that point. In measuring instruments it is generally convenient to have linear transfer functions because the sensitivity is constant, and the magnitude of the physical phenomenon being measured is obtained by multiplying the electric output by the reciprocal of the sensitivity, which is also constant.

The transfer function of Figure 2.2b shows a great sensitivity for large input values, but it is almost insensitive to small input values. Figure 2.2c, in contrast, shows a sensitive transfer function for small input values, but the sensitivity decreases as input increases. The curve of Figure 2.2d is insensitive for middle input values.

In Section (1.6) it was explained that it is desired to have great signal to noise ratios (*S/N*). Thus, for a given noise it is good to have the maximum possible signal. In low sensitive areas of the transference, sensors produce small electrical signals. So in these regions, electrical noise is more likely to interfere with measurements. When the sensor works in the high sensitive part of the curve it produces better data quality because for the same change in the measurand the output change is greater than in the low sensitive part. Therefore the *S/N* ratio increases, and the undesirable noise perturbs less the desired signal. Therefore, from the *S/N* ratio point of view, in a nonlinear transference the low sensitive part should be avoided.

The same concept of a transfer function is also applied to a complete instrument. In this case it is the relation between its output and input, the input being the measurand, and the output, the data measured, recorded or transmitted. As it was mentioned in the description of a generic instrument, an instrument has several stages connected in series, so the complete transfer function of the instrument can be considered as a set of partial transfer functions, also connected in series (Fig. 2.3). In this figure two main transfer functions can be identified, one corresponding to the sensor and the other to the electronics. Figure 2.3 shows how the transfer function of the electronics may be tailored to compensate the sensor's nonlinearities, giving as a result a linear transference for the complete instrument. The electronics include both analog and digital parts; the linearization process may be performed in either part or both.

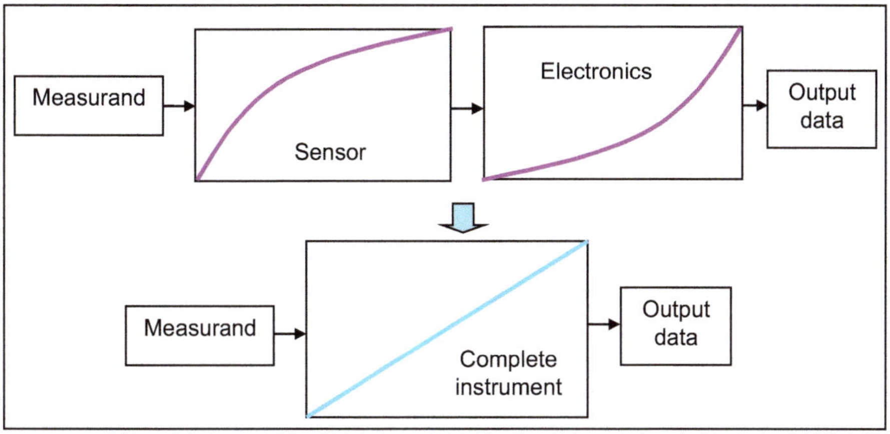

Fig. 2.3: In the upper part, the electronics' transference is designed to compensate the non-linear sensor's transference, giving as a result a complete linear transference (lower part).

2.2.1 Range and Dynamic Range

The input **range** is the difference between the maximum (X_{max}) and minimum (X_{min}) values of the measurand that can be measured by the instrument. The dynamic range (D_{range}) is the ratio between the largest and the smallest measurable signal.

$$Range = X_{max} - X_{min}; \quad D_{range} = \frac{X_{max}}{X_{min}} \tag{2.1}$$

For example, assume that the conductivity of pure water is 0.5 μS/cm and that of seawater is about 50 mS/cm, then if we want to have one conductivity meter to measure from pure water to seawater it should have a dynamic range of 100,000 (also found in manufacturer's brochures as 1:100,000). It is very difficult to cover such a range keeping the measuring error low. So with the purpose of having low errors, conductivity meters are manufactured to measure in a much limited dynamic range, for example 1:1000. Other kinds of instruments have lower dynamic ranges. For example, some electromagnetic current meters can be used only in a dynamic range 1:50, if errors below 0.5 % are desired; vortex flowmeters are insensitive to low velocities, and can be used in the dynamic range 1:10.

Range and dynamic range have been defined for the measurand (input signal) but these definitions also apply for the output of instruments.

2.2.2 Hysteresis

A simple example will be useful to introduce the concept of hysteresis. Some synthetic elastic materials stretched by a force do not return to their initial length when the force is released. Then if a two-dimensional graph is constructed using several pairs of length and force points, two curves will be observed, one when the force is increasing and another when the force is decreasing. It is said that the rubber does not follow Hooke's law because it has some hysteresis.

Likewise, there are sensors whose transfer functions depend not only on the existing input signal but on the previous states of the input (Carr & Brown, 1998). This means that there is not a unique curve (transfer function) but it depends on input signals increasing or decreasing. Thus, the transfer function tends to form a loop as in Figure 2.4. The **hysteresis** (H) is evaluated by means of the maximum variation (h_{max}) observed in the output of a transfer function produced by upscale (increasing) and downscale (decreasing) input variations. It is defined as a percentage of the input range (Eq. 2.2).

$$H\,(\%) = \frac{h_{max}}{Range} \times 100 \tag{2.2}$$

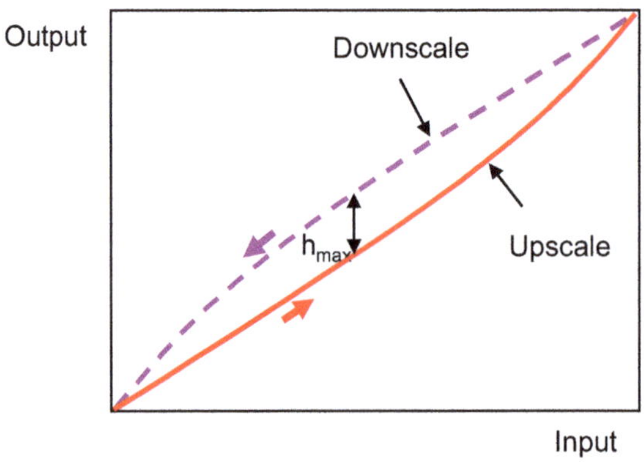

Fig. 2.4: Hysteresis. Upscale and downscale transferences; h_{max} is the maximum variation between both.

2.2.3 Calibration

The word **calibration** may be used in slightly different ways depending on the context. In the vocabulary used by quality control professionals, calibration is the periodically repeated process for verifying the capability and performance of a measuring instrument by comparison to traceable measurement standards. The calibration process assures that the instrument remains capable of making measurements according to its performance specifications. It consists in comparing a given instrument with a measurement system or device having higher specifications and known uncertainty. This comparison allows any variation in the accuracy of the instrument under test to be detected.

It is easily understood that a new instrument could need an initial calibration just after being manufactured, so the word calibration is also used to call the process of alignment or adjustment of an instrument when it is either manufactured or repaired. In order to determine the transfer function of a just manufactured or repaired instrument, other instrument of better quality or a standard instrument has to be used.

For the case in which the sensitivity is variable (nonlinear transference), it may be necessary to determine the transfer curve point by point, which is often a slow and laborious work. This is another reason why sensors with linear transfer functions are more convenient; it is (theoretically) possible to obtain their transfer functions with only two calibration points.

2.2.4 Linearity

The linearity of a sensor (or an instrument) expresses how much its actual transfer curve deviates from a straight line. It is evaluated by the percentage of non linearity as defined by

$$NL\,(\%) = \frac{D_{max}}{Range} \times 100 \tag{2.3}$$

where NL (%) is the percentage of nonlinearity, D_{max} is the maximum output deviation from the straight line and **Range** is the total output range = $O_{max} - O_{min}$.

The percentage of non linearity depends on how the linear approximation of the curve is done (Carr & Brown, 1998; http://www.sensorland.com). From Figure 2.5 the maximum output deviations *D1* and *D2* are different, thus nonlinearity substantially depends on how the straight line is drawn. When the **linearity** of an instrument is specified by its manufacturer, a statement on how the straight line is superimposed on the transfer curve should be made. The user of instruments should be aware that manufacturers specify the possible minimum nonlinearity to show the goodness of their instruments and, in some cases they forget to explain what method they used to draw the straight line. Users should ask manufactures how they specify linearity. Otherwise the information becomes meaningless.

One usual way to make a linear approximation of the transference curve is by means of a straight line whose slope is calculated from the two end (or terminal) points of the transfer function, as shown in Figure 2.5. It is known as the Terminal Straight Line (TSL) method. In this method, the manufacturer calibrates a zero point and a full scale point. The maximum output deviation is *D1*; nonlinearity errors are the smallest around the end points and the highest somewhere in between (Fig. 2.5).

Another way to calculate the linear approximation is by using the Least Squares Best Fit Straight Line (BFSL) method. It is a statistical method that requires knowledge of several calibration points over the whole working range of the sensor. The input and output values at each point are then used to calculate the slope of the straight line which provides the closest possible best fit to all data points on the curve and the intercept. Figure 2.5 shows that the maximum output deviation *D2* is smaller than in the TSL method.

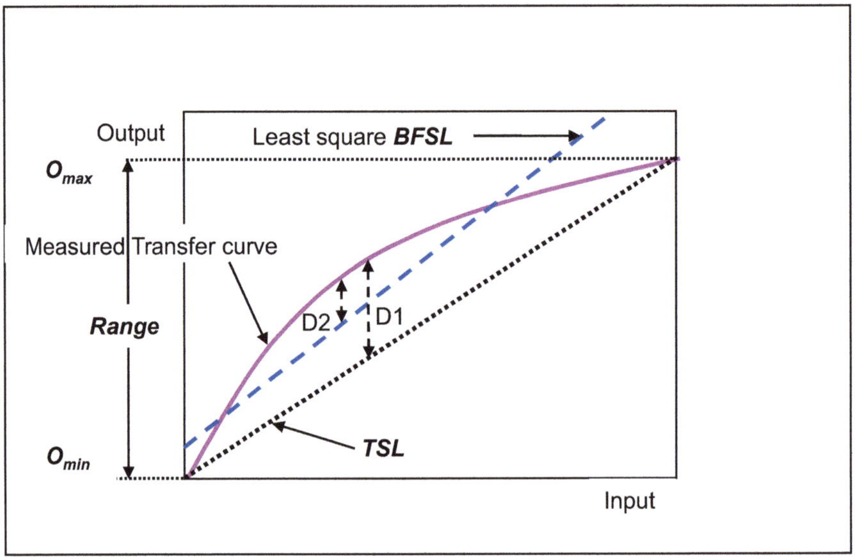

Fig. 2.5: Non-linear transfer curve with TSL and BFSL linear approximations. $D1 \neq D2$ shows that the definition of linearity depends on the linear approximation method.

2.2.5 Offset and Gain Errors

Due to various causes such as daily usage, shocks, vibration, etc., an instrument could suffer changes in its transference, and some measurement error may thus arise when using it. In an instrument with a linear transference, two kinds of changes from the initial calibration can be found; these departures can produce some measuring errors which are known as **offset** and **gain errors**.

The **offset error** of a sensor (or an instrument) is defined as the output that exists when the input is zero. It could be thought as a shift of the whole transfer function by a constant amount. In order to measure the offset a zero input must be forced; for example, in a flowmeter the instrument must be removed from the flow or the flow must be blocked. Then, the output must be read and if the offset were invariant with time, the offset error could be removed by subtracting it from the actual transfer function (Fig. 2.6).

If the slope of the actual transference differs from the ideal (or calibrated) transference it is said that the instrument has a **gain error** (Fig. 2.6).

So far we have allotted the departure from the ideal transfer curve of equipments to their use, but a new instrument also has some offset and gain errors because there is a limit in the quality of the manufacturing processes of sensors (and instruments), which generates a transfer function somewhat different from the ideally desired. In general, manufacturers specify the ideal transfer function as if all sensors were made equal, but at the same time, they also specify the offset and gain errors as a percentage of the total output range or as a percentage of the reading.

Furthermore, if the instrument were subjected to aging, manufacturers should specify the percentage change of **offset** and **gain** per unit time. This happens with some sensors whose transfer characteristics are based on a sensitive material whose properties vary with time in a well known way, and so the change in the transference can be predicted. This change is expressed, for example, as a given gain change of -1% per year for a sensor that is loosing its sensitivity.

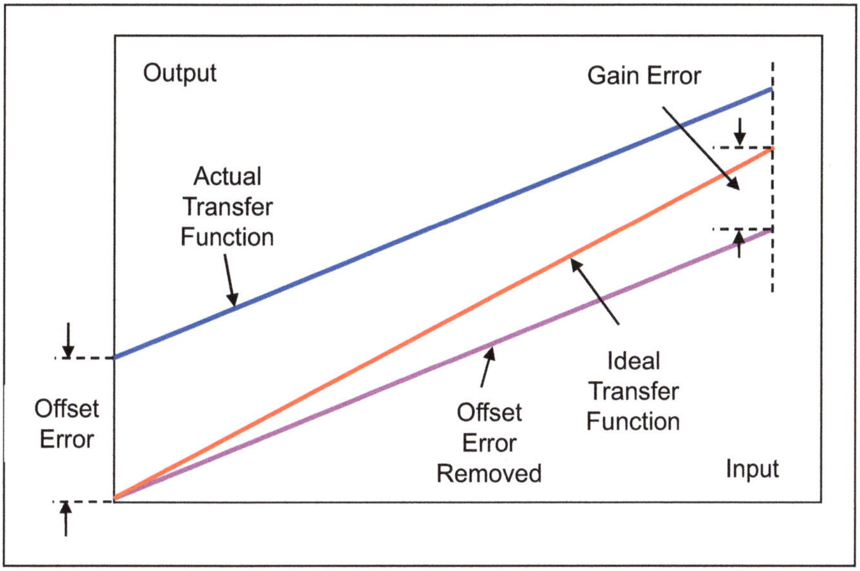

Fig. 2.6: Actual transfer function with offset and gain errors; ideal transfer function; and transfer function with offset error removed. At null and full input, offset and gain errors are indicated.

2.2.6 Drift

Real sensors are not time-invariant systems. Once offset and gain errors are eliminated by some kind of electronic compensation, the transfer function could drift due to environmental factors such as temperature and humidity. The manufacturers usually specify the **zero drift** and the **span drift** as a function of the factor that modifies the transference. For example, for an instrument that measures water table from zero to 10 m a **zero drift** = - 0.1% /°C of full scale, means that for an increase of 10 °C in the instrument's temperature, the measured value will decrease 10 cm, which is not negligible for some environmental studies.

2.2.7 An Example of Sensor Specifications

As an example, the following parts of the whole specifications of a commercial pressure sensor were chosen:

Accuracy .... 1.0% Full Scale
Non-Linearity and Hysteresis........ 1.0% Full Scale
Non-Repeatability.......................... ± 0.15% Full Scale

Accuracies stated are expected for Best Fit Straight Line for all errors, including linearity, hysteresis & non-repeatability **(Note 1).**

Temperature, Operating .... 50° to 145°C
Temperature, Compensated .... 15° to 70°C
Temperature Effects
Zero drift... 0.02% Full scale/ °C
Span drift.. 0.04% Reading/ °C **(Note 2).**

Special Calibration
10 point (5 up/ 5 down) 20% increments @ 20°C
20 point (10 up/ 10 down) 10% increments @ 20°C **(Note 3).**

Comments to the specifications

Note 1: This manufacturer does not specify each error, but gives figures that include all of them. Also, the manufacturer tells us which was the method used to linearize the transfer function (Best Fit Straight Line).

Note 2: The sensor can be used between -50° and 145 °C, but specifications are met only for the compensation range between 15° and 70°C. Temperature changes affect the transfer function. The zero drift is specified as a percentage of the full range, and the span drift as a percentage of the measured value (reading).

Suppose that the full range of this sensor is 10 m of water column (wc) and that it will be used to measure the height of a water column in a tank in an environment whose temperature is limited to the interval between 15 and 70 °C. Let us assume that we calibrate the sensor at 15 °C (offset and gain adjusted). If when performing the measurements the temperature of the water is 70°C, the zero drift for the total change in temperature (70°C – 15°C = 55°C) will be

$$\textit{zero drift} = \textit{0.02\% Full Scale} / {}^{\circ}\mathrm{C} = \frac{0.02 \times 10\ \mathrm{m} / {}^{\circ}\mathrm{C}}{100} \times 55^{\circ}\mathrm{C} = 0.11\ \mathrm{m}$$

If the instrument is measuring at a height of 5 m (half of the total range) the span drift will be:

$$\textit{span drift} = \textit{0.04\% Reading} / {}^{\circ}\mathrm{C} = \frac{0.04 \times 5\ \mathrm{m} / {}^{\circ}\mathrm{C}}{100} \times 55^{\circ}\mathrm{C} = 0.11\ \mathrm{m}$$

Then, the total drift would be 0.22 m over a range of 5 m, which is a total change due to drift of 4.4 % for a change in sensor temperature of 55 °C. Obviously it seems very difficult to find environmental measurements where the fluid changes its temperature in such large amount; but it could be quite reasonable in an industrial process.

This somewhat unrealistic example was chosen to show the importance of temperature calibration. If the temperature of the water whose height has to be measured would range from 15 to 70 °C, it would be much better to calibrate the pressure sensor at half the range = 42.5 °C. In so doing, the change due to drift will be ± 2.2 %.

It should be noted that if the tank had a small column of water of, say, 1 m the zero drift would be still 0.11 m and the span drift would reduce to 0.022, the total drift being 0.132 m. But the total change due to drift is 13.2 % of the measured value. Therefore, an important conclusion is that to minimize the error due to zero drift, the sensor has to be selected in such a way so that it does not measure in the lower part of its range.

This is a good example to show that a sensor has to be selected according to the desired measuring range with the purpose of minimizing errors and, also, to keep a good signal to noise relation. Whenever possible, sensors should be selected to measure in the upper part of their ranges. It is obvious that a sensor able to measure 100 m water column will measure 1 m too, but the price paid is that noise and errors can make the measurements useless.

Note 3: This manufacturer offers two kinds of calibration (at different costs): the 10 points and the 20 points. The first means that the Best Fit Straight Line is obtained from 5 points increasing input values and 5 points decreasing input values. Obviously, the 20 points calibration (10 increasing and 10 decreasing) helps to better know sensor errors due to non-linearity, but the price of the sensor will be higher.

2.3 Spatial Characteristics of Sensors

2.3.1 The Decibel

In order to understand some properties of sensors it is necessary to introduce a useful way of comparing two quantities; e.g. in some cases it is convenient to compare them on a logarithmic scale. This idea was first extensively used for telephone power line (*P*) calculations. The unit of this logarithmic scale is called the decibel (dB). Given two power values P_2 and P_1 the number (*N*) of decibels is defined by

$$N = 10\ log\ \frac{P_2}{P_1} \tag{2.4}$$

A negative value of *N* means that $P_2 < P_1$.

This way of comparison between two power values was then extended to other fields such as sensor theory, in which it is used to compare two signal amplitudes (*A*). The relative change in signal amplitude of sensors is commonly measured in decibels. Because *P* is proportional to the square of the voltage amplitude (*A2*), when comparing two amplitudes the number (*N*) of decibels by which *A2* exceeds *A1* is defined by

$$N = 10\ log \left(\frac{A_2}{A_1}\right)^2 = 20\ log\ \frac{A_2}{A_1} \tag{2.5}$$

Since *N* compares two quantities, *A1* and *A2*, in a relative way, in order to get the absolute value of one of them it is necessary to know the other. For example, when it is said that the voltage gain of an amplifier is 80 dB it means that two voltage **amplitudes** are compared and Eq. (2.5) must be used. From this equation it is found that 80 dB corresponds to an amplification of 10,000 times, but the absolute value of the output voltage remains unknown. If the reference were one millivolt (1 mV) the voltage output would be 10 V.

Obviously, the decibel may be used to give absolute values when a reference value is adopted. Sometimes a suffix is added to the decibel symbol to indicate which reference has been used for comparison. That is the case of dBm, which indicates that one milliwatt (1 mW) has been use as the reference value. A sound amplifier having a 40 dBm gain will have an output of 10 W. Note that because dBm indicates that the reference is a unit of *power*, Eq. (2.4) must be used.

Briefly, when the instrument specifications are expressed in dB, to know absolute values it is required first to recognize if quantities compared are *amplitudes* or *powers*, and second, what is the *reference* value.

Some useful approximated relationships are displayed in Table 2.1.

Table 2.1: The decibel. Approximated relationships for power and amplitude

Ratio	Log_{10}	Power (dB)	Amplitude (dB)
1	0	0	0
1.414	0.1505	1.5	3
2	0.301	3	6
3.16	0.4997	5	10
5	0.6989	7	14
10	1	10	20
100	2	20	40
1000	3	30	60
10000	4	40	80
100000	5	50	100
1000000	6	60	120

2.3.2 Sensor Directivity

A very important property of a sensor is its directivity. Sensors were previously associated with devices that perceive a physical stimulus giving as a result a signal (as explained above, in most devices the output is an electric signal). So far nothing was said about the relative position between the sensor and the direction of the stimulus. A device is said to be **omnidirectional** if it produces the same output signal regardless of the direction where the stimulus comes from. Figure 2.7 shows the voltage output signal of an omnidirectional sensor as a function of the arriving direction in polar coordinates (scale must be multiplied by 10 to cover 360 °). The radial axis is the output voltage (7 V in this case). Few practical sensors are omnidirectional; most of them produce a higher output signal when the stimulus arrives from some preferential direction. This property is often used to know where the signal comes from.

One way to specify the spatial sensitivity of a sensor is to plot its directivity pattern. Usually, the three-dimensional pattern is plotted in two planes containing the axes of the device (generally the horizontal and vertical planes); plots are typically in polar coordinates (Figs. 2.8 and 2.9). The polar diagram of Figure 2.8 shows the main lobe (upwards) and the two secondary lobes (downwards). The main lobe gives the direction of the maximum sensitivity of the sensor, or the maximum radiation angle in the case of an emitting transducer (e.g., a radar antenna). Commonly the main lobe is pointed towards the stimulus of interest. It is desired that the difference expressed in dB between the main and the secondary lobes be as large as possible, because the secondary lobes could catch signals from undesirable stimuli, thus masking or compromising the desired measurement.

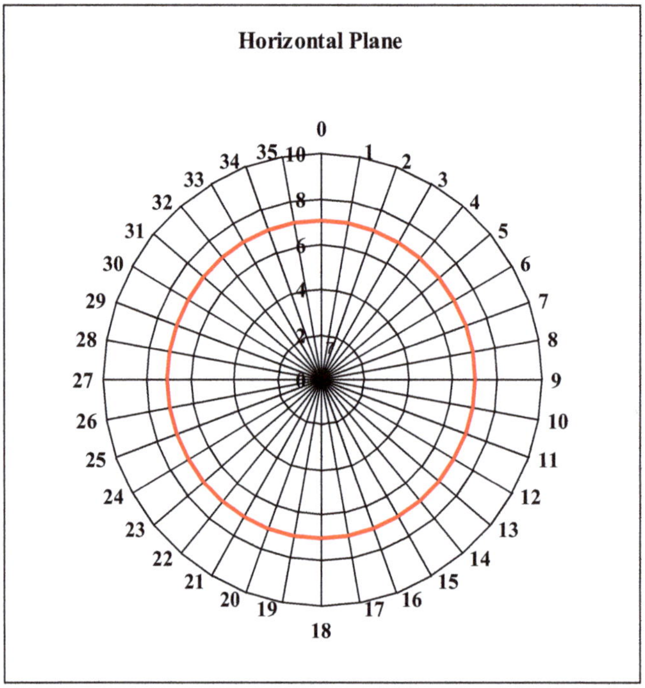

Fig. 2.7: The output voltage of a sensor as a function of the direction the stimulus comes from represented in the horizontal plane. The scale shown on the perimeter must be multiplied by 10.

In the examples shown in Figures 2.8 and 2.9 the horizontal pattern is more directive than the vertical one. A measure of the directivity of sensors is given by its **beamwidth.** It is a measure of the angle between two points of the lobe at some level below the maximum sensitivity angle. Some of the most commonly used levels are -3 dB, -6 dB and -10 dB but manufacturers can define their own.

Generally, two powers are compared in defining the **beamwidth** and so Eq. (2.4) must be used. The level used to define the beamwidth must be specified together with the beamwidth. For example, taking the horizontal pattern of Figure 2.8, the maximum level is $P1 = 10$ and choosing the reference level of -3 dB ($P2/P1 \approx 0.5$), on the left side of the main lobe this level corresponds to about 335 ° (or -25 °), and on the right side it is about 25 °. Then the beamwidth for -3 dB is 50°.

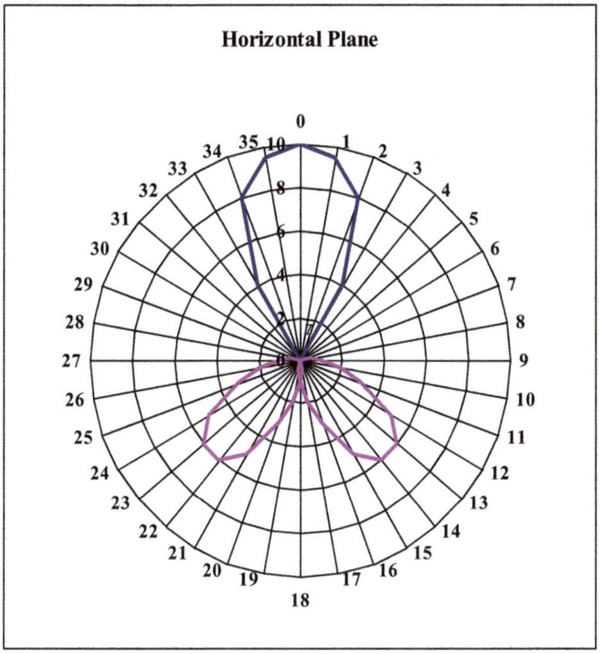

Fig. 2.8: Horizontal plane beamwidth for -3 dB ($P_2/P_1 \approx 0.5$), is - 25 ° + 25 °.

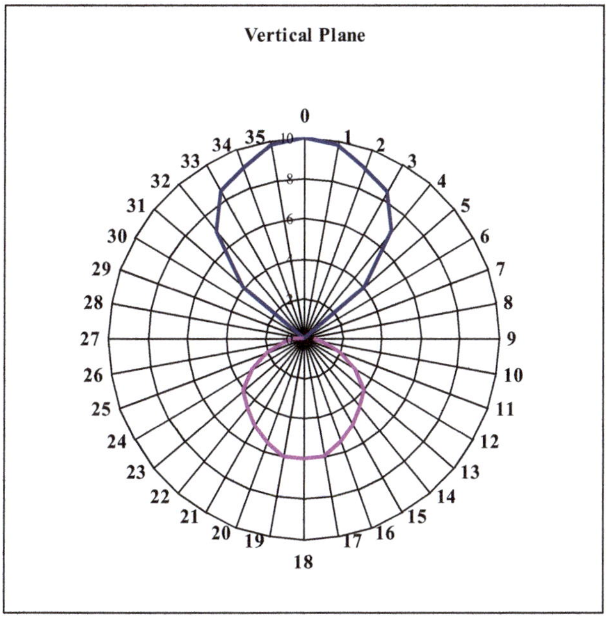

Fig. 2.9: Vertical plane directivity pattern.

2.3.3 Spatial Averaging

The concept of spatial averaging is introduced here to make readers note that measured values do not correspond to an infinitesimal point on the space but to some volume. This volume should be known and compared with the spatial signal we want to evaluate. Suppose that an acoustic sensor for measuring waves is placed on a coastal structure at a height R above sea level with its main lobe pointing down. Assuming that the measurement zone is the intersection of the beamwidth with the sea surface, the area of the sea from which the sensor would take the information is the base of the cone illustrated in Figure 2.10,

$$S = \pi r^2 = \pi \left(R \tan\left(\frac{\alpha}{2}\right)\right)^2 \tag{2.6}$$

where r is the base radius of the cone and α is the vertex angle. For a sensor with a conical beamwidth of $\alpha = 30\,°$ (for a -3 dB level in both the horizontal and vertical planes) and placed at $R = 10$ m, $r = 2.68$ m and the surface of the cone base is $S = 22.6$ m^2. This means that the electrical signal generated at the sensor output will contain information from the stimulus contained in that area. The sensor is thus performing a spatial averaging of the wave information. Waves whose wavelengths (λ) are shorter than the diameter of the cone's base ($\lambda < 2r = 5.36$ m), will be averaged giving a reduced output.

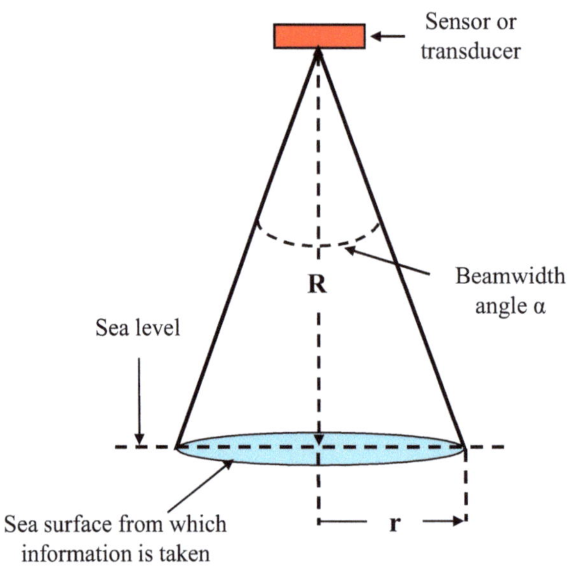

Fig. 2.10: The beamwidth of and acoustic wave sensor takes information of the sea surface averaged over the base of the cone.

If higher spatial resolution is needed (e.g., if it is desired to measure waves of short wavelength), the height of the sensor must be decreased, or a sensor with a narrower beamwidth adopted. For example, to measure the information contained in an area $S < 1\ m^2 (\lambda < 2r \approx 1.13\ m)$ the sensor should be within 2.1 m height above sea level. By the way, in some locations where the tidal range is significant, the wave spatial averaging of such an acoustic wave meter will be a function of the tidal level, which could be inconvenient for some applications.

This phenomenon of spatial averaging occurs for most sensors. Therefore, to recognize whether the sensor used is actually measuring the desired phenomenon, it is important to assess its spatial averaging.

2.4 Time and Frequency Characteristics of Sensors and Systems

2.4.1 Introduction

So far we have analyzed the **static transfer function** of sensors and instruments. We will now turn our attention to the **dynamic transfer function**, which describes how sensors' inputs and outputs are linked when the input signals vary in time. Although this topic is included in this chapter devoted to sensors and transducers, the concepts to be developed are very general and applicable to instruments and systems.

All devices have a finite capacity to reproduce at the output any time-varying signal applied to the input. This means that the time variations of measurands are modified somewhat by sensors and instruments, and users should evaluate if these modifications are significant for the phenomena that they are studying.

Manufacturers specify the capacity of the sensor or instrument to reproduce the measurand dynamic behaviors in a variety of ways. They define different parameters that account for these characteristics: frequency response, bandwidth, time constant, rise time, etc. It is important for the users to understand the meaning of these specifications to evaluate whether any particular instrument satisfies their measuring requirements. As can be intuited from their names some specifications refer to the frequency domain and some to the time domain.

A clear understanding of the above concepts will be useful to users in order to check whether they are losing part of the desired measurand information due to dynamic limitations of sensors or instruments. Also, in those cases that manufacturers do not provide such parameters to account for the instrument dynamic response, these concepts would enable users to perform their own experiments to estimate them. These tests could be as simply as applying a step excitation to the instrument input. This kind of check could become healthy routines before field installation of instruments when some doubts exist about the dynamic instrument performance or one of the following situations occur: factory specifications are not available; the instrument has been used by a prolonged time without maintenance; the instrument has been a prolonged time in storage without use; or the instrument has undergone a deep repair.

2.4.2 Frequency Content of Signals

The **period** (T) of a signal is the elapsed time between two repeating events. The **frequency** (f) is the number of occurrences of the events per unit time, e.g. the frequency of the heart beat is about one beat per second = 1 Hz. Thus, frequency is the reciprocal of period; $f = 1/T$. The SI unit for period is the second (s) and for frequency is the hertz (Hz), the Hz being 1/s.

Most waveforms of practical use in engineering can be analyzed by studying their energy content at each frequency. If the waveform is periodic (i.e. if there is a positive nonzero value of T such that the value of the function at time $t + T$ is equal to its value at time t), it can be described by a series of sine and cosine terms whose frequencies are integer multiples of a fundamental frequency (f_0), which is the inverse of the (fundamental) period (T_0) for the waveform to repeat itself. This series is called a **Fourier series.** The Fourier theorem states that any periodic signal, no matter how complex it is, may be seen as a combination of a number of pure sine and cosine terms with harmonically related frequencies.

If the wave is not periodic, the fundamental period tends to infinite ($T_0 \to \infty$). The fundamental frequency ($f_0 = 1/T_0$) is therefore infinitesimal and the frequencies of successive terms of the Fourier series differ in an infinitesimal amount instead of in a finite amount. The Fourier series becomes thus the **Fourier integral.** It turns out that any waveform can be decomposed or analyzed into only sine and cosine functions of different amplitudes and frequencies.

A fundamental concept already intuitively introduced should be clear to researchers: **all sensors, and then instruments, deliver a signal that is a modified version of the real signal they want to measure.** The measuring process changes the temporal (and frequency) characteristics of the signal, sometimes in a seamlessly, and therefore acceptable way, and others in an appreciable and inconvenient way. This undesirable trimming of the signal is due to an unwanted filtering which restricts the frequency content of the input signal. This happens because real devices cannot respond to very fast or very slow stimuli. Figure 2.11 shows how a signal could be distorted due to unwanted filtering. The amount of distortion and attenuation of the parameters being measured depends on the relation between **signal frequency content** and **instrument frequency response.**

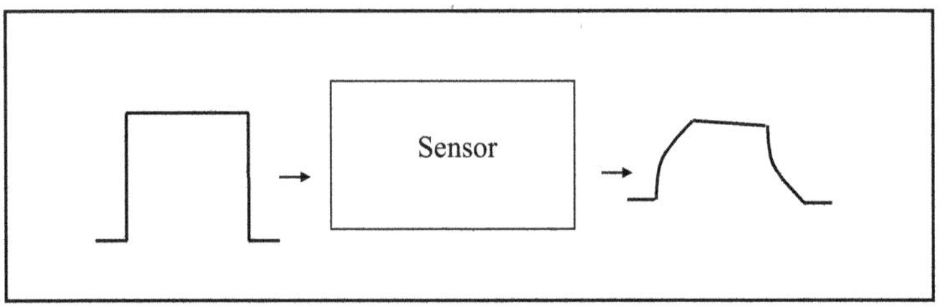

Fig. 2.11: The input signal (left) is modified by the sensor frequency response giving a different signal at the output (right).

2.4.3 Frequency Response

Frequency response is one of the ways in which manufactures inform the users how instruments perform dynamically. The **frequency response** of a device describes its ability to reproduce at the output any signal applied to the input. The **frequency response** of a system is a measure of the magnitude and phase of the output as a function of frequency, in comparison to the input.

In simple terms, when a sine wave of given frequency is injected at the input of a linear system, it will output that same frequency with a certain magnitude and a certain phase angle relative to the input. To obtain the **frequency response** of a system, sinusoidal inputs [$x(t)$] of known amplitudes may be applied to the system, and the corresponding output amplitudes and phases (relative to the inputs) [$y(t)$] must be measured over a range of frequencies,

$$x(t) = X \sin 2\pi f_0 t; \qquad y(t) = Y \sin(2\pi f_0 t + \phi) \tag{2.7}$$

The corresponding gain $G(f_0)$ and phase $\varphi(f_0)$ of the frequency response for the frequency f_0 are defined as

$$G(f_0) = \frac{Y}{X}; \qquad \varphi(f_0) = \phi \tag{2.8}$$

The following example shows how a real instrument can distort part of the information contained in the measurand. The approximate transfer function, gain and phase as function of f (*frequency response*) of an old commercial wave meter is shown in Figures 2.12a and 2.12b. These figures are also referred to as the **Bode plot** of the instrument response.

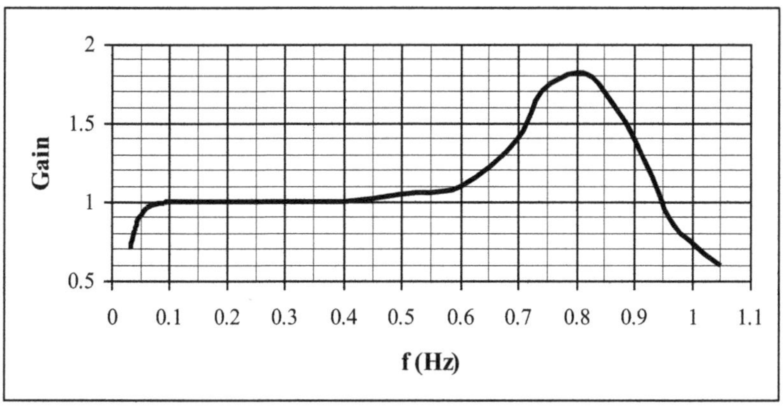

Fig. 2.12a: Wavemeter gain as a function of frequency. As modified from Datawell (1980).

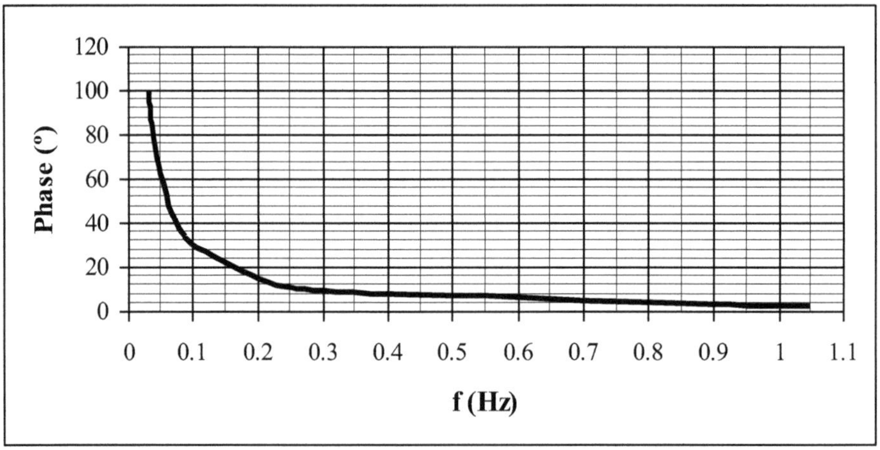

Fig. 2.12b: Wavemeter phase output relative to the input, as a function of frequency. As modified from Datawell (1980).

Let us see how the instrument modifies the recorded data. Waves of frequencies $0.1 < f < 0.4$ Hz (or period $2.5 < T < 10$ s) are measured without attenuation ($G = 1$) and with a small change of phase. Long waves of $f = 0.05$ Hz ($T = 20$ s) are attenuated by 10 %; short waves between $0.4 < f < 0.6$ Hz (or period 1.67 s $< T <$ 2.5 s) are overestimated up to 10 %. Waves of $T = 1.25$ s are overestimated up to 80 %.

For most studies this instrument is adequate, but if the study requires to measure outside the range from 0.1 to 0.6 Hz perhaps another instrument best fitted to the desired range could be found. Then, before using an instrument the frequency range of the phenomenon to be measured should be compared to the transfer function of the instrument.

Manufacturers of the wavemeter whose Bode plots were presented did a good job specifying their instrument and allow users to better know how the measuring system influences the recorded data. Unfortunately, many times, manufacturers do not provide a complete Bode plot of their devices.

In some particular sensors and instruments whose transferences are more flat and regular than that shown in Figures 2.12a and 2.12b, manufacturers give the plot of the gain as a function of frequency, $G(f)$ and phase shifts are inferred from some equations accepted as standard. The transference region in which gain is constant permits the useful frequency range of the instrument to be identified, also known as the **bandwidth** of the instrument.

2.4.4 Bandwidth

Bandwidth is used to characterize electronic filters, sensors, antennas, audio amplifiers, instruments, communication channels, etc. It describes their ability to allow the passage of a signal without significant attenuation and distortion. Figure 2.13 shows the gain of a given device together with its bandwidth. The frequency band of a circuit or sensor in which the output signal is attenuated less than -3 dB is called the **bandwidth.**

In Figure 2.13 there is a range of frequencies over which the output $G(f)$ is almost constant (G_0) and then all the input information is translated to the output without amplitude distortion. At frequencies f_1 and f_2 the output falls to $G(f_1) = G(f_2) = 0.707\ G_0$; this drop in signal level corresponds to -3 dB. Frequencies f_1 and f_2 are called the lower and upper frequency points, respectively, and the bandwidth is defined as $B = f_2 - f_1$. For many practical cases $f_1 \approx 0$, so $B = f_2$.

The -3 dB points are also referred to as the half-power points because power is proportional to the square of signal amplitude, then a drop to 0.707 in amplitude means a drop in power to $P \equiv (0.707)^2 \approx 0.5$.

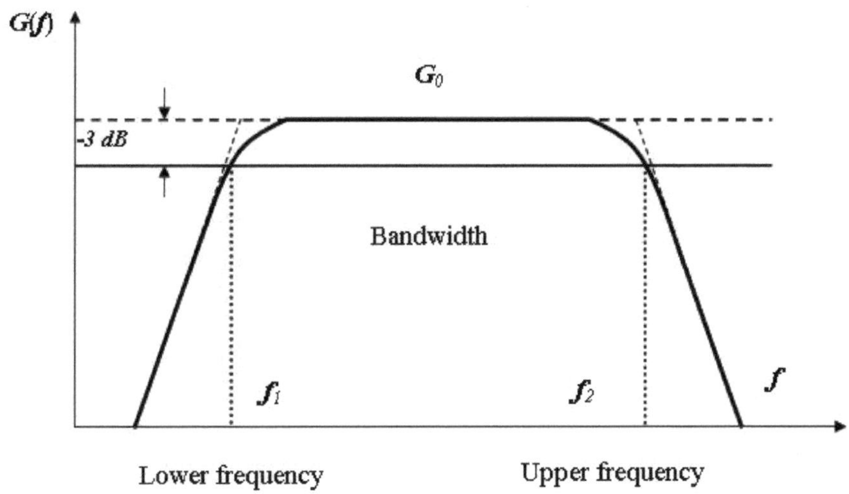

Fig. 2.13: Gain as a function of frequency showing a constant gain (G_0) over a certain range of frequencies. Lower and upper frequency points (f_1 and f_2, respectively) and the bandwidth are also shown.

There is a simple equation for the *bandwidth* of a first order linear time invariant (LTI) system. This simplification is very useful because in nature there are many first order LTI systems, among them thermal, hydraulic, mechanical, biological, radioactive phenomenon may be treated as LTI systems. A similar behavior can be found in electronic circuits, viscoelastic materials, humidity meters, etc.

Out of the almost constant (G_0) region (central region of the transference of Figure 2.13 where the information is not modified) it can be demonstrated (Millman & Halkias, 1967) that for a sinusoidal forcing function, the corresponding gain ($G(f)$) and phase ($\Phi(f)$) of the frequency response of a first order LTI system are:

$$G_{UP}(f) = \frac{1}{\sqrt{1+(N_2)^2}}; \qquad \Phi_{UP}(f) = \arctan N_2; \qquad N_2 = \frac{f}{f_2}$$
$$G_{DOWN}(f) = \frac{1}{\sqrt{1+(N_1)^2}}; \qquad \Phi_{DOWN}(f) = -\arctan N_1; \quad N_1 = \frac{f_1}{f} \tag{2.9}$$

where G_{UP} and Φ_{UP} are valid for high frequencies and G_{DOWN} and Φ_{DOWN} for low frequencies.

The gain and phase given by Eq. (2.9) are plotted as a function of $\boldsymbol{N} = \boldsymbol{N_1} = \boldsymbol{N_2}$ in Figures 2.14a and 2.14b. For $\mathbf{N} = 1$ the amplitude of the output is as expected from the definition of the upper and lower frequencies $G_{UP} = G_{DOWN} = 0.707$ times the input; and the phase is $\Phi_{UP} = +45°$ and $\Phi_{DOWN} = -45°$.

Typically it is desired that the input signal passes the system with minimum attenuation, and for this to happen, as a rule of thumb, the upper frequency of the system bandwidth should be one decade up the maximum frequency of the signal input and the lower frequency should be one decade down the minimum frequency of the input. For this condition, the relations of frequency should be $f/f_2 = f_1/f = 0.1$ resulting $G = 0.995$ (see Table 2.2 following) or, in other words, the attenuation would be only 0.5%.

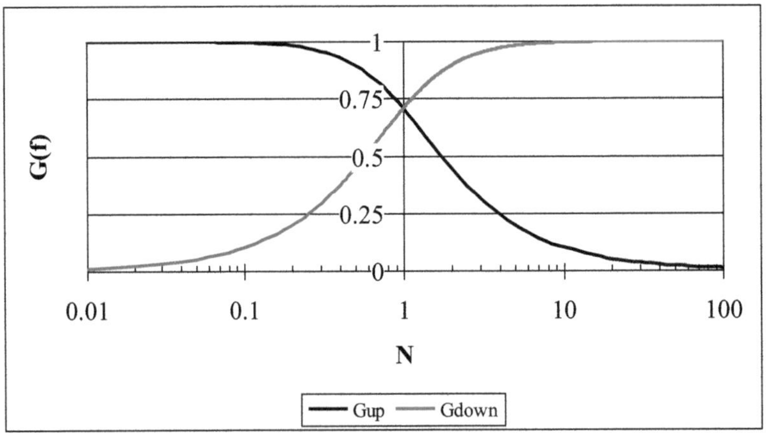

Fig. 2.14a: G_{UP} and G_{DOWN} are the gains in the upper and lower frequencies, respectively, as a function of $N = N_1 = N_2$.

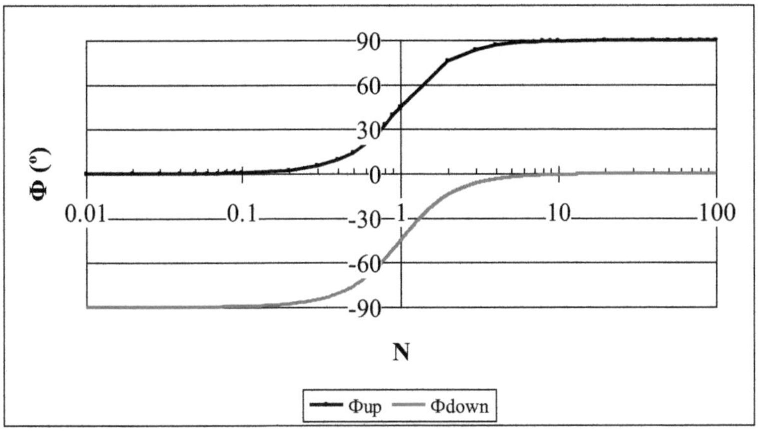

Fig. 2.14b: Φ_{UP} and Φ_{DOWN} are the phases in the upper and lower frequencies, respectively, as a function of $N = N_1 = N_2$.

Another less demanding rule of thumb states that the output signals attenuated and distorted by the system are still reasonably good if the frequency components of the input signal are included in the *bandwidth* of the device. That is to say that it is accepted that the measurand information content, close to the upper and lower frequencies, be somehow attenuated, because the output at f_2 and f_1 is 0.707 of the input. This means that a researcher who uses in the entirely bandwidth an instrument whose frequency response correspond to that of a first order LTI system will have the lower and higher frequency components attenuated about 30%.

Some points of Gain and Phase are shown in Table 2.2.

Table 2.2: Gain and Phase

N2	Gain	Phase (°)	N1	Gain	Phase (°)
0.1	0.99503719	0.5729387	10	0.99503719	- 0.5729387
0.2	0.98058068	2.29061004	9	0.98058068	- 2.29061004
0.3	0.95782629	5.14276456	8	0.95782629	- 5.14276456
0.4	0.92847669	9.09027692	7	0.92847669	- 9.09027692
0.5	0.89442719	14.0362435	6	0.89442719	- 14.0362435
0.6	0.85749293	19.7988764	5	0.85749293	- 19.7988764
0.7	0.81923192	26.104854	4	0.81923192	- 26.104854
0.8	0.78086881	32.6192431	3	0.78086881	- 32.6192431
0.9	0.74329415	39.0074726	2	0.74329415	- 39.0074726
1	0.70710678	45	1	0.70710678	- 45

Higher orders of **LTI** systems are also described by simple equations similar to Eq. (2.9) but it is left to the interested readers to find them in references such as (Millman & Halkias, 1967).

2.4.5 Time Constant

Hitherto the discussion has been centered in how manufacturers specify the capacity of instruments to reproduce the measurand dynamic behavior based on characteristics defined in the frequency domain. Because instruments record data as a function of time it would be interesting to find some parameter in the time domain that could give us an idea of how instruments modify the dynamic characteristics of the measurand. It will be even better if this time measured parameter could be related to the bandwidth of the system. This is difficult to do with complex transferences such as that of the wavemeter of Figures 2.12a and 2.12b, but simple approximations could be proposed for some LTI systems.

Now we will analyze the **time response** of systems to understand how much measuring systems distort the information contained in the measurand. For first-order LTI systems the **time constant** is a quality indicator of the dynamic response of the system; it is a way to characterize its dynamic response (http://www.sensorland.com). Equation (2.10a) represents a first-order LTI system. Recall that a linear time invariant (LTI) system is a system that can be described by a linear differential equation with constant coefficients,

$$\frac{dx(t)}{dt} + k\,x(t) = U(t) \tag{2.10a}$$

where $U(t)$ is the forcing function, or system input, and $x(t)$ is the system response, or system output.

As an example of a mechanical first-order LTI system we could consider the force $U(t)$ required to move a block of mass m at a velocity $x(t)$ sliding on an oil layer (Fig. 2.15a). The sliding viscous friction force is $f = b\,x(t)$, where b is the viscous friction coefficient. Then the differential equation is:

$$m\frac{dx(t)}{dt} + b\,x(t) = U(t) \tag{2.10b}$$

If $U(t)$ is the step function, the solution to Eq. (2.10a) is (Millman & Taub, 1965):

$$x(t) = C_0 + C_1\,e^{-t/\tau} \tag{2.11}$$

where τ is known as the time constant of the system and has units of time.

If $C_0 = 0$ the result is a decaying exponential, and when $C_0 = -C_1$ the solution can be written as

$$x(t) = -C_1(1 - e^{-t/\tau}) \tag{2.12}$$

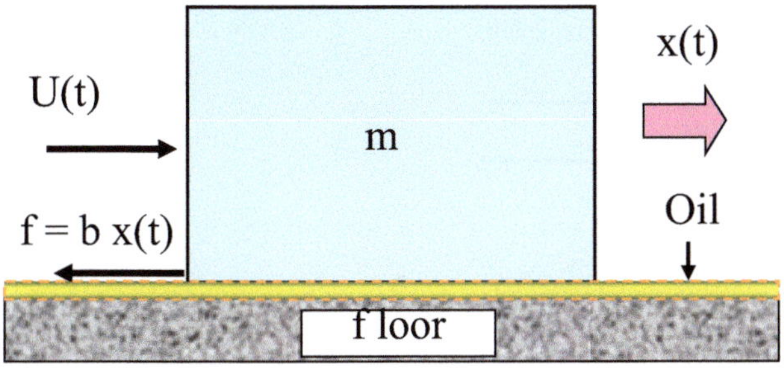

Fig. 2.15a: A block of mass *m* pushed with force *U(t)* and sliding at a velocity *x(t)* on an oil layer. It experiences a viscous friction force proportional to the velocity. The friction coefficient is b.

Figure 2.15b shows the functions $y_1 = x(t)/C_1$ for $C_0 = 0$, and $y_2 = - x(t)/C_1$ for $C_0 = - C_1$, both as functions of $z = t/\tau$.

For a system excitation with a step-like shape, the system response will rise or fall exponentially to approximately 63 % of its final (asymptotic) value when $t = \tau$. Table 2.3 presents some points of the y_1 and y_2 curves.

The **time constant** of a system was found to be a useful tool to evaluate the system frequency performance, because, as will be shown below it can be related to the system bandwidth. Using a step-like shape input signal, the **time constant** is easy to measure for some first order linear networks and could be used to estimate the bandwidth.

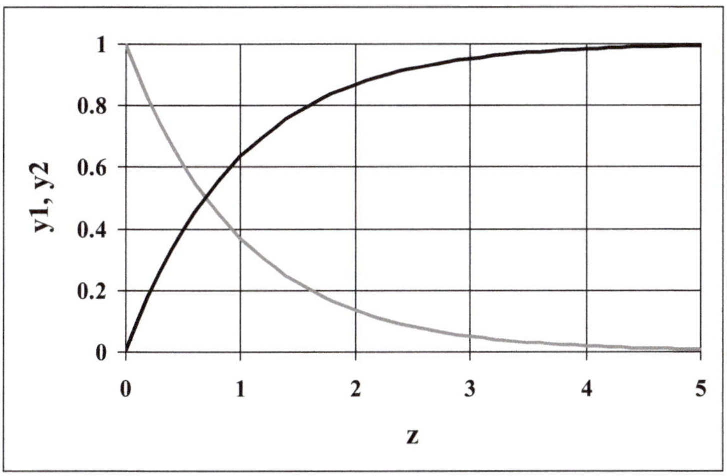

Fig. 2.15b: Solutions to Eq. (2.10a) when *U(t)* is the step function. $y_1 = x(t)/C_1$ and $y_2 = - x(t)/C_1$ as functions of $z = t/\tau$.

Table 2.3: Some points of the y_1 and y_2 curves (Fig. 2.15b)

z	y_1	y_2
0.5	0.607	0.393
1	0.368	0.632
2	0.135	0.865
3	0.050	0.950
4	0.018	0.982
5	0.007	0.993

2.4.6 Rise Time and Fall Time

Before showing how to relate frequency and time indicators of dynamic response quality, another tool will be introduced to evaluate the time response of a system: the **rise time**, which for special cases can be related to the **time constant**. When an input step function is applied to a system, the **rise time** (*rt*) refers to the time required for the output to change from low to high specified values; typically, these values are 10% and 90% of the step height (*H*). Conversely, the **fall time** is the time required for the output to change from 90 to 10 %.

$$rt = t\,(h_2) - t\,(h_1) = t\,(0.9H) - t\,(0.1H)$$

It is simple to establish a relation between the rise time and the time constant for any first-order LTI system. It was found (Millman & Halkias, 1967) that for this case the $h_1 = 0.1$ H is reached after $t = 0.105\,\tau$ and the $h_2 = 0.9$ H after $t = 2.302\,\tau$; thus,

$$rt = (2.302 - 0.105)\,\tau \approx 2.2\,\tau \tag{2.13}$$

2.4.7 Time Constant and Bandwidth Relation

This relation is well known in electronic circuits and can be found in the literature (Millman & Halkias, 1967; http://en.wikipedia.org/wiki/Time_constant), and then it will not be developed here. It can be demonstrated (Millman & Halkias, 1967) that the upper frequency of a first-order LTI system bandwidth is related to τ by

$$f_2 = \frac{1}{2\pi\tau} \tag{2.14}$$

A very interesting conclusion can be drawn from this equation: by applying a step input to a first-order LTI system, its time constant can be estimated and the upper frequency of the **bandwidth** can be determined by Eq. (2.14). Thus, for a first-order

LTI system the time and frequency response can be linked. It means that measuring the time response, some information on the frequency response can be obtained.

Figure 2.16 illustrates the time response of a temperature sensor. It was immersed in a bucket with soil at T = 20 °C until the thermal equilibrium was reached; then the soil temperature was suddenly changed (as a step-like shape input signal) to 46°C and the temperature change registered until the final temperature was reached. Because the time constant is the elapsed time since the change in temperature was initiated, until the 63.2 % of the total change was reached (Tab. 2.3), the temperature value to calculate the time constant is $T(\tau)$ = 20 °C + 0.632 × (46 – 20) °C = 36.432 °C, which correspond to t_1 = 14 s. Since the temperature change was initiated at t_0 = 4 s, the time constant is $\tau = t_1 - t_0$ = 10 s.

Once the time constant is estimated, the upper frequency of the bandwidth can be calculated from Eq. (2.14), the result being f_2 = 0.016 Hz. When using this sensor in a measuring application, Table 2.2 can be used to know the attenuation and phase change for a given frequency of the input temperature.

For example, applying the rule of thumb which says that to have low attenuation, the upper frequency of the system bandwidth should be one decade up the maximum frequency of the signal input, a phenomenon with a maximum frequency of 0.0016 Hz (period T = 625 s), measured with this temperature sensor, will result attenuated less than 0.5 %. In other words, sinusoidal temperature changes of about 10 minutes of period would be attenuated less than 0.5 %, but sinusoidal changes of about 1 minute would be attenuated about 30% (because 1 minute is approximately the period of the upper frequency f_2 = 0.016 Hz).

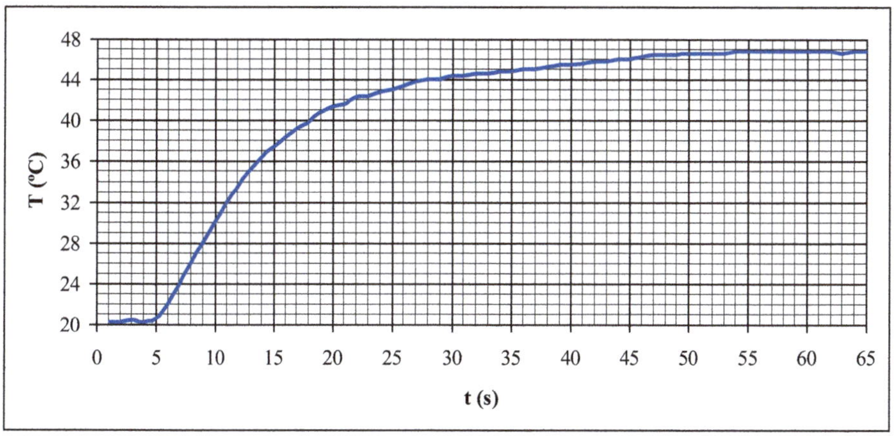

Fig. 2.16: Time response of a temperature sensor when a temperature step of 26 °C is applied to it. The delay from the beginning of the step until the temperature of the sensor reaches 63.2 % of the step amplitude is the time constant of the sensor.

2.4.8 Rise Time and Bandwidth Relation

Multiplying Eq. (2.13) by Eq. (2.14) we get,

$$rt\,f_2 = \frac{2.2\,\tau}{2\pi\tau} = 0.35 \tag{2.15}$$

The resulting product is a dimensionless constant that is often used to evaluate the bandwidth of a first-order LTI system by measuring the rise time.

2.4.9 Measuring the Rise Time of a Phenomenon by Means of an Instrument

The rise time of a phenomenon (*d*) that is to be measured by an instrument will be overestimated due to the own instrument rise time (*i*) according to Eq. (2.16), where *m* is the measured rise time (Walter, 2004).

$$m = \sqrt{(i)^2 + (d)^2} \tag{2.16}$$

A practical rule is to use an instrument whose rise time is 1/3 to 1/5 the rise time of the measured signal. In these cases errors introduced by the instrument are 5.5 and 2 % respectively. Therefore, if a steep slope of the phenomenon should be measured, it should be very important to verify that the rise time of the instrument meets Eq. (2.16).

2.4.10 Summary

Manufacturers inform the time and frequency characteristics of systems, instruments and sensors in diverse ways. When the information is reported in the frequency domain, the **Frequency response or Bode plots** permit the input-output relationships for any kind of systems to be described. **Bandwidth, upper and lower frequency and gain** allow the manufacturers to inform the dynamic properties of LTI systems.

When the information has to be reported in the time domain, the **rise and fall time** may be used for describing any type of systems. When the system is a first-order LTI one it is possible to compare the **rise and fall time** with the increasing and decreasing **time constants** of the system. Also, for these cases, **rise time** and **time constant** may be related to **bandwidth.**

Users should be aware of all these definitions to be able to understand the potential and limitations of the instruments they are using. They should confront the expected dynamic behavior of the measurand to the dynamic response of the systems to assure that the instrument does not significantly modify the parameter under study.

2.4.11 Examples to Help Fix Previous Concepts

Some examples extracted from real cases and somewhat modified will be presented to clarify the previously developed concepts These examples may be considered somewhat unrealistic but they were chosen in order to stress how in some cases the dynamic responses of the instruments could distort the measurand.

Example 1 - Figures 2.12a and 2.12b show the gain and phase (**frequency response**) as a function of **f** for a sea wavemeter whose sensor is an accelerometer; these figures are referred to as the **Bode plot** of the instrument response.

It can be draw from Figure 2.12a that this particular device has a maximum transference at $f \approx 0.8$ Hz, being G (0.8 Hz) ≈ 1.8 and φ (0.8 Hz) ≈ 9°. This means that a measurand (sea wave) of such frequency (0.8 Hz) will be measured by the instrument 80 % higher and with a phase lag of 9°. Wave energy is directly related to the square of the wave amplitude. Thus, because $A^2 = 1.8^2 = 3.24$, energy of this frequency component would be estimated with a considerable error.

Wave amplitudes whose frequencies are between 0.1 and 0.4 Hz are not modified, but phases are; for example φ (0.1 Hz) ≈ 30°. This behavior implies that if the study of the ocean is devoted to obtain wave energy between 0.1 and 0.4 Hz, because energy is related to the amplitude, results will be adequate. But if we are interested in the shape of the wave it will be distorted compared to the input shape due to the phase shifts.

For example, using the instrument between 0.1 and 0.4 Hz, where amplitudes are not modified, an artificial wave was synthesized by adding four sine waves of frequencies 0.1, 0.2, 0.3 and 0.4 Hz of the same amplitude and zero phase shift (Fig. 2.17, Input); the same signals were shifted in phase 30, 15, 10 and 9° respectively, which is approximately the phase lag indicated in the Phase Bode plot, and summed up (Fig. 2.17, Output). Clearly the shape has been drastically modified just due to the phase shifts. Perhaps in sea wave studies it could not be of importance to know the shape of the waves, but for other kinds of instruments it could be of great concern.

Suppose that two instruments from different manufacturers are used to analyze the traveling time of a wave between two points, one without phase shifts and other with a phase distortion similar to that of Figures 2.12a and 2.12b. Let us assume that such time is calculated by means of a correlation between the waves measured at both points. Since correlations are based on shape information, the calculated time could have very important errors because the instruments delay frequency input in different ways.

Finally, below 0.1 Hz the signal of the wavemeter results attenuated and the energy would be underestimated.

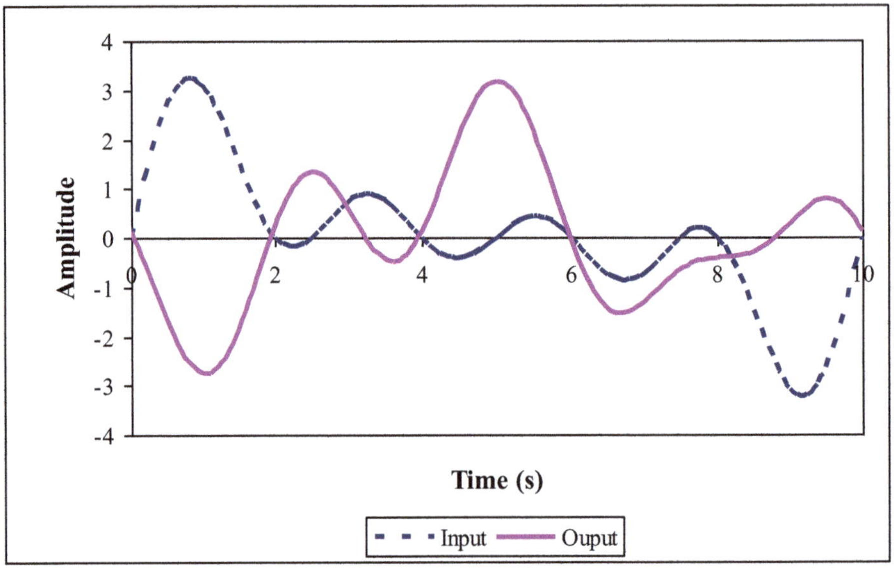

Fig. 2.17: An artificial wave was synthesized by adding four sine waves of frequencies 0.1, 0.2, 0.3 and 0.4 Hz and applied to the wavemeter dynamic transfer function of Figures 2.12a and 2.12b. Output shows how the shape of the wave is distorted due to the phase shift introduced by the meter.

Example 2 - What is the price paid when frequency components of the measurand equals the **bandwidth** of the instrument?

It has already been stated that some manufacturers specify the frequency response of instruments by means of their bandwidths. It was shown that measurand frequencies close to the upper and lower frequency of the transference will appear attenuated and shifted at the output.

Usually when users are buying an instrument they only verify that the bandwidth of the phenomenon being measured matches the bandwidth of the instrument. Let us see what happens with a signal whose bandwidth matches the instrument bandwidth. Suppose the instrument has f_1 =10 Hz, f_2 = 1000 Hz and the input signal (measurand signal) is the sum of three sine waves whose frequencies are: 10 Hz, 100 Hz and 1000 Hz. Figure 2.18 shows the input and output signals as function of time. Again, it is shown how the instrument could distort the shape of the measurand even if the bandwidth of the instrument matches the bandwidth of the phenomenon being measured. Perhaps for many applications this distortion is not a trouble, but for others it could be inconvenient. Users should be aware on how instrument characteristics are reflected in the recorded data.

Example 3 - This example illustrate how to evaluate whether an instrument of unknown or suspected frequency response can be used for a specific study. Suppose that a laboratory has a submergible pressure sensor with unknown specifications and it is desired to know whether it could be used to measure waves with periods between 1 and 20 s.

The first step is to estimate the sensor's frequency response. It can be easily done if the sensor could be treated as a first-order LTI system.

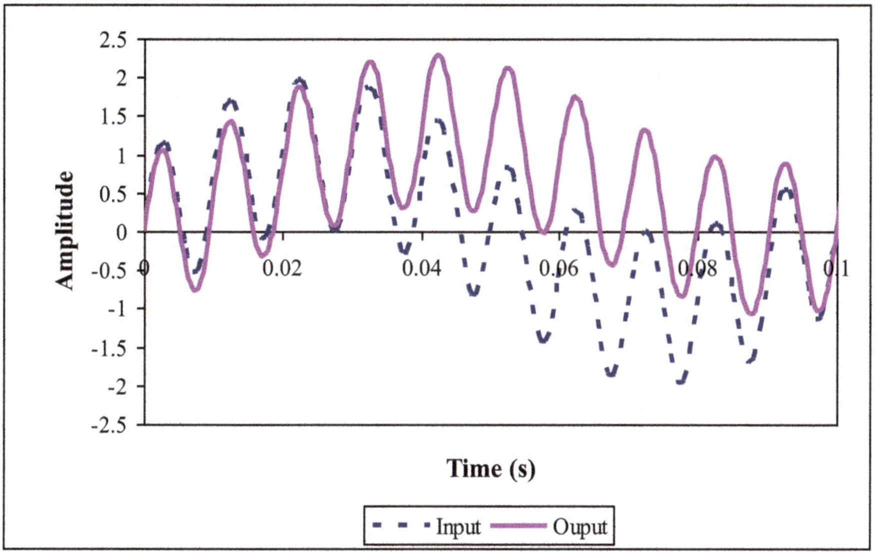

Fig. 2.18: An instrument has lower frequency f_1 =10 Hz and upper frequency f_2 = 1000 Hz, and the measurand signal is the sum of three sine waves whose frequencies are: 10 Hz, 100 Hz and 1000 Hz. Output signal is different than the input due to attenuation and phase shift introduced by the instrument.

In general, pressure sensors used in environmental sciences are able to measure the still water level; in other words, the lower frequency of the bandwidth is f_1 = 0. Then, if correctly installed, the sensor will measure long waves without attenuation.

In contrast, pressure sensors do have an upper -3 dB gain frequency and the higher frequency content of the measurand may result attenuated. Although in this case f_2 is unknown, some test can be done to estimate it. We have pointed out that the time constant gives information about the bandwidth of the instrument. Therefore a test to measure the time constant should be performed. For this purpose a pressure step input should be applied to the sensor.

In order to apply a step of pressure to the sensor it can be placed inside a pressurized balloon. Then the electrical output of the pressure sensor has to be connected to a data logger. After starting data recording the balloon has to be popped. Data have to be continuously recorded until the new pressure output becomes constant. For many practical purposes this sudden change in pressure behaves like a negative step that allows the time constant to be measured. Proceeding as was done with Figure 2.16 the time constant of this hypothetical test was found to be $\tau = 10$ ms, then

$$f_2 = \frac{1}{2\pi\tau} = \frac{1000}{20\pi} Hz \approx 16 Hz$$

Following the rule of thumb that states that for the input signal to pass the system with only 0.5% attenuation, the maximum frequency of the signal input should be one decade below the upper frequency of the bandwidth, this sensor could be used to measure from the still water level up to 1.6 Hz ($T = 0.625$ s). Therefore, it has been confirmed that this sensor can be used to measure waves with periods between 1 and 20 s if 0.5% of attenuation in the higher frequencies is accepted.

A more realistic and complete experiment to obtain time and frequency characteristics of pressure sensors are presented in Section (11.3). The ideas behind this experiment can be extended to other kind of sensors.

Example 4 - This example will show how far an instrument can follow a rapid change in the measurand without appreciably distorting the data. In this case the measurand is the wind speed and the instrument is an anemometer. It is desired to know how the time response of the instrument affects the measurement of fast changes in wind speed.

In general, mechanical anemometers are not first-order LTI systems because they do not respond in the same way to increasing and decreasing wind speeds. But in order to approximately know the response of an anemometer an increasing step of wind was applied to it and its output recorded. From the time data series the time required to reach 63.2 % of the final value was estimated and hence the time constant was established as $\tau = 2.5$ s.

Question: What is the minimum wind rise time (steep slope of the wind speed) that this instrument could measure if an error of 10 % is accepted?

Suppose that this anemometer behaves as a first-order LTI system, then following the naming of Eq. (2.16), the instrument rise time is $i \approx 2.2\,\tau = 2.2 \times 2.5$ s $= 5.5$ s. Since the maximum value admitted for the error of the measured rise time (m) is only 10 % larger than the value of the unknown rise time to be measured (d), we have $m \leq 1.1\, d$.

The minimum d for this condition may be calculated from Eq. (2.16),

$(m)^2 = (i)^2 + (d)^2$ and because $m \leq 1.1\, d$; $(1.21 - 1)d^2 = i^2$;

$$\text{then } d \leq \sqrt{\frac{i^2}{0.21}} = 12\,s \tag{2.17}$$

Therefore, if the error in the slope of the wind speed must be smaller than or equal to 10 %, the minimum wind rise time that can be measured with this anemometer is 12 s. This means that for a change in wind speed that jumps from 10 to 90 % in less than 12 s, the instrument will overestimate the wind rise time in more than 10%.

For example, if $d = 8$ s, from Eq. (2.16) $m = 9.7$ s. It means that a wind rise time of 8 s will be measured as 9.7 s. The percentage of error can then be calculated as

$$e\,(\%) = \frac{(9.7-8)}{8} \times 100 \approx 21\%$$

2.5 Filters

Some general concepts on filters and filtering will be introduced in this chapter because as it was mentioned in the topic dedicated to time and frequency characteristics of sensors and systems, all stages of instruments are to some extent filters.

Electronic filters are networks specifically designed to process signals in a frequency-dependent manner (http://www.analog.com). Figure 2.19 shows the gain as a function of frequency for four ideal types of filters. The frequency response of a filter defines the filter properties. A **high pass filter** (HPF) is a filter that prevents passing low frequency inputs to the output. A **low pass filter** (LPF) allows passing low frequencies, but beyond some frequency the output decreases to zero.

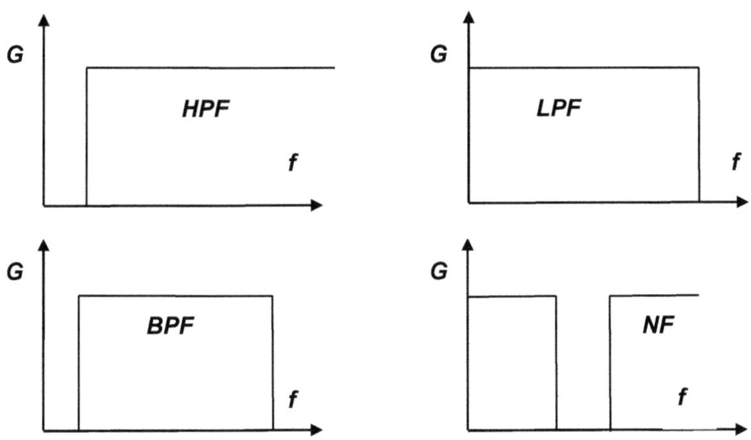

Fig. 2.19: HPF (high pass filter), LPF (low pass filter), BPF (band pass filter), NF (notch filter).

Connecting in series a low pass filter and a high pass filter a **band pass filter** (BPF) is obtained. BPF filter out low and high frequencies, keeping the central frequencies. A filter with the opposite characteristics than the BPF is known as a **notch filter** (NF). This last kind of filter is used to reject a particular range of frequencies. In the case of noise due to the AC power supply (for example $f = 50$ Hz) the notch is centered at this frequency.

Sensor manufacturing has technological limitations that make them inherently frequency selective and therefore acting similarly to filters. All devices produced by humans have limited time responses, or in other words, they have always a maximum frequency from which the sensor output begins to decrease until it becomes zero. This gives to sensors and then instruments an intrinsic low pass filter characteristic.

Because sensors are the first link of several elements in series that comprise the overall measuring instrument, then the sensor characteristics strongly influence the characteristic of the entire instrument. In general, all the information lost or distorted by the sensor cannot be recovered in the following steps. Digital signal processing helps to rescue signal buried in noise, but cannot substitute the quality of a sensor or save their weakness. Signal processing efforts should be made starting from a good sensor, thus the same effort will produce better results. The knowledge of the frequency response of the sensor allows knowledge of the "window" through which the sensor perceives the phenomenon being measured. Therefore, it is very important to select the adequate sensor for the desired measurand. The sensor should be selected to have a frequency response that allows passing the signals of interest without attenuation or distortion. A linear sensor that does not distort the signal must have constant gain and a linear phase change as a function of frequency, in all the measuring range of interest.

If the phenomenon to be studied is completely unknown, the frequency characteristics of the measurand could not be specified and this lack of information could lead to an inappropriate sensor or instrument. In this case, users should be aware that due to filtering the acquired information may be a limited version of the desired signal.

2.5.1 Noise Reduction by Filtering

As explained above, filtering is the property of a device to select some frequencies and reject others. It was mentioned that it is undesirable that a sensor filters out the input signal, but it would be desirable that some kind of filtering were used to reduce noise. Noise filtering is the process of removing undesirable noisy frequencies from signals. Filtering is a selective process that permits only useful information to be stored in, or transmitted by, the instrument, thus reducing the need for memory space or communication bandwidth.

The procedure of reducing the noise frequency components of a signal requires knowledge of the frequency characteristics of the wanted signals and the undesirable noise. By selectively rejecting some frequency bands, it is possible to extract the wanted signal from the noise, improving the signal to noise ratio. When the frequency content of the signal and noise are very different filtering is usually a simple solution. On the other hand, if both frequency of noise and input signal match, filtering can be extremely difficult and part of the input signal may be lost.

There are physical and mathematical filters; the first are analog circuits and are in general used in the first stages of an instrument or system. The second are applied after the signal has been converted from analog to digital. Most instruments incorporate microprocessors and **digital signal processing** (DSP) for a very quick mathematical manipulation of readings, thus making it possible that digital filtering of the signal be carried out on board the instrument.

Nowadays, some instruments allow users to decide how much filtering to perform before storing data. Some modern instruments have the capability of allowing users to select different input signal bandwidth, that is, permit tailoring the filter to their needs. In these cases, it is useful to identify the maximum frequency signal needed and keep the measuring bandwidth as low as possible. In this way high frequency noise is impeded from entering the system.

On the other hand, filtering modifies the data permanently, preventing a later signal processing on the original data. Designers have always a compromise on how much on board filtering will be applied, and users should make sure that filtering is not clipping the information they want.

When signals are low frequency and have random noise, averaging a large number of readings could produce good estimate of the true value. Present instruments use this technique to improve the data quality. Also, they usually calculate the standard deviation, which gives information on the amount of noise that accompanies the signal. The larger the standard deviation, the noisier is the signal. Some instruments verify that the standard deviation is below some specified level. If not, the data is discarded.

Several instruments record the signal to noise ratio and present this data to the users. This information can be employed to validate the data or not. In general, the manufacturers know and provide to the users the level of the minimum signal to noise ratio required to consider a data input as reliable.

A simple example of digital filtering follows: assume that air temperature data is needed for agricultural purposes and a sample of the soil temperature every ten minutes is enough. Then, if a data logger with a maximum sampling rate of one sample per second is available, a good solution to decrease random noise could be taking one sample every second and averaging 600 samples. This would give a good estimate of the soil temperature measured in the last 10 minutes. This averaged result will be then recorded by the data logger.

If the standard deviation (STD) is also recorded, and it was found in normal functioning conditions (for example during laboratory tests or tests under controlled conditions) that STD = 0.5 °C. This value could be used to validate field data, for example, disregarding data whose STD > 1 °C.

Let us see a different example in order to show how the dynamic of the measurand and the instrument performance condition the filtering process. If it is desired to study how a fan refrigerates a small metal piece, perhaps at least one temperature measure every one second will be required. Then, if the same data logger were used (with maximum sampling rate of one sample per second), the averaging procedure to reduce noise described above could not be used. Therefore, the same data logger would produce more accurate readings for agricultural purpose than for metallurgical applications.

From these examples we can draw a general conclusion: it is possible to reduce errors by averaging noise **if the dynamics of the phenomenon is sufficiently slow with respect to the data acquisition rate.**

2.5.2 Filter Delay

In the previous example random noise was reduced by averaging 600 samples over 10 minutes; at the end of that period one value representing all these samples is recorded. Thus, averaging introduced a lag of 10 minutes to the data being measured. In the agricultural application, this delay in knowing the data is irrelevant, but could not be the case in other applications.

Assume a tide gauge placed at the entrance of a harbor which is used to judge the request of ships to enter or leave the port. As it happens in shallow water ports small changes in the tide may be important in allowing or rejecting the ship request. Then, to minimize the influence of waves on the sea surface measurement it could be desired to filter the data by averaging several samples of the sea surface over a period of time. The longer the period the lesser the wave influence, but the longer the delay in having the information.

Let us assume that the tide gauge averages during 20 minutes, then, the averaged data obtained is not the tide at the moment the data was calculated but the average over the previous 20 minutes. But, in the last 20 minutes the tide changed, and depending on the tidal cycle and weather conditions, it could change an amount not acceptable for the application the data is intended for.

Summarizing, filtering that allows knowing better the mean values of a phenomenon by averaging random noise or undesirable high frequencies, may introduce an unacceptable lag on the needed information.

2.5.3 Spatial Filtering

An interesting signal processing method that can be applied to some measurement cases is the spatial filtering of the signal. The concept of spatial filtering is the same discussed in Section (2.3.3). In this case spatial filtering is a wanted fact in opposition to the unwanted filtering shown before.

In order to explain this idea, the previous case of a tide gauge in a harbor is taken again. Let us assume that instead of having one tide gauge in the entrance of the port; a number of level meters are installed, distributed in a certain way (the allocation has to be studied according to the port shape, bathymetry, etc). All level meters are connected to a computer that can process the information collected by them.

Suppose that these level meters allow a Low Pass Filter of one minute to be applied to the level signal (it could be done for example by sampling ten times by second and averaging 600 samples), thus these level meters will introduce a delay of only one minute, but the output level measured by each instrument will vary ostensibly due to long waves.

If the spatial distribution of the tide gauges is such that, at the same instant, different gauges measure the sea surface level at different phases of the waves, then, at the same time, some meters will report the wave crest, other the wave trough, while others will report the mean value of the sea surface. Then the average of the tide gauge outputs will tend to compensate the surface changes introduced by waves. This average of several levels measured in a certain area can be called a spatial filtering of the tide; it has a great advantage with respect to the time average presented previously for the harbor application. Because the average of several measures performed by the tide gauges is calculated in a short time, then in our example the tide is known with only a one minute delay. This is an interesting fact in such cases where a lag is not admitted and the phenomenon has a spatial distribution that allows the allocation of several sensors which measure the same measurand with different phases (Cavalieri & Curiotto, 1979).

2.6 Summary

The front ends of instruments are usually known as sensors and they are devices that take information from a physical or chemical phenomenon and create or modify an electrical signal. It can be found that they are named in diverse ways such as transducer, sensor, active transducer, passive transducer and detector. Also, devices that are actuated by a form of power and supply another one are called transducer, active transducer and actuator. Most of the time, all these devices will be mentioned in the future chapters as sensor or transducer.

Sensors and transducers are characterized by their transfer function (or transference) which is a curve or equation representing the relationship between the input and the output. A transfer function may be obtained with constant inputs (static transference) or varying inputs (dynamic transference).

Several parameters are used by manufacturers for specifying the static transfer such as: range, dynamic range, hysteresis, calibration curve, linearity, offset and gain errors, drift, etc.

It is a little harder to specify the dynamic transference of a sensor or instrument and for representing them the frequency response, expressed as a diagram of amplitude relations and phase relations, also known as Bode plot is used. This is a very detailed way of expressing a transference that can be used with any kind of device. The drawback in obtaining it is that devices have to be excited with each frequency in the complete frequency range of interest and their outputs related in amplitude and phase to the respective inputs.

A simplified way to characterize the dynamic transference of first order linear time invariant (LTI) systems is to specify their bandwidth. In these cases the frequency response in amplitude and phase are linked by known equations, thus, by knowing the upper and lower frequencies the transference can be obtained. As a rule of thumb, when instruments can be treated as LTI systems and the upper frequency of the system is one decade up the maximum frequency of the signal input and the lower frequency is one decade down the minimum frequency of the input, the amplitude attenuation at the output is less than 0.5 % and the phase shift less than 0.6 °.

In order to characterize the dynamic behavior of sensors and instruments in the time domain, some parameters like time constant, rise time and fall time have been defined. For LTI systems it is possible to easily relate the time constant and bandwidth, thus, by means of a test in the time domain which provides the time constant, some frequency characteristics can be inferred.

Sensors and instruments are natural filters of the signal we want to measure because they discriminate how the different frequencies comprising the input signal are allowed to pass to the output. The instrument designer may also build some filters to let some frequencies pass or be blocked with the purpose of reducing noise. They are known as high pass, low pass, band pass and notch filters.

References

Carr J. J., & Brown, J. M. (1998). *Introduction to Biomedical Equipment Technology*. Upper Saddle River, NJ: Prentice Hall.

Cavalieri L., & Curiotto, S. (1979). A fast-response shallow-water tide gauge. *Il Nuovo Cimento*, serie *1*, vol *2-C*, 273-287. Datawell (1980). http://rpsmetocean.com/products_services/pdfs/oceanographic/Waverider%20f.pdf

Millman, J., & Taub, H. (1965). *Pulse, Digital and Switching Waveforms*. New York: McGraw-Hill Book Company Inc.

Millman, J., & Halkias, C. C. (1967). *Electronic Devices and Circuits*. Tokyo (Japan): McGraw-Hill Book Company Inc.

Walter, P. L. (2004). *Shock and Blast Measurement - Rise Time Capability of Measurement Systems*. Engineering Faculty, Texas Christian University, Fort Worth TX (USA) – PCB Piezotronics, Inc., Depew, NY (USA).

http://www.merriam-webster.com/dictionary

http://digital.ni.com/public.nsf/allkb/084702CE98679BB886256CA3006752D7

http://www.sensorland.com/HowPage026.html

http://en.wikipedia.org/wiki/Time_constant

http://www.analog.com/library/analogdialogue/archives/43-09/EDCh%208%20filter.pdf

3 Review of Concepts

3.1 Introduction

In some countries people attending post graduate studies in environmental sciences have different backgrounds. Some of them come from harder sciences than others. Then it is difficult to determine what level of knowledge should be taken for granted that the students already have. This chapter attempts to save possible background differences, making available to all readers the tools that will be needed to understand how sensors and instruments perform.

For some students this chapter will be useful only as a remainder of subjects already studied, but it could be a necessary introduction for others. The material presented here can be found in many different books, but we considered that this information is somewhat disperse and could be of some help to have it all together in the same source, avoiding a continuous pilgrimage by diverse textbooks.

In this chapter several subjects with little connection between them are introduced with the only objective that they will be useful to understand the material developed in the following chapters. Each subject is directly linked to one or more subjects presented at later stages.

The analysis of each kind of instrument developed in the next chapters makes reference to the previous material needed to understand it. Rather than reading this entire chapter, readers interested only in some particular instruments will find it more convenient to go straight to the pages where the instrument is described and see what previous material is needed and, if they are not familiar with it, then reading just what is necessary to satisfy their requirements.

3.2 Waves

3.2.1 Introduction

Because many instruments base their working principles on the propagation of sound and electromagnetic waves, it is convenient to remember briefly some important properties of waves. This review begins with sound waves because they are easy to understand for those not yet introduced in wave theory. The subject is presented in a descriptive way without using complicated equations in order that it can be understood by those readers with little background knowledge on waves.

Among the most applied phenomena of wave propagation used in instruments devoted to the measurement of environmental and hydraulic parameters is the **Doppler effect.** For this reason some effort is dedicated to explain it. This effect is also known as the **Fizeau effect** in the theory of electromagnetic waves and used in radars and the Global Positioning System (GPS).

Once concepts are made clear for sound waves they are directly extended to electromagnetic waves, and the most remarkable differences between sound and electromagnetic waves are presented.

We believe that to grasp the material presented in the following chapters a deep understanding of electromagnetic wave theory is not required and it is optional to the readers. Electromagnetic wave theory is a subject so vast that any attempt to introduce this topic within the limits of a book dedicated to instrumentation keeping certain academic rigor, would certainly be unsuccessful. Many good books with a deeper insight into this matter are available for those readers wishing to review them (Skilling, 1974).

3.2.2 What is a Wave?

A wave may be defined as a disturbance that carries energy across or through a material at a given velocity (c) and without any significant overall transport of the material itself. Whereas mechanical waves require a medium in which to propagate, electromagnetic waves can propagate through vacuum. We must distinguish between transverse and longitudinal waves. In transverse waves the displacement of the medium is perpendicular to the direction in which the wave travels. A vibrating string is an example of a transverse wave because all points on the string move up and down but the wave (disturbance) moves along the length of the string. In longitudinal waves the disturbance takes place parallel to the wave propagation. As will be seen below, sound waves are longitudinal waves. Water waves, however, are neither completely transverse nor completely longitudinal, but a combination of both. A water particle moves in an almost closed circular path, the orbital motion having both transverse and longitudinal components. We have used the adverb 'almost' because real water waves do exhibit a small net component of forward motion in the direction of wave travel. This net forward motion is called **wave drift**.

3.2.3 Wave Properties

The **wavelength** is determined by considering the distance between consecutive corresponding points of the same phase, such as maximum displacements or zero crossings. In other words, it is the distance over which the wave's shape repeats. The **wavelength** (λ) and **amplitude** (A) of a sinusoidal wave are shown in Figure 3.1.

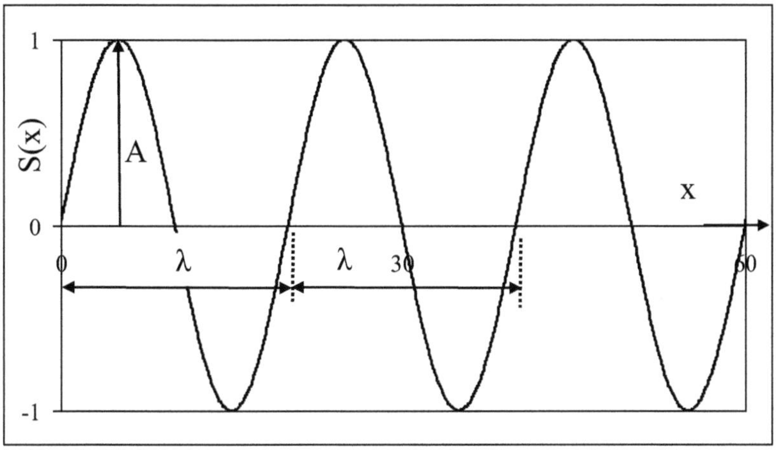

Fig. 3.1: A sinusoidal wave as a function of *x*. The amplitude (*A*) and the wavelength (*λ*) are shown.

3.3 Harmonic Sound Waves

These waves are generated by a mechanical source vibrating with a simple harmonic motion, like a tuning-fork. The vibrating focus makes nearby air molecules oscillate about their equilibrium positions.

These molecules collide with their neighbors, spreading the sound wave. The displacement of the molecules $s(x, t)$ as function of space x and time t is given by

$$s(x,t) = s_0 \sin(kx - \omega t); \quad \omega = \frac{2\pi}{T} = 2\pi f; \quad k = \frac{2\pi}{\lambda} \tag{3.1}$$

where s_0 is the maximum displacement of a molecule about its equilibrium position (amplitude); l is the wavelength (distance between two successive peaks); f is the number of peaks that pass through a point per unit time, called the frequency; T is the period ($T = 1/f$) and k is the wave number.

For a sinusoidal wave there is a relationship between frequency (f), wavelength (λ) and velocity (c),

$$c = \lambda f \tag{3.2}$$

Because of the motion of particles, there are regions in the medium where the particles are compressed together and other regions where they are spread apart. These regions are known as **compressions** and **rarefactions**, respectively. The compressions are regions of high pressure whereas the rarefactions are regions of low pressure. Sound waves produce fluctuations of pressure above and below the atmospheric pressure. The pressure wave has the same waveform as the displacement one, but is delayed by $\pi/2$. Sound waves are longitudinal waves because they transfer energy in the same direction as the pressure and displacement disturbances.

Particle compressions and rarefactions are shown at the top of Figure 3.2, whereas their associated displacement and pressure waveforms are illustrated at the bottom.

The speed of sound (c) is the speed at which the wave propagates. It depends on the properties of the medium, e.g. for air $c_a = 343$ m/s and for water $c_w = 1500$ m/s. Generally speaking, the human hearing frequency range is from 20 Hz to 20 kHz. Below the lower limit the phenomenon is known as infrasound, and over the upper limit it is called ultrasound. Ultrasound is used in many instruments, from biomedical applications to plastic welding.

So far we have described one-dimensional waves, i.e. waves that propagate in a straight line. Figure 3.3 shows two-dimensional waves on the surface of a water body. The waves were generated by a drop impacting on the surface. The concentric circles are called wave fronts. Most sound sources produce three-dimensional waves, their wave fronts being concentric spherical surfaces.

When a point source radiates energy uniformly in all directions, the energy at a distance r from the source is distributed uniformly over a spherical shell of radius r and surface $S = 4p\, r^2$. If the power emitted by the source is P, the power per unit area at a distance r (P_r) decreases with the square of the distance to the source,

$$P_r = \frac{P}{4\pi r^2} \tag{3.3}$$

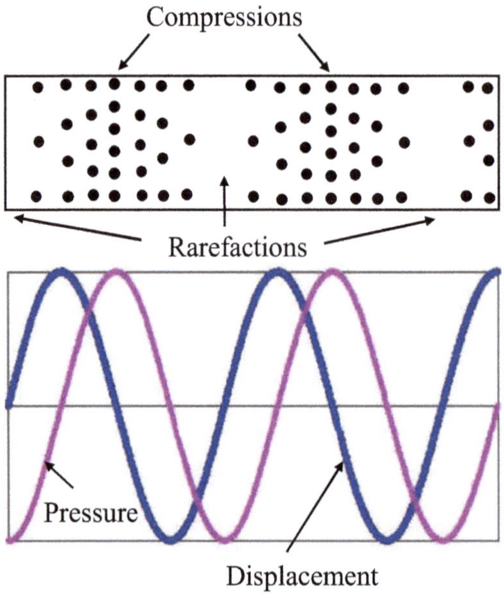

Fig. 3.2: (Top): Compression and rarefaction zones for a wave traveling through the air. (Bottom): The associated displacement and pressure waveforms.

Power received at a distance *d* from the source is four times smaller than that received at a distance *d*/2. This is one of the reasons why signals from more distant sources are notably more affected by noise.

Fig. 3.3: Two-dimensional waves propagating on the surface of a water body.

3.3.1 The Doppler Effect

The Austrian physicist Christian Andreas Doppler (1803-1853) described the effect that bears his name around 1842. The best known example of this effect is the frequency difference of a moving train horn as perceived by an observer standing alongside the train track. When the train is approaching the observer he/she perceives a sharper tone than when the train moves away.

In Figure 3.4 a walker moves towards a fixed sound source. The frequency that he perceives is higher than that when he remains stationary, because when moving towards the source he meets a greater number of wave fronts, in the same period of time.

Figure 3.5 shows the effect produced by a sound source moving towards a fixed observer. Wave fronts are closer to each other with respect to a fixed source due to the movement of the source. Then the observer perceives that the time between arrivals of successive wave fronts is less than when the source is fixed. In other words, the observer perceives a higher frequency.

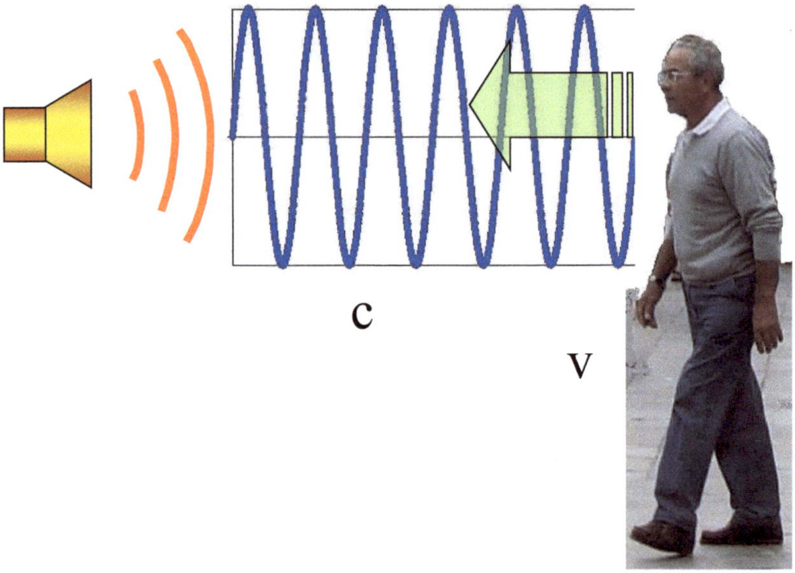

Fig. 3.4: The Doppler effect. The observer walks towards a fixed sound source and perceives a higher frequency.

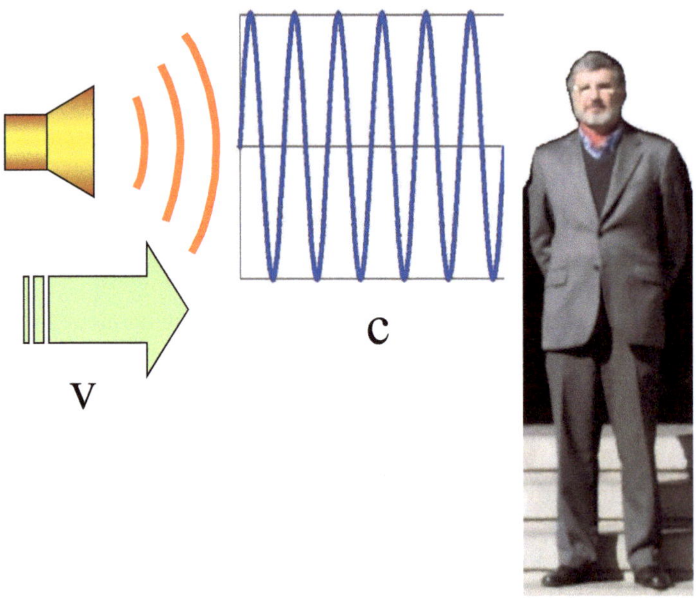

Fig. 3.5: The Doppler effect. A moving sound source moves towards a fixed observer, who perceives a higher frequency

The frequency f_0 of the sound source is fixed and it is a characteristic of the source; f_0 is the number of oscillations per second of a mechanical device that makes air particles vibrate, as in the case of a tuning-fork.

When the observer moves at a speed v toward the sound source, the speed of the wave fronts from the observer's point of view is $v' = c + v$, then the apparent frequency f' perceived by the observer is:

$$f' = \frac{v'}{\lambda} = \frac{c+v}{\lambda} = \frac{c}{\lambda} + \frac{v}{\lambda} = f_0 + \frac{v}{\lambda} = f_0\left(1 + \frac{v}{\lambda f_0}\right) = f_0\left(1 + \frac{v}{c}\right) \tag{3.4}$$

On the other hand, if the observer moves away from the sound source with the same speed, the apparent frequency is:

$$f' = f_0\left(1 - \frac{v}{c}\right) \tag{3.5}$$

A similar analysis for a moving source and a stationary observer can be done. A general relationship between the observed (f) and the emitted (f_0) frequencies which considers the velocity of the receiver relative to the medium v_r and the velocity of the source relative to the medium v_s is given by

$$f = \left(\frac{c + v_r}{c + v_s}\right) f_0 \tag{3.6}$$

In Eq. (3.6) $\mathbf{v}_r$ is considered positive if the receiver is moving towards the source, and v_s positive if the source is moving away from the receiver.

Assume that a fixed sound source is emitting with frequency f_0 to clouds that are moving horizontally with velocity $\mathbf{v}^*$ (Fig. 3.6). The velocity component in the direction of wave propagation is $v = v^* \cos\theta$. When the sound wave impinges the clouds, the cloud's particles will vibrate with a higher frequency than that of the source because clouds are moving towards the source (as in the case of a moving observer).

The clouds reflecting the sound act as a moving sound source. Thus, clouds are simultaneously receiver and emitter. Therefore, the velocities of the receiver and the emitter (source) relative to the medium are the same, $\mathbf{v}_r = \mathbf{v}_s = \mathbf{v}$. They are positive in the numerator and negative in the denominator of Eq. (3.6),

$$f = \left(\frac{c + v}{c - v}\right) f_0 \tag{3.7}$$

The total frequency shift $\Delta f = f - f_0$ is computed taking into account that $c >> v$.

$$\Delta f = \frac{2\,v\,f_0}{c - v} \approx \frac{2\,v\,f_0}{c}; \quad v = \frac{\Delta f\,c}{2 f_0} \tag{3.8}$$

A moving object irradiated by a fixed emitting source returns a signal shifted in frequency. Equation (3.8) allows the velocity of the object to be known if the emitted frequency and shift are measured. This is the working principle of radars.

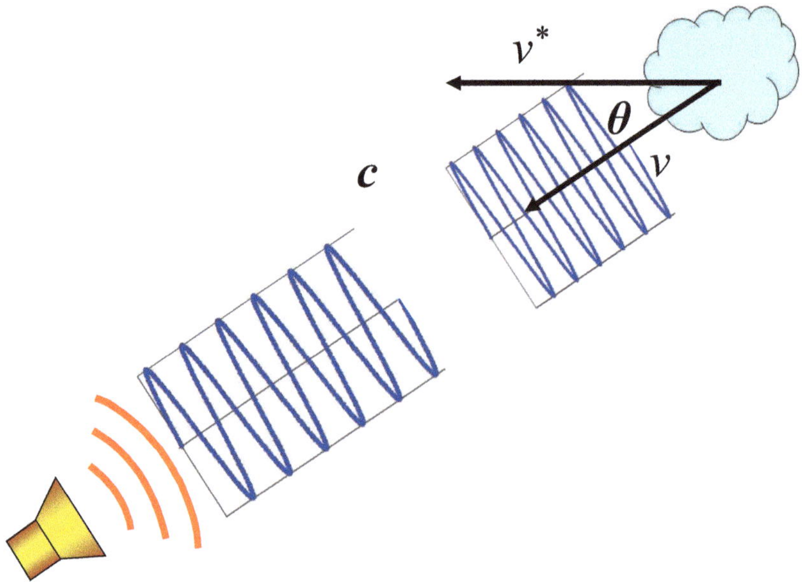

Fig. 3.6: A fixed sound source emits towards a cloud moving with velocity **v***. The cloud returns a frequency shifted signal.

3.4 Electromagnetic Waves

Only a few ideas on the theory of electromagnetic waves, necessary to understand the working principles of some measuring methods and instruments, will be introduced here. Then in this book the approach to this issue is merely conceptual. Those readers willing to follow a more rigorous and thorough introduction could easily find very good classical literature (Skilling, 1974).

Most of the concepts needed to explain how some instruments work have been introduced in the previous sections on harmonic sound waves. Therefore, only the most remarkable differences between sound waves and electromagnetic waves, needed to make clear the operation of the instruments, will be explained.

Waves in water and sound waves have been naturally introduced because everyone has experienced them, and they are easily perceived by the human being. Mechanical waves travel through a medium due to molecular interactions that transfer energy from one molecule to the next. They cannot travel in vacuum because there are not molecules to transmit energy. A big difference is that electromagnetic waves do not need a medium to propagate; they can travel through solids, air and also vacuum.

Electromagnetic waves are formed by the oscillation of electric and magnetic fields, and their propagation velocity is equal to the velocity of light (3×10^8 m/s). These fields are perpendicular to one another and both are perpendicular to the travel direction of the wave. This is the reason why electromagnetic waves are called transverse waves. Electric and magnetic fields are produced in a solid like a wire, a coil or an antenna and come off the solid and go free as electromagnetic waves.

The definitions of frequency and wavelength defined for mechanical waves are valid for electromagnetic waves. Equation (3.2) also applies if the speed at which the wave travels in a medium is replaced by the velocity of light.

The same phenomenon described previously as the Doppler effect was independently discovered on electromagnetic waves by Hippolyte Louis Fizeau (1819-1896), a few years after Doppler. With the aim of applying Eq. (3.8) to electromagnetic waves, the speed of sound has to be changed by the speed of light. Many instruments and weapons are based on this equation.

3.5 Useful Concepts on Wave Propagation

3.5.1 The Bragg Law

Many research methods in earth sciences are based upon the capture of backscattered signals that reach a receiver device in such a way that they can interfere constructively. The physical foundation of these methods is to be found in the well known *Bragg's scattering law*, so named after William Henry Bragg (1862-1942) and his son William Lawrence Bragg (1890-1971) who, in studying how X-rays are scattered by a crystal lattice, found the conditions to be met for the constructive interference of the scattered wavelets resulting from the incidence of an X-ray wave upon the atoms of the lattice. Two points are worth noting in the scattering of X rays by a crystal lattice, as well as in the scattering of electromagnetic and acoustic waves in general. First, the dimensions of the scattering sources are smaller than the wavelength of the incident wave and second, the distance between scattering centers is comparable to the incident wavelength. For a better understanding of Bragg's law we will first describe the scattering of X rays by a crystal.

A crystal consists of a regular array of atoms that can scatter any electromagnetic waves falling upon them. When a monochromatic X-ray wave front falls upon a crystal the atoms of the crystal lattice become the source of scattered wavelets, or secondary wave fronts. The scattering of these secondary wave fronts is generally multidirectional, and the interference between them, destructive, thus the backscatter signal is weak and useless. However, due to the regular arrangement of the atoms in the crystal lattice, there will be directions in which the scattered wavelets will interfere constructively and are strong enough to be amplified and analyzed.

Figure 3.7 shows a section of a crystal with three successive layers of atoms separated by a distance d (layer spacing). An X-ray beam strikes the crystal from the left making an angle θ with the crystal surface, and is scattered by the first two layers of atoms. The physicists Bragg proved that if two conditions are met, the scattered

X rays will interfere constructively. The first condition is that the angle between the emergent X-ray beam and the crystal surface must also be θ. As this is exactly the law of light reflection in geometrical optics, Bragg scattering is also called, incorrectly, Bragg reflection. The second condition refers to the X rays scattered by the lower layers of atoms. Consider the wavelets scattered by the first and second layers. The second condition says that for constructive interference to take place, the paths followed by both scattered wavelets must differ in an integral multiple of the X-ray wavelength. If this condition is fulfilled, the scattered rays are said to be in phase.

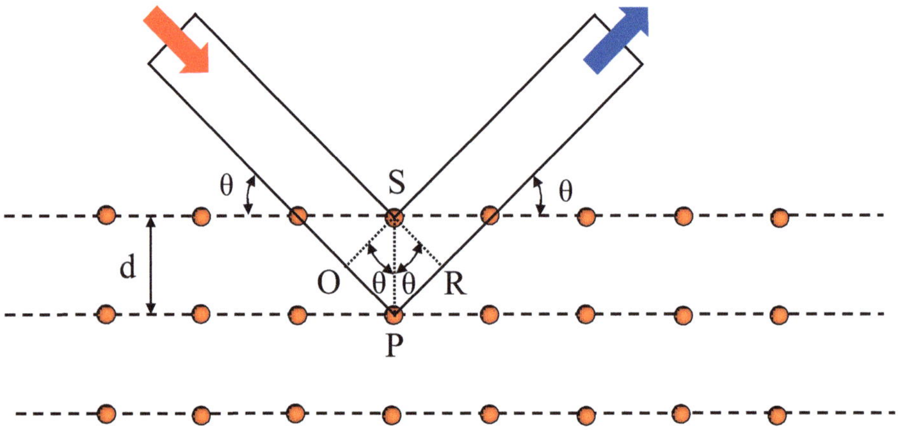

Fig. 3.7: Schematic of X-ray scattering by the atoms of a crystal lattice

Turning the attention again to Figure 3.7, it is observed that the ray scattered by the atoms in the second layer travels an additional distance farther than the ray scattered by the atoms in the surface layer. To determine this additional distance let us draw the lines SO and SR perpendicular to the directions of the incident and scattered rays, respectively. Each of these lines makes the same angle θ with the line SP = d as the incident and scattered rays with the crystal surface. It follows from Figure 3.7 that the ray scattered by the second layer travels an extra distance equal to OP + PR with respect to the ray scattered by the surface layer. As OP = PR = $d \sin\theta$, this extra distance is thus OP + PR = $2\,d \sin\theta$. Then the second Bragg condition states that the wavelets scattered by the first two layers of the crystal will interfere constructively if the additional distance $2\,d \sin\theta$ is an integral multiple n of the X-ray wavelength, i.e. if

$$2\,d\sin\theta = n\,\lambda \tag{3.9}$$

Although the above explanation refers to the scattering of an X-ray beam striking the surface of a crystal, it is also valid for those electromagnetic and acoustic waves captured by receiver devices after being backscattered by a given object. The backscattered signal carries information about the scattering objects. Thus, energy could be deliberately impinged upon an object in order to receive some information embedded in the backscattered portion of the impinging energy. The Bragg scattering law gives the conditions for the backscattered signals to arrive at the reception device in phase, so as to interfere constructively, making the signal useful for processing and analysis.

3.5.2 Array Signal Processing

A sensor array is a group of sensors located at different spatial positions which are disposed to measure a propagating wavefield. The array permits the field energy distribution in time at diverse places to be known. An array of sensors samples the propagating field spatially. The collection of these distributed data may be combined to improve the quality of the information about the wavefield. The clever combination of these data is usually referred to as *array signal processing.*

Propagating wavefronts contain information about the sources that produce them and about the objects that reflect or scatter the wave energy. Signal processing methods allow some desired signals to be selectively acquired, rejecting others of no interest.

The array signal processing is a powerful tool that may enhance the signal-to-noise ratio beyond that of a single sensor's output. It may be used to track the source signal as they move in space, or permits the number of sources and their locations to be determined (Johnson & Dudgeon, 1993).

As was introduced in Section (2.3), individual sensors and antennas may have an omnidirectional pattern or a beamwidth with some directivity which lets the user receive a preferential propagating field with respect to direction. If information from another direction is desired, the receiver must be pointed to such direction.

Propagating signals vary in time and space. A sensor's array allows choosing the signals according to their propagating direction (Johnson & Dudgeon, 1993). Arrays are used to collect the data required to apply the information process called *beamforming.*

Beamforming is a kind of spatial filtering which retains signals propagating in one direction, attenuating those propagating in other directions. It is a spatial filtering operation that reduces noise and interference. Beamforming is also a generic name for a wide variety of processing algorithms which have the capability of selecting signals in a particular direction. The most adequate algorithm should be adopted according to the characteristics of the problem (plane wave, spherical wave, near field, far field, etc.)

3.5.3 Beamforming

Processing signals from several transducers (or antennas) arranged in a certain spatial allocation permits the signal that comes from a preferential spatial direction to be selected. The process of combining signals arriving from different transducers to reinforce some signal coming from a particular direction and rejecting the others is the goal of beamforming.

Beamforming can be achieved as a result of a physical action, e.g. by shaping and moving the transducers in the array, by an electrical process (analogically controlling the phase and amplitude of the individual transducer's outputs), and mathematically by digital signal processing, the last one being the most used nowadays.

Beamforming is used either for receiving or transmitting signals. A very simple and didactic acoustic example will be schematically described to introduce the concepts, but a deep introduction to the subject requires a certain degree of mathematical manipulation which is beyond this introductory book. With some precautions, the example could be extended from acoustic to electromagnetic waves.

Let us introduce a linear array of transducers (Fig. 3.8), each of them receiving the signal from a scattering source; assume that the source is a particle, previously irradiated by an acoustic pulse, which scatters the received energy in different directions. Because the distances from the source to each transducer (d1, d2, d3 and d4) are not the same, the scattered signals will suffer different delays to arrive at the transducers and a different degree of attenuation.

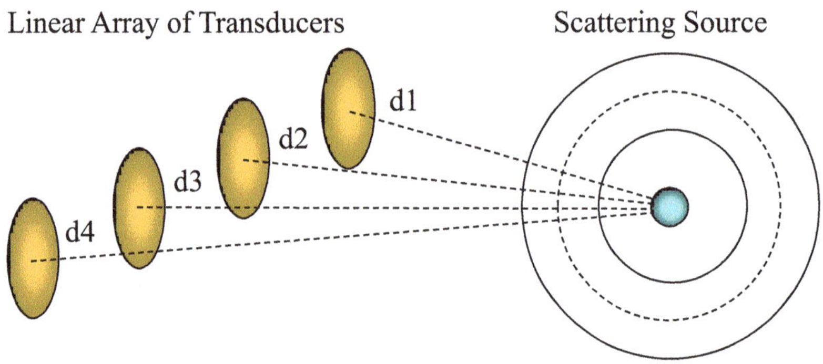

Fig. 3.8: A linear array of transducers receives energy scattered by a particle. The distances between the particle and the transducers are d1, d2, d3 and d4.

If the spatial position of the source, the propagation speed of the wave and the attenuation coefficient of the medium are known, the delay and attenuation introduced in the signals arriving at the transducers could be estimated. Therefore, if a correction in time and amplitude is made for each signal and then summed up (Fig. 3.9), the resulting signal will have information from four transducers, improving the signal to noise ratio with respect to that of a single transducer.

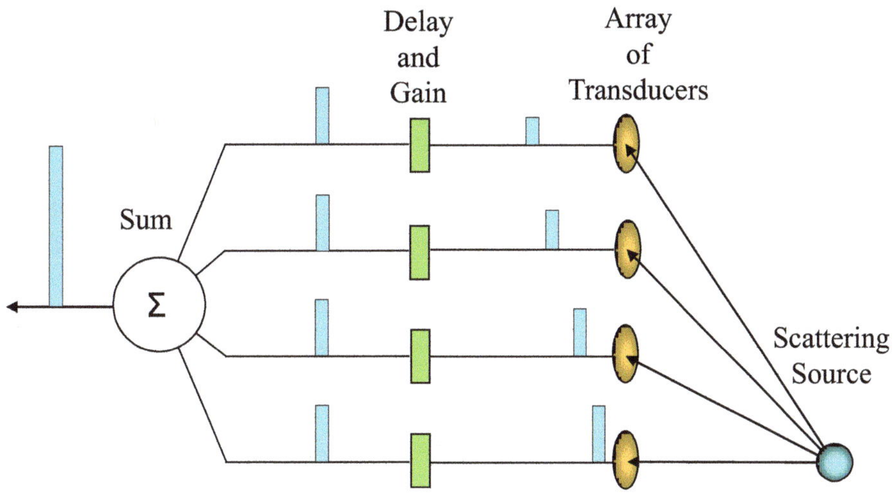

Fig. 3.9: Each transducer receives the scattered signal with different amplitude and at a different time. Because the attenuation and delay of the propagating signal is known, they may be compensated by the Delay and Gain circuits.

We want to underline that this upgrading is achieved for the signal coming from that particular source. In this way, if a source in another position is transmitting an undesired signal, the Delay and Gain circuit that was "tuned" for the first source would result in unsuitable for the unwanted one. This shows the spatial selectivity of the array.

The beamforming in the transmitting case can be explained in a similar way. Figure 3.10 is a schematic representation of a signal generator whose output is passed through a Delay and Gain circuit before arriving at the transducers. In this case it is desired that the wave fronts of each individual transducer be summed up in such a way that the resulting wave front propagates in a particular direction

The pulse arrives first at the upper transducer and generates a propagating wave front. When the pulse arrives at the second transducer, the first wave front produced by the first transducer had already moved away. The same happens with the next transducers. In this way the individual wave fronts are summed up at different distances from the transducers and the resulting wave front will have a particular direction that depends on the delays.

The attenuation of the propagating wave produced by each individual transducer, which produce the resulting wave front, depends on the distance traveled from the transducers to the target we want to irradiate. If this distance is long all the transducers will irradiate the target with approximately the same energy, because the differences in attenuation due to the different path lengths will be negligible. But if the target is at a distance comparable to the separation between transducers, the transducer's gains could be individually adjusted to place a similar amount of energy with each transducer.

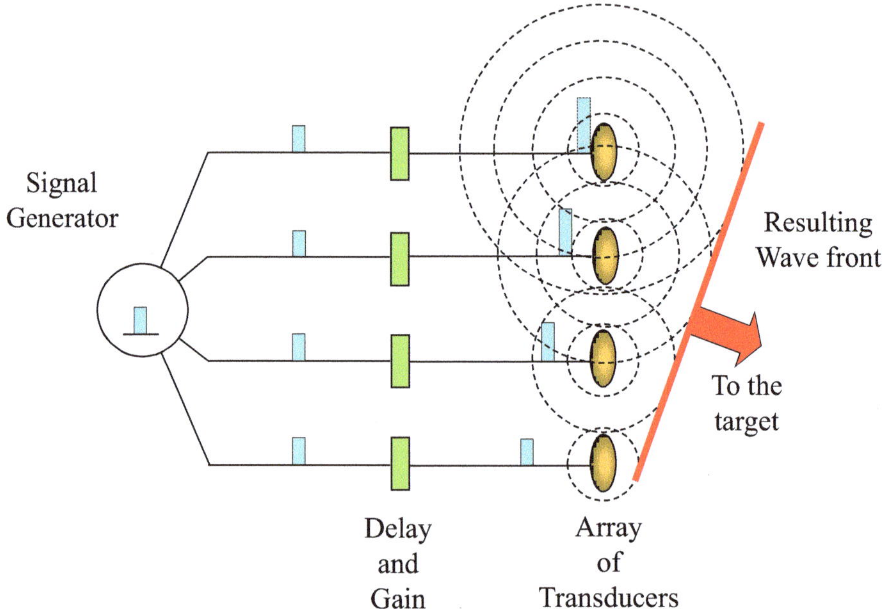

Fig. 3.10: The signal produced by the Signal Generator is delayed and amplified in a differential way such that the wavefronts are summed in a resulting wavefront with the desired particular direction.

As an example of the second case, assume an ultrasound array of five transducers; the size of the array is comparable to the distance where it is desired to concentrate the energy. This case is schematically depicted in Figure 3.11, where the upper and lower transducers should be excited first and with the maximum gain; the opposite (last excited and minimum gain) should be done with the central transducer. Thus, the set of transducers constitute a *phased array* (Fig. 3.11) where each transducer contributes a similar amount of energy.

The methods described above are frequently used in processing environmental data and in some remote sensing applications where the beamforming is referred to as phased-array technology and will be presented later.

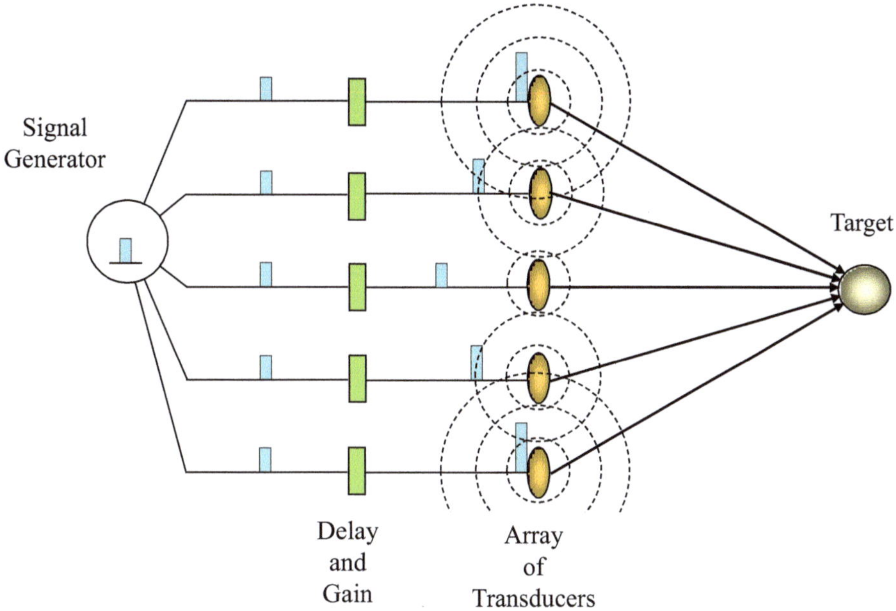

Fig. 3.11: Attenuation and phase of the Signal Generator is processed (in amplitude and phase) to focus the energy on the target.

3.6 Signal Conversion

3.6.1 Signals

In general, a signal is a sound, gesture, object, image, electromagnetic wave, voltage, etc, that conveys information, notice or warning. More specifically related to instrumentation, a signal is a detectable physical quantity (such as a voltage, current, magnetic field strength or sound) by which information can be transmitted, stored or processed.

3.6.2 Analog and Digital Signals

The purpose of this section is to introduce some concepts on how instruments process information. The process starts with an analog signal and ends with the information stored in a digital memory. A deep insight into the mathematics involved in this process is avoided because it is beyond an introductory book. Those readers interested in the subject can find it in books on signal analysis (Smith, 2003). Only the main ideas supported by some drawings will be presented to explain the transformations that the signal undergoes in this process. Users of Instruments should be aware on how these alterations introduce certain modification on the signal information. **The data recorded by instruments is therefore a modified version of the actual input signal.** The user of an instrument should evaluate whether the modifications are acceptable to the study being carried out, or conversely, too altered, making data useless.

The signal processing performed within instruments introduces successive modifications on the signal information content. The goal of presenting some fundamental concepts on this subject is that users be able to evaluate whether the altered version of the signal recorded by the instrument, adequately preserves the real phenomenon that they are trying to measure. Some examples will be given at the end.

3.6.3 Analog Signals

An analog signal is a kind of signal that is continuously variable. The word analog indicates something that is mathematically represented as a set of continuous values. It can be assumed that there exist an infinite number of values between two arbitrary levels of the signal. For this reason it is said that analog signals have a theoretically infinite resolution. The analog signal is a continuous function; its slope is never infinite, which would imply an instantaneous change of value.

A set of continuous values of some physical phenomena such as light, sound, pressure, or temperature is an analog signal. Sensors transform the information contained in the phenomenon into an electrical signal which is later on conditioned in the first stages of instruments (Fig. 1.1). Because an analog signal has infinite values, it would require recording an infinite amount of numbers to store the complete signal information which is obviously unpractical.

How could this amount of numbers be reduced without loosing useful information? What is the minimum amount of numbers needed to keep the signal information? Some answers to these questions will be developed below.

3.6.4 Sampling an Analog Signal

Intuitively, given a signal which varies slowly it is possible to reconstruct it from a few samples. As the signal varies faster and faster its high frequency content increases and it becomes necessary to increase the number of samples to keep its information content. For example, with the intention of recording how the temperature of a river changes due to sun radiation a few samples per hour would be enough because the mass of water is quite large, but to know how a small puddle is heated by the same amount of energy requires several samples per hour, because its temperature change is faster than that of the river.

The minimum amount of samples required to keep the signal information content unchanged is given by the sampling theorem. It states that in order to recover a signal *w*(t) from a number of samples equally spaced in time, it is necessary to sample *w*(t) at a rate equal to or greater than twice the highest frequency component of the signal. This sampling frequency is known as the Nyquist frequency (NSC, 1980). For example, a sea wave having a maximum frequency of 3 Hz requires a sampling rate higher than 6 Hz (six samples per second) to preserve and recover the waveform exactly. It is convenient to clarify that this is a theoretical limit and that in practice more samples are needed because, in general, not all the theoretical assumptions of the theorem are met.

The first step to convert analog signals to digital ones is to take analog samples that represent the original signal but discretized in time. At the end of the first step of the signal process, samples of the analog signal are pulses of a given width (*w*), separated by a period of time equal to the sampling period (*T*) (Fig. 3.12). The amplitude of these pulses is equal to the amplitude of the signal at the moment it was sampled. Now the resulting signal is pulsed. Also it is observed that if *w* is long, the amplitude of the signal will vary during sampling; for this reason it is convenient that *w* be short.

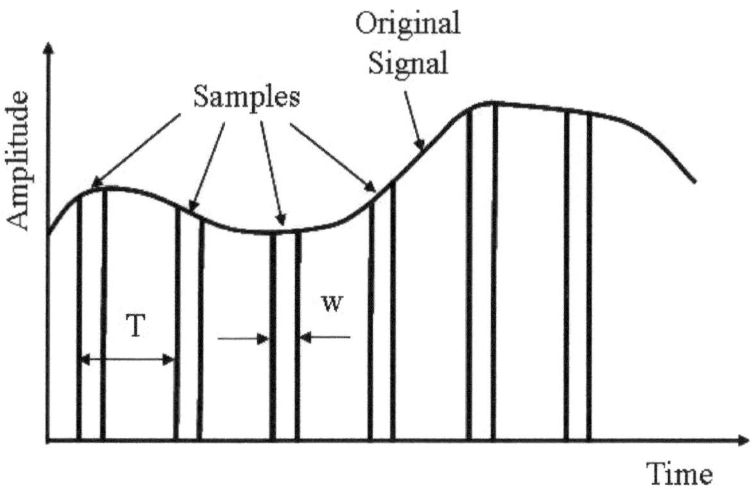

Fig. 3.12: The analog original signal is sampled at periods *T* during a sampling time w, which results in a pulsed signal.

Because we want a single value to represent the sample, then, in general, the average value of the original signal amplitude over the period *w* is taken as the representative amount. The sampling theorem states that samples should be taken in an infinitesimal time, so they should have an infinitesimal width. Actually, real samples are always finite and they may have a *w* of milliseconds, microseconds or nanoseconds, depending on the signal frequency content and measurement requirements.

Even when these samples do not satisfy the theoretical assumption of infinitesimal width required by the theorem, for low frequencies signals, as are most of the signals encountered in environmental sciences and industrial world; this theoretical limitation does not modify the signal information content appreciably.

If it is attempted to recover a signal from their samples when it was sampled at a rate below twice its highest frequency component, a completely different low frequency signal will appear which is known as **aliasing**. This phenomenon is visualized in Figure 3.13; the higher frequency signal (in blue-dashed line) is the signal of interest sampled at equal intervals; the black bars with a dot at the end show the sampling points. In this example the sampling rate is about 1.4 times the highest frequency of the signal. If a sinusoid were reconstructed from the samples, the lower frequency sinusoid (in red-continuous line) would be obtained. This artifact is known as aliasing.

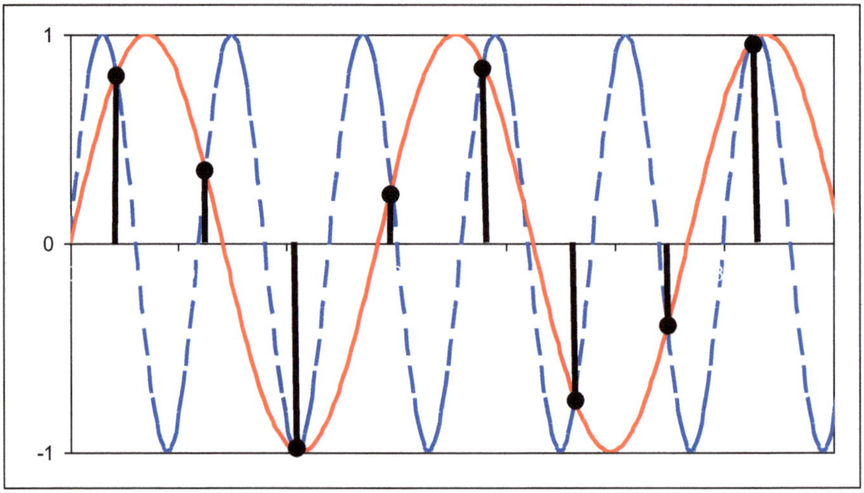

Fig. 3.13: The blue (dashed line) signal is sampled with fewer samples per second than required by the sampling theorem. The black bars with the dots symbolize the acquired samples. When reconstructing a sinusoidal signal from the samples, the red (continuous line) signal will be obtained which is of lower frequency than the original one.

The application of the theorem to real cases hides a problem: previously to adequately sample a signal, it is necessary to know the maximum frequency component so as to sample, at least, at twice this frequency. Most actual signals contain a wide spectrum of frequency components. Then to have enough samples to recover the signal from the samples, the sampling rate could be impractically high.

In order to solve this problem, the desired input signal bandwidth has to be limited by filtering before sampling. The filter to limit the frequency content of the input signal is called anti-aliasing filter. It is a low pass filter which allows passing the lower spectral content of the signal. When we have a previous knowledge of the frequency content of the signal, it is a relatively easy task to choose the filter bandwidth but if the signal is unknown, the filtering process could cut away useful information.

3.6.5 Spatial Aliasing

So far the sampling problem has been analyzed in the time and frequency domains, and the aliasing effect was avoided by means of an anti-aliasing filter. The spatial filtering concepts were introduced in Section (2.3), and above in Section (3.5.2). In this way the concept of sampling of a signal was extended from time to space. Now an example of how aliasing could affect data collection in space will be given. This example is a recreation of a real world case.

Some students committed to collecting data, measured soil properties in a country land. They took samples every 2.5 m approximately on a 100 m line (Fig. 3.14) and found a regular distribution of the soil porosity about every eight to ten meters. They were excited about this discover but were not able to find an explanation of such amazing result.

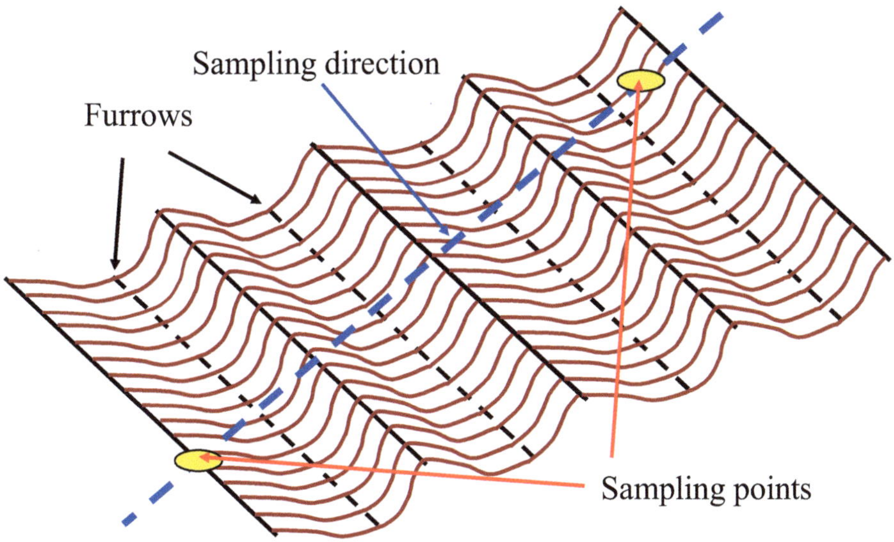

Fig. 3.14: Surface porosity is modified at regular intervals. When regularly sampled at a rate less than two samples per furrow, aliasing could appear in the results.

They disregarded that this regularity could be due to plowing because at the time they collected the samples the land was not planted and they thought that because the length of the regular distribution of the porosity was much longer than a typical furrow there could not be a relation.

After some reading they realized the existence of spatial aliasing and that a higher frequency periodical regularity could give rise to another of lesser frequency when sampled below the Nyquist frequency. Further research determined that the field had been plowed at a regular distance of about 1 m some months before.

3.6.6 Analog to Digital Conversion

Once the necessary samples per unit time to reconstruct the signal were taken, they must be converted from discrete pulsed signals to numbers able to be processed on a computer. This step of the process is known as analog to digital conversion and could be subdivided into two operations: quantizing and coding.

Figure 3.15 shows the quantization process of converting samples, represented by dots, into a set of discrete output levels for a 3-bit resolution analog to digital converter (ADC). At this point a slight digression is necessary. Figures 3.12 and 3.13 show a signal discretized in time, now the signal is discretized again, but in amplitude.

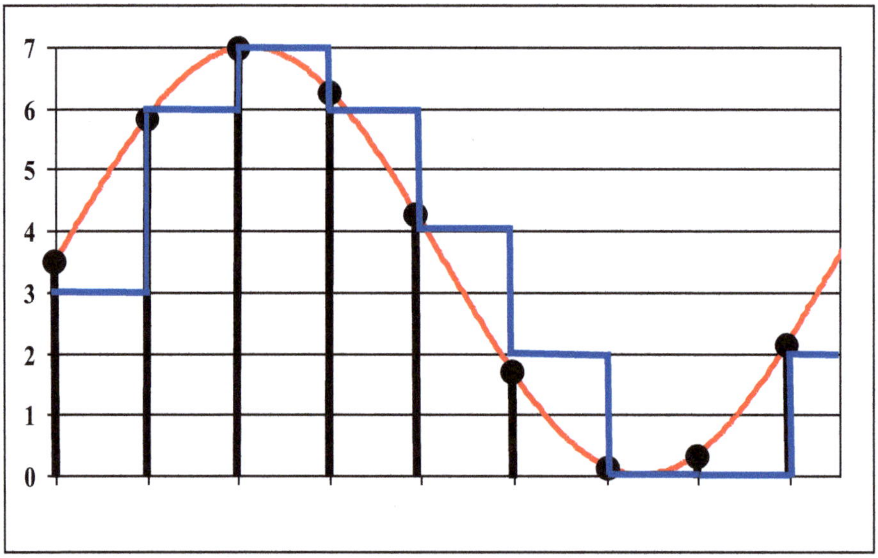

Fig. 3.15: The analog signal is the red (continuous sinusoid); the bars with dots are the acquired samples. Numbers from 0 to 7 are the values that the samples adopted in the quantization process. The steps in blue are the individual values assigned to each sample.

Using a 3-bit resolution converter eight discrete levels (N) can be defined in a binary code as shown in Table 3.1. The first stage of the ADC compares the amplitude of the samples from the analog signal to digital levels (quantization process). The digital level closer to the sampled value is adopted by the converter to represent the sample. This gives the resulting signal the stepwise appearance. Once a level was adopted the converter assigns a binary number to each level (coding).

Table 3.1: Levels in a 3-bit resolution converter

Decimal number	**0**	**1**	**2**	**3**	**4**	**5**	**6**	**7**
Binary number	000	001	010	011	100	101	110	111

The representation of the signal depicted in Figure 3.15 in decimal and binary number series is shown in Table 3.2 (see the ordinate axis in Figure 3.15 and Table 3.1). The decimal numbers correspond to the quantized levels adopted in the coding process. The binary numbers are the actual way in which information is stored in digital memories.

Table 3.2: Representation of the signal of Figure 3.15 in decimal and binary codes

Decimal number	**3**	**6**	**7**	**6**	**4**	**2**	**0**	**0**	**2**
Binary number	011	110	111	110	100	010	000	000	010

As shown in Figure 3.15, the digital signal does not represent the original signal very closely because eight levels are not enough to represent it. However, increasing the resolution (the amount of bits) the representation improves, for example with 10 bits the number of levels increases to 1024. The size of the minimum level change is called quantization step **q** or least significant bit (LSB). Devices that perform the sampling, quantization and coding processes are called analog to digital converters (ADC).

The amplitude uncertainty in digitizing a signal is inherent to the amount of bits of the ADC, and is referred to as **quantization error** which is equal to ± LSB/2. The amount of bits sets the resolution of the converter, which determines the resolution of the instrument where it is used. The maximum quantization error is $\pm q/2$, defined in Eq. (3.10). The number of levels (N)), may be calculated as follows:

$$\text{Number of levels} = N = 2^{\text{number of bits}}; \quad q = LSB = \frac{1}{N} \tag{3.10}$$

Then for a 10-bit ADC the maximum quantization error is $\pm q/2 < \pm 0.0005$ or ± 0.05 %. Nowadays it is common to find instruments with ADC of 16 and 24 bits. Table 3.3 presents examples of the amount of bits and the number of levels for the most used ADC.

Table 3.3: Amount of bits and number of levels

Bits	8	10	12	16	24
N	256	1024	4096	65536	16777216

Byte is a very frequently used term in digital language to denote a digital word of 8 bits, and it is symbolized by the letter *B*. In old digital storage media, a byte corresponded to the minimum memory unit. For example, it is usual to say that an image occupies 1.5 MB, which means 1.5 mega Bytes that is equal to 1,500 k Bytes = 1,500,000 × 8 bits. Table 3.4 shows the relation between bytes, bits and levels.

Table 3.4: Bytes, bits and levels

Bytes	1	2	3	6
Bits	8	16	24	48
N	2^8	2^{16}	2^{24}	2^{48}

3.6.7 Application Example

Resolution should not be a problem when buying an instrument because manufacturers account for it. A different situation arises when researchers want to acquire data using sensors and a data logger. In this case, the user should verify how the resolution could affect the data being recorded.

3.6.8 Data Logger

Let us describe a data logger in order to later introduce an example of the concepts previously explained. A data logger is an electronic device that receives an analog voltage and converts it to digital information stored in a memory (Fig. 3.16). The first electronic stage (*A*) permits coupling the voltage input signal to the data logger without disturbing the signal. The second stage is an ADC, just explained. A data storage medium (*DS*) follows that consists, generally, in a non-volatile memory. Finally, there is a communication port (*CP*) required to extract the data and to program the data logger functions (such as the amount of samples per second to be acquired). This port is accessed using an external computer (*C*). An internal or external battery supplies the energy needed by the circuits.

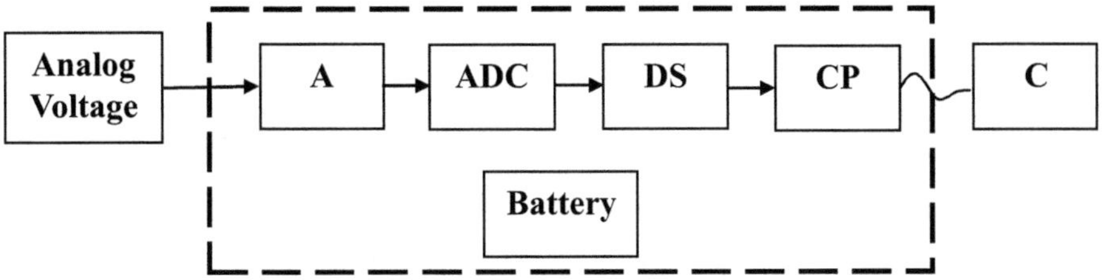

Fig. 3.16: Schematic of a data logger.

Let us introduce an example of a quite real case in which it is desired to record the soil temperature in a beach. Suppose that the researcher has at his disposal an electronic analog thermometer composed of a temperature sensor and an amplifier whose voltage output from 0 to 1 V corresponds to a temperature range from 0 to 100 °C. The data logger available at the laboratory has an analog input of 10 V and a resolution of 10 bits. Both instruments and their static transferences are shown in Figure 3.17. Both transferences are considered linear. K_1 and K_2 represent the individual sensitivities and K the total sensitivity of the whole set.

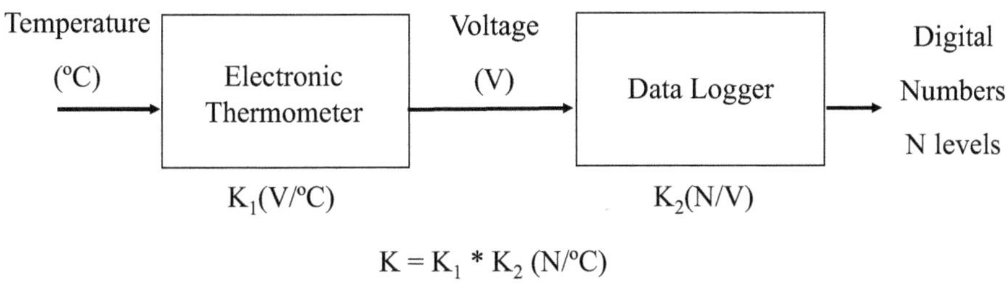

Fig. 3.17: An electronic thermometer produces a voltage output proportional to the input temperature, its sensitivity being K_1. The data logger converts input voltage to digital numbers with sensitivity K_2. The total sensitivity is the product of both.

Question 1: What is the combined temperature quantization error of the set formed by the electronic thermometer and the data logger?

Because a 10-bit data logger has 1024 output levels, the transference of the data logger is K_2 = 1024 / 10 V ≈ 100 /V. The constant transference of the thermometer is K_1 = 1 V/100 °C = 0.01 V/°C. Then the sensitivity of the whole system is K = 1/ °C and this means that **one level** represents 1 °C. Because the maximum quantization error is ± LSB/2 the maximum quantization error turns out to be ±0.5 °C. This means that the temperature of the beach will be known with an uncertainty of ±0.5 °C, which would not be enough for some temperature studies.

Question 2: How could the resolution be improved if another set of instruments could be selected?
A first improvement can be achieved by using a thermometer better suited for the particular beach where temperature has to be measured. It is unlikely that the temperature of a beach could reach 100 °C. For example, a sensor with a range from 0 to 50 °C seems more reasonable. Combined with an amplifier with a 10 V output it gives $K_1 = 10$ V/50 °C = 0.2 V/°C.

For the same data logger the total sensitivity (K) increased and a new $K = 20$ / °C is obtained, resulting $\pm\, q/2 = \pm 0.025$ °C. Thus, selecting a sensor with a temperature range closer to the measurand expected range and amplifying the voltage signal by ten, all the input range of the data logger is exploited and the resolution is increased 20 times.

If more resolution is needed a data logger with the same analog input (10 V) but a better resolution, 12 bits (number of levels = 4096) could be used. Now $K_2 = 4096$ /10 V ≈ 41/V and $K \approx 80$ / °C; resulting $\pm\, q/2 = \pm 0.00625$ °C.

It is observed that there are different options to improve the resolution of the measuring and recording system. The fist step was simply to select a sensor with a measuring range according to the range of the physical variable to be acquired, this improved the resolution twice. The second step consisted in matching the amplifier output to the data logger input, which improved the resolution 10 times, and the last step was to use a 12 bits data logger which upgraded 4 times the resolution. Thus, with these procedures the total resolution was improved 80 times. It is just an example of a particular solution, but some general conclusion will be drawn below.

Keeping in mind to improve only the resolution, it is valid to use any of the above-mentioned measures, but improving the resolution does not assure the improvement of data quality. The simpler solution that users frequently adopt is to increase the data logger resolution, but this could undermine data quality because noise can make this high resolution useless.

With the purpose of keeping the signal to noise ratio high it is healthier to keep the signal level as high as possible. Thus, **it is a good practice to amplify in the first stages of a measuring system in order to have large signals.** Therefore, in all measuring systems, **selecting the adequate sensor range is the most important step.**

Also, **matching the output of the amplifier to the data logger input** helps in the same direction because the signal is amplified and the complete ADC range of the data logger can be used.

3.7 Electricity

With the purpose of making the following chapters understandable it is needed to deal with some concepts on electricity that play important roles in describing the working principles of instruments. It would not be possible to understand how sensors work without the concepts that follow.

3.7.1 Electrical Energy

Voltage in an electrical circuit is the potential energy per unit charge able to move electrons in a circuit. The amount of charge per unit time displaced in a circuit is called the electric current. In the International System of Units (SI), the unit for voltage is the volt (V) and for the current is the ampere (A). There are mainly two different kinds of electric current, **direct current** (DC) and **alternating current** (AC).

3.7.2 DC

This is the current supplied by batteries, solar panels, thermoelectric devices and electrostatic generators. It is called direct current because electrons flow in only one direction in the electric circuit. In general, DC voltage sources are of low voltage and they are used to supply portable equipments such as cell phones, notebooks, lanterns and most autonomous instrumentation. Typically, DC sources are used to provide energy to instruments used in environmental sciences.

The most common battery voltages are 12, 6 and 3 V. The amount of energy stored in a battery is expressed in ampere - hour (Ah). It is the product of a current and a time. It is the current that the battery can deliver without dropping the voltage, expressed in amperes by the time it can do it, expressed in hours. For example, the batteries used for car ignition are in general 12 V and have approximately 80 Ah. If it were required to power an instrument with a current consumption of 0.5 A one car battery should last about 160 h.

Batteries undergo an aging process which decreases the amount of energy that can be stored in them. The energy that a battery can supply depends on its initial charge, the room temperature, the battery age and the rate of current discharge. Therefore, for field instruments it is convenient to take a conservative attitude at the time of calculating the autonomy of batteries. In general, losing data is more expensive than supplementary batteries.

3.7.3 AC

This is the voltage find in most houses to power electrical appliances. It has a sinusoidal waveform with a frequency of 50 or 60 Hz and amplitude of 110 or 220 V. In a circuit supplied with AC voltage, electrons flow in both directions. This kind of energy is used to supply most electrical motors, home tools and industrial machinery. In general it is used to cover most medium and high power requirements.

3.7.4 DC Ohm's Law and the Resistance Concept

Ohm's law states the relationship between voltage (V) and current (I) in a conductor. The current through conductors is proportional to the voltage applied to them.

$$I = GV \quad \text{or} \quad V = RI; \quad G = \frac{1}{R} \tag{3.11}$$

The constant of proportionality G is called the *conductance* and R is called the *resistance*. The unit for R is the ohm, whose symbol is the Greek capital letter omega (Ω), and for G is the siemens, symbolized by the letter S. V is the voltage across a conducting element and I is the current flowing through it.

3.7.5 Resistor

R (or **G**) is a property of the physical conducting element called "resistor", which depends on its shape and the material it is made of. A piece of conductive material with length l and cross section A is shown in Figure 3.18 and its resistance given by Eq. (3.12a). In words, the resistance is directly proportional to the length of the resistor and inversely proportional to its cross section. The constant of proportionality ρ is an intrinsic property of the material called **resistivity**. Equation (3.12a) is equivalent to Eq. (3.12b), where σ is the **conductivity**, σ being $1/\rho$.

$$R = \frac{\rho l}{A} \tag{3.12a}$$

$$G = \frac{A}{\rho l} = \frac{A\sigma}{l} \tag{3.12b}$$

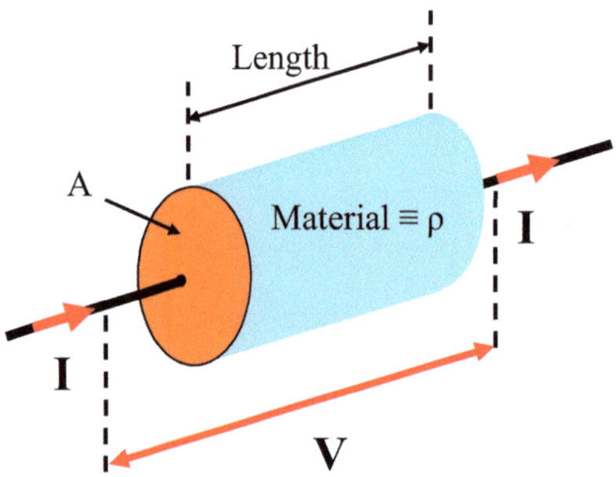

Fig. 3.18: A resistor has a resistance that depends on its geometrical properties and on the material it is made of.

For those readers unfamiliar with electricity it could be useful to use a hydraulic analogy to explain Ohm's law, keeping in mind that this analogy works only to some limited extent.

Assume a water tank filled with water placed on the roof of a house and connected through a pipe to a faucet. The difference of pressure between the tank and the faucet can be thought of as the voltage (*V*). The flow through the faucet is the analogy of the electric current (*I*). The hydraulic resistance introduced by the pipe to the passage of water could be thought of as the electrical resistance (*R*). Note that the hydraulic resistance is directly proportional to pipe length and inversely proportional to the pipe cross sectional area (the same happens in the case of the electrical resistance).

Ohm's law may be found written in any of the following ways:

$$I = \frac{V}{R}; \quad V = RI; \quad R = \frac{V}{I}; \quad I = GV; \quad V = \frac{I}{G}; \quad G = \frac{I}{V} \tag{3.13}$$

This law is also valid for AC voltages applied to resistors. When an AC sinusoidal voltage is applied to a resistor the current will be also a sinusoid and both waveforms (voltage and current) will be in phase.

3.7.6 Resistors in Series and Parallel

When two resistors are placed in a circuit such that the current passing by them is the same, it is said that they are in series (Fig. 3.19a). When the voltage is the same on both resistors, it is said that they are in parallel (Fig. 3.19b).

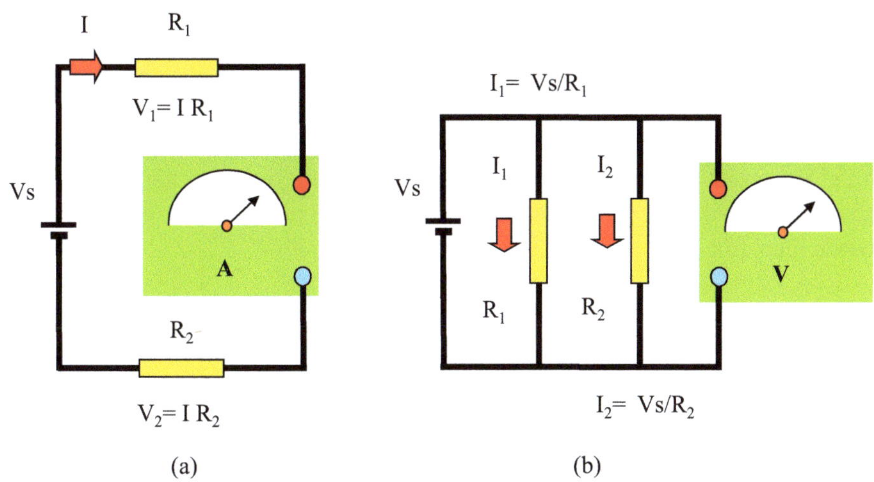

Fig. 3.19: (a) Two resistors in series: the current is the same for both. (b) Two resistors in parallel: the voltage is the same for both.

From the point of view of the voltage supply, two resistors in series ($R1$ and $R2$), can be replaced by one (Rs) whose value is the sum of the two. Then, if one is much greater than the other, the current in the circuit will be governed by the greatest.

From the viewpoint of the voltage supply, two resistors in parallel ($R1$ and $R2$) can be replaced by one (Rp) with the value shown in Eq. (3.14). In this case the value of Rp is always less than the smaller of the two resistors. The current in the circuit will be governed by the smaller of the two resistors.

$$R_s = R_1 + R_2; \qquad R_p = \frac{R_1 R_2}{R_1 + R_2} \tag{3.14}$$

These concepts explained for resistance are also valid for impedances in series and parallel. The concept of impedance is defined following in 3.7.11.

3.7.7 Electric Power

From Figure 3.18, the electric power (P) dissipated on the resistor R through which a current I flows and a voltage V is applied is the product of the two last and is calculated by Eq. (3.15). The unit for power is the watt (W).

$$P = VI = \frac{V^2}{R} = I^2 R \tag{3.15}$$

3.7.8 Capacitance and Capacitors

Capacitance is the ability of a body to store charge in an electric field (**E**). To place this charge in the body requires doing work, and an electrical potential energy is accumulated in the **capacitance**, and this energy can be later recovered. The device used to store charges is called **capacitor** and the simplest capacitor is formed by two conductive plates of similar size facing each other (Fig. 3.20). Assuming that the sides of the plates are much larger than their separation, the capacitance is directly proportional to the area (A) of the plates and inversely proportional to the distance between them (d):

$$C = \frac{A\varepsilon}{d} \tag{3.16}$$

where ε is the permittivity of the dielectric material separating both plates (dry air is a dielectric material).

The charge accumulated on the capacitor's plates is represented by Q and the voltage between plates by V. The capacitance (C) is the amount of charge accumulated per unit voltage, $C = Q/V$, Q being measured in coulombs and V in volts, thus the unit of capacitance is coulomb / volt, which adopts the name of **farad** (symbolized by F). Unfortunately, the symbol for coulomb is C and some attention should be paid to avoid confusing it with the symbol for capacitance.

The energy stored in a capacitance is expressed by Eq. (3.17), and the unit is the **joule** (J).

$$U = \frac{1}{2} C V^2 = \frac{1}{2} \frac{Q^2}{C} \tag{3.17}$$

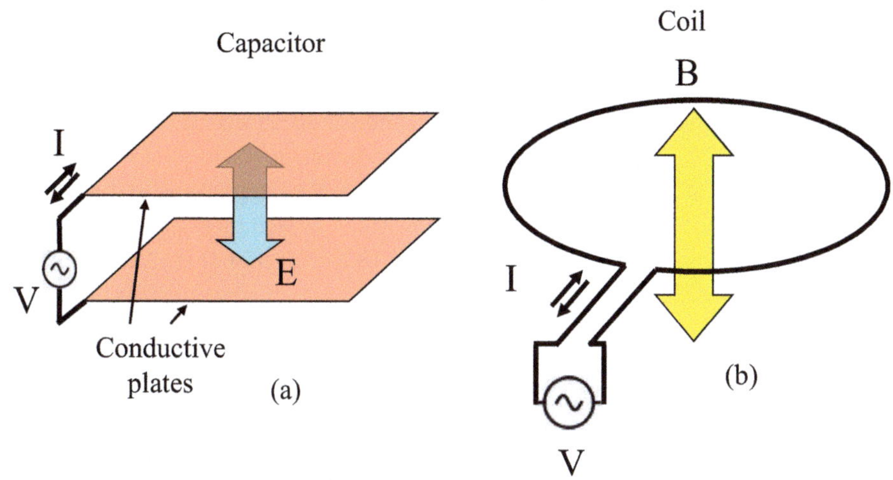

Fig. 3.20: (a) Capacitance. (b) Inductance.

3.7.9 Inductance and Inductors

Inductance is the ability of a device to store energy in a magnetic field (**B**). To create the magnetic field a current is required doing work. Thus potential energy is stored in the field. This energy can be recovered when the circuit opens and the current creating the magnetic field is interrupted.

A device which produces a magnetic field is the inductance (*L*) that can be materialized by an **inductor** or coil (Fig. 3.20b). It is basically a winding of several turns (*N*) of conductive material (usually cooper). The unit used for inductance is the henry (H). Sometimes a core of magnetic material is used inside the coil to increase the energy stored in the magnetic field. The magnetic material confines and guides the magnetic field, e.g. by placing an iron core inside a hollow coil the inductance can increase several times.

There are many approximate equations to calculate *L* which take into account the coil length (*l*) and area (*A*), and the material of the core. In all of them, the inductance is directly proportional to the square of the number of turns. For a long coil the inductance is given by

$$L = \frac{\mu N^2 A}{l} \tag{3.18}$$

where μ is the magnetic permeability of the coil core. The energy stored in an inductance is given by

$$U = \frac{1}{2} L\, I^2 \tag{3.19}$$

3.7.10 AC Ohm's Law

This law states the relation between AC current and voltage in a circuit which contains not only resistance but also capacitance and/or inductance. As noted above, if an AC sinusoidal voltage is applied to resistors the current will be in phase with the voltage. This does not happen with storing energy elements as capacitances and inductances.

When an AC sinusoidal voltage is applied to a capacitor (Fig. 3.20a) charges create an alternating electric field between plates. Charge movement gives origin to a current. This current is sinusoidal but unlike what happens with resistors, it is not in phase with voltage but V lags I by 90°.

When AC sinusoidal voltage is applied to an inductance (Fig. 3.20b) the current through the coil creates an alternating magnetic field, the current is sinusoidal and V leads I by 90°.

Inductance and capacitance oppose the passage of an alternating current in different ways. Capacitance presents a greater opposition to low frequency currents than to high frequency ones. DC current (frequency = 0) is blocked by capacitances. This opposition to the current flow, which is a function of the current frequency content, is called the capacitive **reactance** (X_C).

Inductance has the contrary behavior to capacitance, increasing its opposition to current flow as frequency increases. In this case it is called **inductive reactance** (X_L) and formulas for calculating both reactances are given by

$$X_C = -j\frac{1}{2\pi\, f\, C}; \quad X_L = j 2\pi\, f\, L \tag{3.20}$$

where f is the frequency of the AC voltage and j is the imaginary unit. The phase of the voltage and current is taken into account by j.

3.7.11 Impedance

The magnitude Z, called the **impedance,** is formed by any combination of resistances and reactances. The phase between V and I depends on the values of R, L and C and on how they are placed in the circuit (whether they are in series or in parallel). As an example, a circuit formed by one resistance, one capacitance and one inductance connected in series has the impedance

$$Z = R - j\,\frac{1}{2\pi\, f\, C} + j\, 2\pi\, f\, L \tag{3.21}$$

The more general form of AC Ohm's law is thus

$$I = \frac{V}{Z} \tag{3.22}$$

Summarizing, the **impedance** is the term that symbolizes the ratio of voltage to current in AC circuits. It is a complex number which takes into account the phase between the voltage and the current. If the impedance has only a resistance, the phase between voltage and current equals zero. If the impedance is only composed by one reactance, the phase will be +90° or −90°. When the impedance is composed of resistances and reactances the phase can adopt any value between ±90°.

3.7.12 Faraday's Law

In order to understand how several instruments work (geophone, electromagnetic flow meters, current meters, etc.), it is required to have a conceptual understanding of Faraday's law of induction; with this purpose this law is introduced below.

Figure 3.21a shows a one turn coil connected to a galvanometer. If a magnet moves towards the coil, the galvanometer will deflect while the magnet is in motion, indicating that current flows through the coil. If the magnet is at rest no current flows. When the magnet is removed, the galvanometer deflects in the opposite direction. If the other pole of the magnet moves towards the coil, the experiment will be as before but the deviations of the galvanometer will be opposite to the above. It doesn't matter if the magnet moves towards the coil or the coil towards the magnet, what matters is the relative motion. The current that appears is called induced current and it is produced by an induced **electromotive force** (emf) (ε). (Do not confuse the symbol ε as used to denote an electromotive force with the same symbol when used for representing the permittivity of a dielectric material).

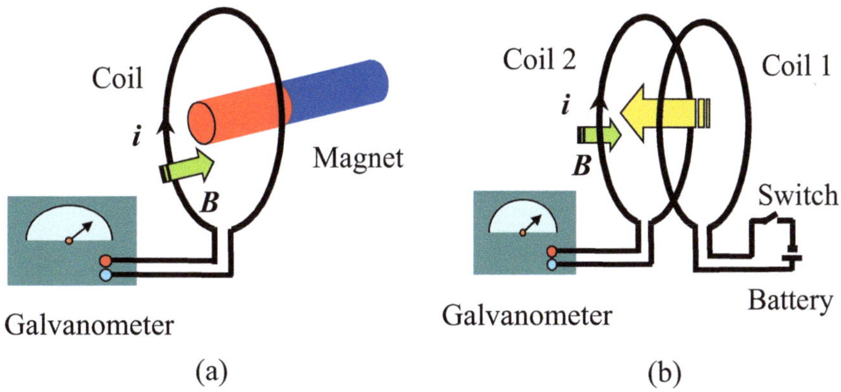

Fig. 3.21: (a) When a magnet moves towards the coil, the galvanometer will deflect indicating that current flows through the coil. (b) A coil through which a current flows behaves as a magnet and it is called an electromagnet.

In Figure 3.21b there are two coils at rest, i.e. there is no relative motion between the coils. When the switch closes the battery generates a current in coil 1 and the galvanometer connected to coil 2 is momentarily deviated. When the switch is kept closed and current in the coil 1 is constant, the galvanometer does not deflect. When the switch is opened (Fig. 3.21b), the change in the magnetic field makes the galvanometer deviate again. The current change in coil 1 produces a magnetic field variation which induces an emf in coil 2.

An immediate conclusion that will be used below is that a coil through which a current flows behaves as a magnet and it is called an electromagnet.

Faraday realized that the induced emf (ε) depends on the variation of the **magnetic flux** (F). Lenz stated that the induced electromotive force produce a current in coil 2 whose magnetic field opposes the original change in magnetic flux. The Faraday – Lenz law states that the induced emf in a circuit equals the negative value of the time rate at which the flux through the circuit changes. The induction law is

$$\varepsilon = -\frac{d\Phi}{dt} \tag{3.23}$$

3.7.13 Generation of Electrical Energy

The above mentioned Faraday-Lenz law is the operating principle of the generation of electrical energy from mechanical energy. If a magnet is mounted on a shaft and a coil is placed in such a way that due to the rotation of the magnet a variation of the magnetic flux appears in the coil, then a voltage is generated at the terminals of the coil (Fig. 3.22). The coil voltage is a pulsed signal; pulses are produced when the magnet approaches the coil. The polarity of the voltage depends on the magnetic pole approaching the coil.

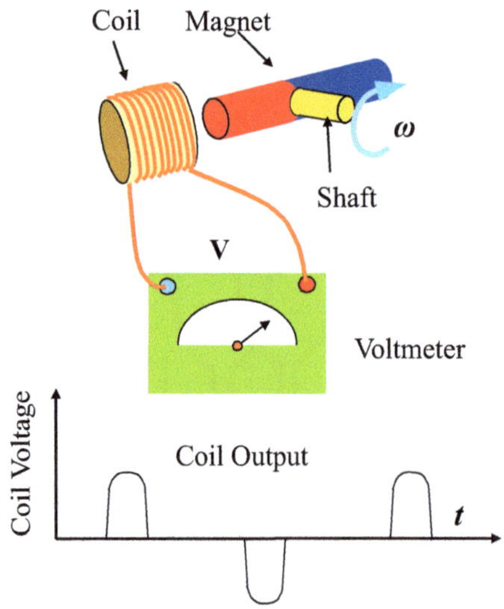

Fig. 3.22: A magnet produces a variable magnetic flux which induces and emf on a coil. Coil output shows pulsed voltage.

This elementary description is quite far from a real electric generator in which, usually, the magnet is replaced by an electromagnet and several coils with magnetic core are employed.

3.7.14 Flux Change Due to a Changing Area

The change of the magnetic field in a coil induces electromotive forces not only in conducting materials but also in non-conducting ones and in empty space.

Figure 3.23 shows a cylindrical metal rod that can slide on a "c-shaped" conductive wire (Kip, 1962). The set forms a one turn coil in a uniform magnetic induction field **B** (also called **magnetic flux density**).

By definition, the magnetic flow F is the integral of the dot product

$$\Phi = \int \mathbf{B} \cdot d\mathbf{A} = BLD \tag{3.24}$$

where $d\mathbf{A}$ is an infinitesimal area (in vector notation), L is the length of the cylindrical rod included in the coil and D the distance.

From Eq. (3.24), F may vary due to a change in **B** as in Figure 3.21a and b, or due to the change of the coil area. In Figure 3.23, **B** and L are constants whereas D is variable, so that moving the rod with speed Ve will produce a flow reduction in the coil due to the decrease in area. This change generates an emf on the coil that originates a current i which produces $\mathbf{B}_1$. By the Lenz law, the induced $\mathbf{B}_1$ will have a sense such as to oppose to the variation of the flow.

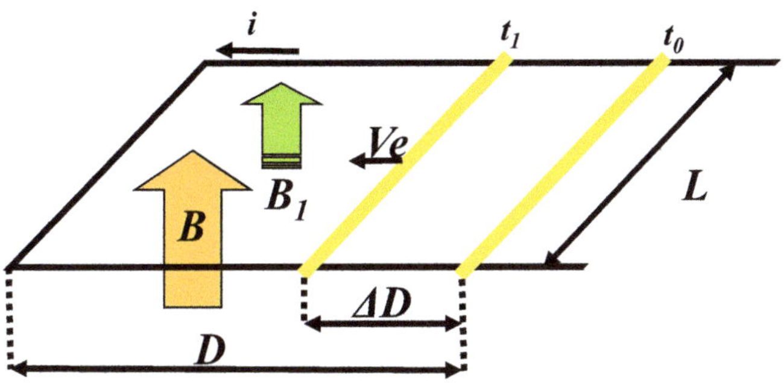

Fig. 3.23: A conductive "c-shaped" wire and a cylindrical metal rod forming a one turn coil are placed in a uniform magnetic field **B**. The rod can slide on the conductive wire varying the coil area, thus the flux Φ changes generating an emf on the coil that originates a current i which produces $\mathbf{B}_1$.

Assuming a constant Ve, the emf (ε) may be calculated as follows,

$$|\varepsilon| = \left| -\frac{d\Phi}{dt} \right| = B\frac{dA}{dt} = BL\frac{\Delta D}{t_1 - t_0} = BLVe \tag{3.25}$$

It can be noted from Eq. (3.25) that the emf generated on the moving rod is independent of the electrical conductivity of the rod, but the current in the coil will decrease with decreasing electrical coil conductivity making current detection difficult. If the rod were made of a non-conducting material the same emf would be induced but current would not flow.

A more formal and general equation expressed in vector form relates the electric field (**E**), the magnetic field (**B**) and coil velocity (**Ve**) (Skilling, 1974).

$$\nabla \times \mathbf{E} = -\frac{\partial \mathbf{B}}{\partial t} + \nabla \times (\mathbf{Ve} \times \mathbf{B}) \tag{3.26}$$

In this equation the symbol **∇** (called 'nabla' after an old Hebrew harp of triangular shape) stands for the vector operator $(\mathbf{i}\,\partial/\partial x + \mathbf{j}\,\partial/\partial y + \mathbf{k}\,\partial/\partial z)$. If it is applied to a vector field via a cross product (×), the result is a new vector called the curl of that vector. The curl of a vector field allows the variation of the field in a direction *normal* to the field to be determined.

The first term of the second member in Eq. (3.26) accounts for the time variation of **B**, whereas the second term accounts for the movement of the circuit in space.

For the sake of completeness, if the vector operator **∇** is applied to a vector field via a dot product (**.**), the result is a scalar called the divergence of the vector. The divergence of a vector field allows the variation of the field *along* the direction of the field to be determined. Finally, if the vector operator is applied to a scalar field, the result is a vector called the gradient of the scalar field. The direction of the gradient is that of maximum change of the scalar field.

3.7.15 Magnetic Force

If an electric charge (Q) moves in a magnetic field (**B**) with velocity **u**, it will experience a force (**F**) given by:

$$\mathbf{F} = Q\,(\mathbf{u} \times \mathbf{B}) \tag{3.27}$$

The magnetic force vector is the result of a vector product of the charge velocity and the magnetic field, thus the force is perpendicular to the plane containing **u** and **B**. The magnetic force is part of the Lorentz force, which also takes into account the electric field.

As forces appear on charges moving in a magnetic field, so do they appear upon a piece of wire placed in a magnetic field through which an electric current flows (charge per unit time). The force on a current element $I\,d\mathbf{l}$ of the wire is given by:

$$d\mathbf{F} = I\,d\mathbf{l} \times \mathbf{B} \tag{3.28}$$

where I is the current flowing through the wire and $d\mathbf{l}$ is a differential length along the wire upon which the force is exerted. This phenomenon is represented in Figure 3.24 for a square loop of wire carrying a current I and placed in a magnetic field **B**.

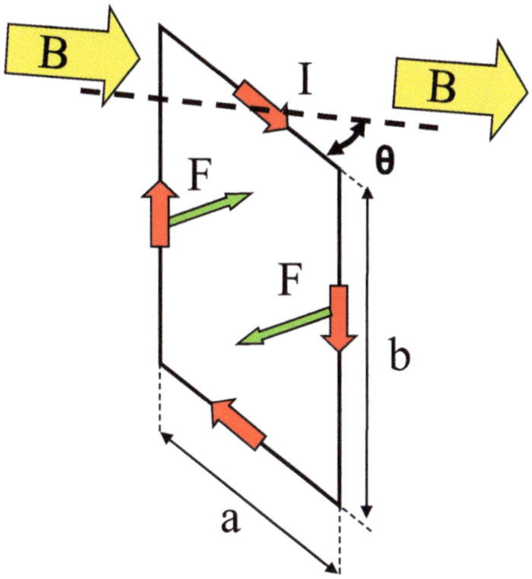

Fig. 3.24: A conductive loop placed in a magnetic field **B** carries a current *I*, appearing a force on the vertical wires. θ is the angle between **B** and the plane of the loop.

Using Eq. (3.28) it is found that in the upper and lower sides of the loop (of length *a*) the forces are respectively pointing up and down, then, considering a rigid loop, there is no net force on the loop. Instead, on the lateral sides of the loop (of length *b*) forces are as indicated in Figure 3.24, which tend to produce a rotation of the loop. Summarizing, if a current flows in a loop placed in a magnetic field, a torque will appear on the loop that will tend to rotate it. This is the operating principle of electrical motors and some acoustic transducers.

3.7.16 Transformers

Transformers are devices often used in electricity and electronics to transfer electrical energy between separated circuits. They are based on the concepts explained before and extended to alternating current in Figure 3.25, where coils 1 and 2 are composed of N_1 and N_2 turns respectively. An alternating voltage (v_1) on coil 1 (primary transformer) originates an alternating current (i_1) that generates a magnetic field $\mathbf{B}_1$ varying in time. The variation of the magnetic flux (F) in coil 1 induces an emf in coil 2 that generates a current i_2, which, in turn, generates voltage a v_2 on the load connected to the secondary. The load may be an impedance or a resistor. In voltage transformers the relation between input and output voltages depends on the turn's number relation which is fixed at the time the transformers are manufactured.

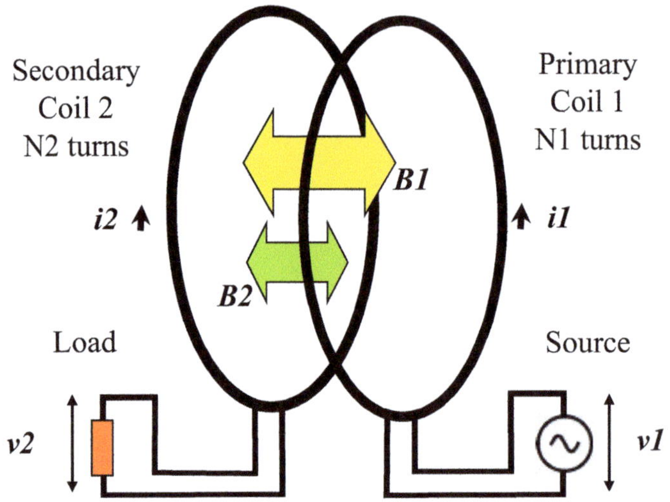

Fig. 3.25: Voltage source v_1 produces a current i_1 in the primary coil which generates a variable $\mathbf{B}_1$. The flux change generates an emf in the secondary coil which produces a current i_2 and a voltage v_2.

Let us now introduce some concepts that will be used later in instruments. As the resistor (or load) in the secondary of a transformer decreases, more current i_2 will flow through the load because the voltage v_2 remains constant (Ohms' law). This means that the power in the transformer's secondary, which is the product of $v_2 i_2$, increases. This increase has to be provided by the primary of the transformer through the magnetic flux. Again, because the primary voltage v_1 is constant, the only way to increase the primary power is by increasing i_1. It is interesting to note that a change in a resistor in the secondary of a transformer produces a change in the primary current. Therefore, by measuring the primary current it would be possible to know the resistance in the secondary. This feature is used in several kinds of sensors.

By means of transformers it is very easy to change the amplitude of alternating voltages and currents. In an ideal transformer the relation between the primary and secondary voltages, currents and turns is

$$\frac{v_1}{v_2} = \frac{N_1}{N_2} = \frac{i_2}{i_1} \tag{3.29}$$

Most transformers use a closed magnetic core because it greatly concentrates the effect of magnetic fields. A real magnetic transformer is depicted in Figure 3.26, where the core and coils are clearly seen. The core increases the transference of energy from the primary to secondary by thousands of times with respect to the coils in air. Transformers are also used to construct sensors, and one of the most widespread sensors is the Linear Variable Differential Transformer explained later on.

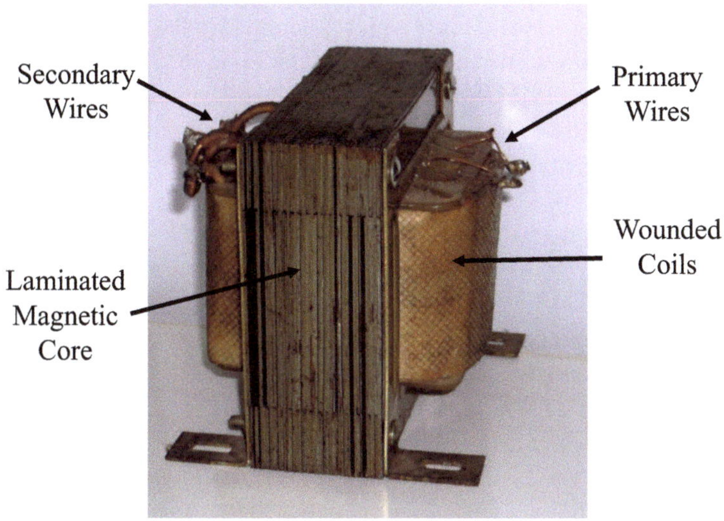

Fig. 3.26: A homemade transformer allows the laminated magnetic core and the coils to be appreciated.

3.7.17 Toroidal Transformer

A toroid is an object with the shape of an O-ring. A toroidal transformer is composed of a core of magnetic material with an annular shape, and the primary and secondary coils wound around it (Fig.3.27).

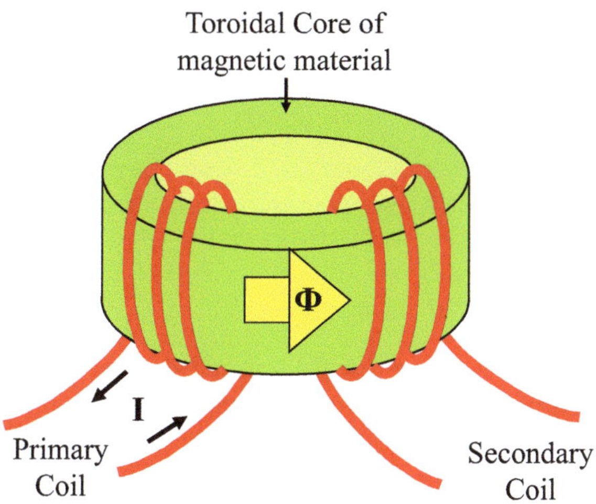

Fig. 3.27: A toroidal transformer.

The current in the primary generates a flux Φ which is confined to the core. A toroidal core is constructed by winding a strip of magnetic material on to a helix to form a circular core. This type of core is more efficient than that shown in Figure 3.26, but it is more difficult to wind the coil on the core. Toroidal transformers are used in instruments such as conductivity meters for liquids.

3.7.18 Hall-Effect Sensor

Figure 3.28 shows the Hall effect on a bar of rectangular cross section made of metal or semiconductor. The bar is placed in a magnetic field **B** and carries a current *I*. A force on the carriers (electrons for a metal bar) will appear (due to the Lorentz magnetic force explained above) deflecting the carriers and generating an electric field **E** in a direction perpendicular to **B** and *I* (Millman & Halkias, 1967; http://www.electronics-tutorials.ws). Output electrodes are placed on the bar to detect a voltage V_H that is proportional to the current and the magnetic field (Eq. (3.30)). The proportionality constant *k* is a function of the bar geometry.

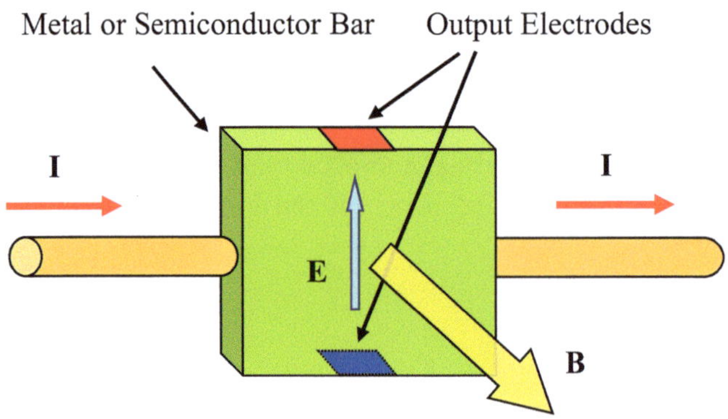

Fig. 3.28: A current *I* flows through a bar made of metal or semiconductor and placed in a magnetic field **B**. Carriers are deflected due to the magnetic force, thus generating an electric field **E**. A voltage can be measured at the output electrodes.

$$V_H = k\,BI \tag{3.30}$$

Then by means of Eq. (3.30), if a known magnetic field is applied on the bar, the current can be measured. Conversely if a known current is supplied, the magnetic field can be measured. Many times these devices are used only as a switch, to open or close a circuit according with the presence or absence of an electric field. In this case they are know as Hall effect switch. Some applications of Hall sensors are in automotive systems for the sensing of position, distance and speed of a shaft.

In environmental instruments it is used with the same purpose. Such is the case of measuring the rotation speed in mechanical anemometers.

3.7.19 Reed Switch and Reed Relay

The reed switch consists of a sealed glass capsule filled with an inert gas containing two ferromagnetic reeds soldered to external contacts (Fig. 3.29). The free ends of both reeds are overlapped but separated by a small gap. When the reed switch is placed in a magnetic field both reeds become opposite magnetic poles attracting each other and closing an electric circuit through the external contacts.

When closed, the contact resistance is of a few milliohms, and when opened, it is on the order of petaohms (10^{15}), then they perform as ideal switches.

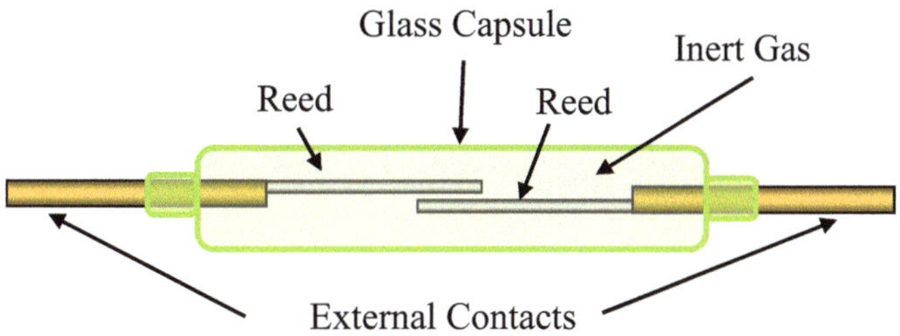

Fig. 3.29: A reed switch.

Reed switches are used in instruments for a variety of applications. A typical one is to turn on and off an electronic circuit housed in a sealed housing (such as underwater equipment). A magnet is placed from outside the housing near a reed switch that is inside the housing. In this way an internal circuit is closed without opening the housing. It is also used to count mechanical events. For this purpose a magnet is attached to a moving piece and a reed switch is placed such that the piece displacement opens and closes the reed switch. This is used in some rain gauges.

When a coil is wound around the glass capsule and a current is made to circulate through the coil creating a magnetic field, the reeds join, so the device can be operated by an electric or electronic circuit rather than a magnet. This is called a reed relay. Because of their high resistance when open, reed relays are used in instruments where high impedance electrodes have to be connected and disconnected, according to a program run by an electronic circuit, such as in automatic analyzer for chemical substances.

3.7.20 Summary of Some Electrical Units

Quantity	Unit	Equivalent
charge	coulomb (C)	ampere second (A s)
current	ampere (A)	coulomb/second (C/s)
voltage	volt (V)	-----
resistance, impedance, reactance	ohm (Ω)	volt/ampere (V/A)
power	watt (W)	$AV = A^2\ \Omega = V^2/\Omega$
energy	joule (J)	-----
inductance	henry (H)	V s/A
capacitance	farad (F)	A s/V
conductance	siemens (S)	$S=1/\Omega$

3.8 Piezoelectricity

3.8.1 Piezoelectric substances

The piezoelectric effect is one of the most used effects to transform mechanical signals into electrical signals and vice-versa. It is the base of many electro-acoustic transducers. The term piezoelectric means electricity obtained from pressure. This effect is observed in some crystals and ceramics and consists in the accumulation of electrical charge as a result of a mechanical stress. The Curie brothers (Pierre and Jacques) were the first in demonstrating this effect. Among other substances, they used crystals of quartz and Rochelle salt. Because piezoelectric materials are used in many sensors and instruments an introduction on this subject seems necessary.

Piezoelectric substances are crystals whose molecules are polarized. Every molecule has one end negatively charged and the other end positively charged, thus forming a dipole. The polar axis is an imaginary line in the molecule that runs through the center of both charges (http://www.electronics-tutorials.ws; www.aurelienr.com; Piezo Systems, 2013).

In some materials piezoelectricity occurs naturally. In others it is generated by a manufacturing process. When some polycrystals are heated under the application of a strong electric field the dipoles line up in nearly the same direction, and if the electric field remains until the substance is cooled the dipoles will stay aligned, becoming a piezoelectric substance.

A piezoelectric material exhibits the following effects:

a) It produces a displacement of its electric charges when a mechanical stress is applied on it.
b) A mechanical deformation is produced on it when an electric field is applied between two opposite faces.

Figure 3.30 shows a simplified bar of piezoelectric material to explain how it is used to manufacture a transducer. The black arrow indicates the polar axis. Two electrodes with conductive wires are represented at the top and the bottom of the piezoelectric bar. The thick arrow indicates compression and extension of the crystal.

At the top center of Figure 3.30 the crystal remains without deformation and without electric charges. At the left, the material is compressed and a voltage of the same polarity as the polar axis appears between the electrodes. At the right the bar is stretched and a voltage of opposite polarity appears.

Conversely, if a voltage is applied to the bar it will be deformed (Fig. 3.30, bottom). A voltage with the opposite polarity to that the polar axis will expand the bar and a voltage with the same polarity will cause the bar to compress.

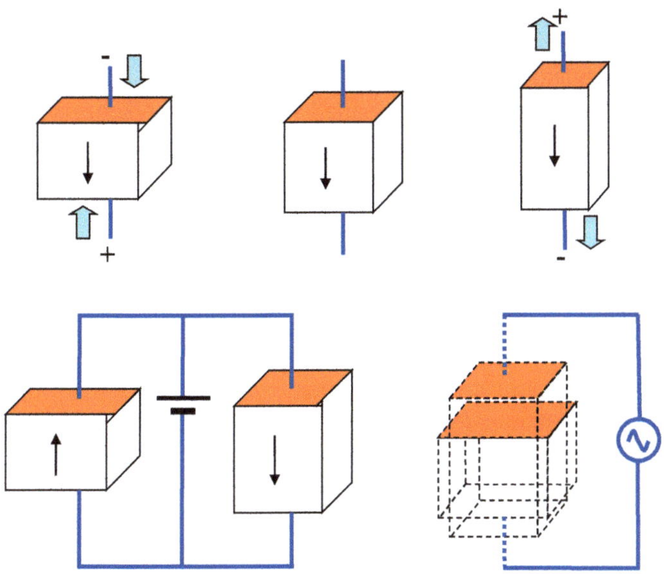

Fig. 3.30: (Top drawings) At the center a piezoelectric bar with electrical contact on top and bottom is shown; the black arrow indicates the polar axis. On the left, the bar is compressed generating a voltage with the same polarity than the polar axis. On the right, the bar is stretched and the voltage polarity changes. (Bottom drawings) A voltage is applied to the bars which compress and stretches them depending on the voltage polarity. Applying an alternating voltage to the bar it vibrates at the voltage frequency; conversely if a vibration compresses and stretches the bar an alternating voltage appears on the electrical contacts.

If an AC electrical signal is applied to the material it will vibrate at the same frequency as the electrical signal. If a mechanical vibration is applied to the material, an electric alternating voltage of the same frequency than the vibration will appear on the electrodes. This property is fundamental to develop mechanical to electrical and electrical to mechanical transducers.

The piezoelectric crystal may have various shapes. Some are crystal bars which bend in different ways at different frequencies. It is said that each shape has its own vibration modes. Several modes have been developed to operate over a wide frequency range from kHz up to MHz.

The thickness of the crystal and the wavelength of the acoustic signal have a particular relationship. If the frequency of the signal corresponds to a wavelength equal to twice the thickness of the crystal the amplitude of the crystal vibration will be maximum. It is said that the crystal vibrates at the resonance frequency. Therefore, the piezoelectric crystal thickness is selected accordingly to the desired range of frequencies in which the transducer will work.

3.9 Magnetostrictive Effect

Transducers based in the magnetostrictive effect are very robust. They can take a great amount of power and can be mechanically overloaded without risk of damage. They are a good alternative to piezoelectric transducers in the low-frequency band and are used in underwater applications, especially in deep-sea measurements.

In 1842, James Joule discovered that ferromagnetic materials change their length when placed in a magnetic field. This property of materials is known as the magnetostrictive effect. In 1856, it was found, conversely, that a change in the stress experienced by a ferromagnetic material produces a corresponding change in its magnetism. This property is known as the Villari effect.

The maximum change in length produced by magnetostriction is 14 parts per million (ppm) for a rod of iron, 30 ppm for nickel and for new special alloys changes can reach 2500 ppm (Bhattacharya, http://www.iitk.ac.in).

If a rod of ferromagnetic material is inserted in a coil, and a current is made flow in the coil, the resultant magnetic field applied to the material will make the length of the rod change from L_1 to L_2 as shown in Figure 3.31a (NUSC, 1990). Conversely, if the rod is stressed by a force **F** while the bar is magnetically biased by **B**, it will produce changes in the rod permeability and a voltage on the coil appears as a result of changes in the magnetic flux (Fig. 3.31b).

Figure 3.32 shows the strain (the change in length per unit length) produced on a nickel rod as a function of the direct current in the coil. The curve is symmetric for positive and negative direct current. Therefore, when an alternating current is applied to the coil the strain is always positive regardless of the direction of the current, so that the mechanical vibration will have twice the frequency of the electrical input. For example, it is frequent to perceive a low frequency hum in transformers connected to a 220 AC power supply. If the power supply is 50 Hz the hum frequency will be 100 Hz. This hum frequency is the result of the strain produced by the alternating current on the iron core of the transformer.

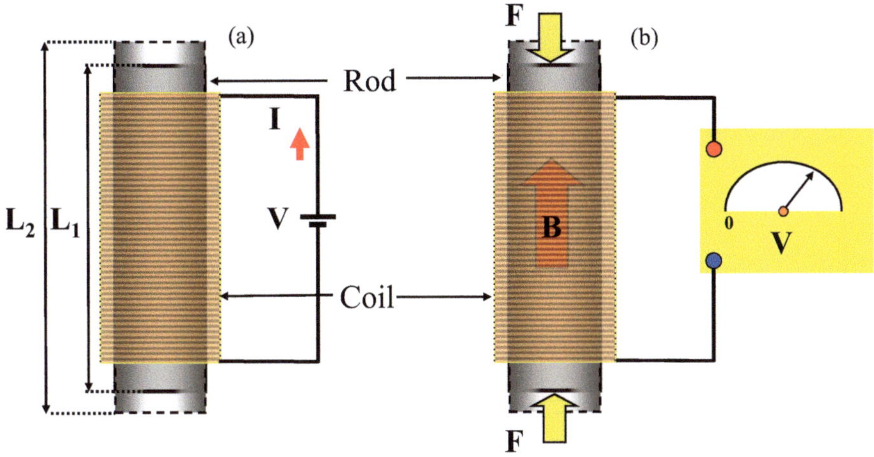

Fig. 3.31: A ferromagnetic cylindrical bar is inserted in a coil. When the coil is connected to a power supply V the magnetic field generated by the coil will change the length of the bar. Conversely, when the bar is biased by a magnetic field **B** and stressed by forces **F**, the permeability of the material changes and the magnetic flux varies producing an induced voltage in the coil.

An alternating current as a function of time (t) and the mechanical response of the rod are illustrated in Figure 3.32. Due to the shape of the transference, for low currents the change in length is small and the transference is not linear.

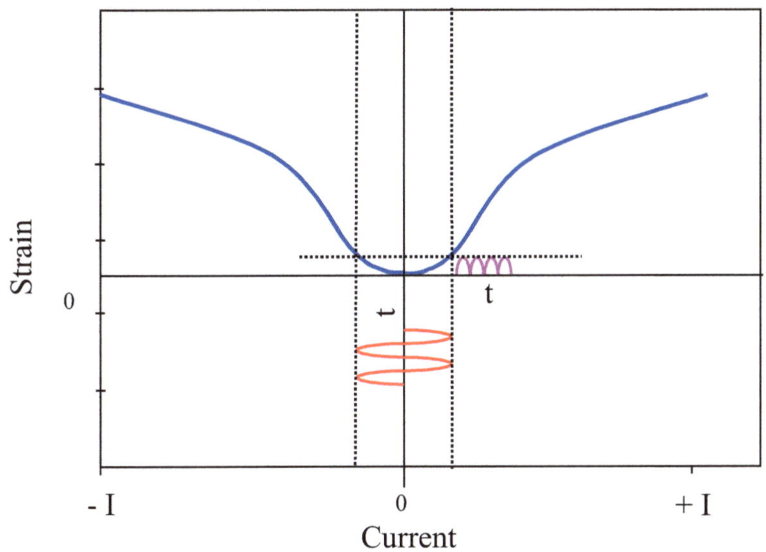

Fig. 3.32: Strain as a result of applied current. An alternating current will produce a strain always positive regardless of the direction of the current.

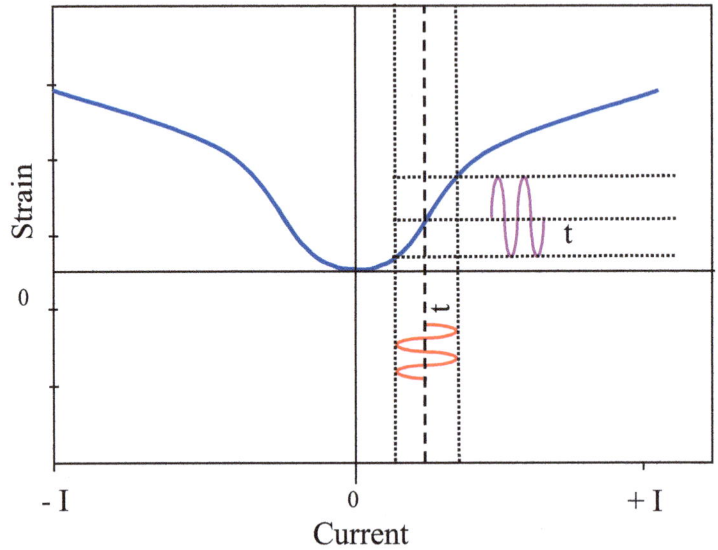

Fig. 3.33: A working point in the linear part of the transferences must be chosen to have the maximum sensitivity and linearity by adding a constant magnetic field to the rod.

For practical applications it is desired to have the maximum sensitivity and linearity of the transference. Then a working point in the linear part of the transferences must be chosen. To place the working point in the linear region a constant magnetic field has to be added to the rod as depicted in Figure 3.33. This can be done by a permanent magnet, by another coil with a DC current or using just one coil where both DC and AC currents can be superimposed.

3.10 The Coriolis Force

3.10.1 Introduction

An application of the Coriolis force concept is found in some mass flowmeters introduced in Chapter 5, and then a discussion of the Coriolis force in this introductory chapter follows.

As it is well known, Newtonian mechanics is valid only in inertial frames of reference. In fact, Newton's first law can be seen as a definition of an inertial frame of reference. The first principle of Newtonian mechanics (principle of inertia) states that there is at least one coordinate system in which if no forces act on a particle, or if the net force acting on the particle is zero, the particle is either at rest or moving with uniform motion in a straight line. In either case the particle is said to be in equilibrium. Such a coordinate system is called an inertial frame of reference, and every other system moving with constant speed with respect to this inertial frame of reference is also an inertial frame. So the principle of inertia postulates the existence of such frames and points out that Newtonian mechanics is valid in these frames of reference.

A coordinate frame that accelerates with respect to an inertial system is not an inertial system. It is a non-inertial coordinate frame and Newton's laws are not valid in it. However, Newton's laws can still be used in such non-inertial coordinate frames if some fictitious forces, the so-called 'inertial forces', are introduced. These inertial forces depend on the acceleration of the non inertial system and are not due to any external driving agent. Newton's third law (action and reaction) is thus not valid for them. One of these inertial forces is the Coriolis force, which acts on bodies moving with respect to rotating systems.

3.10.2 The Coriolis Force

In 1835, the French mathematician and engineer Gustav Gaspard Coriolis (1792-1843) showed that when Newton' s ordinary equations of motion are intended to be used in a rotational reference system, it is necessary to add a force whose direction depends on the direction of rotation of the system; this force is known as the Coriolis force.

Figure 3.34a shows a platform rotating with an angular velocity $\boldsymbol{\omega}$ where, at time t_0, a person standing at the axis of rotation A throws an object with velocity **V** (relative to the rotating platform) to someone standing at B_1, Δt being the time of flight of the object. An observer who is outside the platform will see that the object describes a straight trajectory that ends at B_2 (Fig. 3.34b), but the person who was at B_1 moved during the flight time of the object, so he will perceive that the object deviates describing a curved path (Fig. 3.34c). For the observer on the platform, which is not aware of its displacement, it is as if a force (**Fc**) acting on the object deflects it. This force is described by Eq. (3.31) and is known as the Coriolis force.

$$\mathbf{F}_C = -2\,m\,(\boldsymbol{\omega} \times \mathbf{V}) = -2\,m\,\omega\,V \sin\theta\,\mathbf{n} = -2\,m\,\omega\,V\mathbf{n} \qquad (3.31)$$

In Eq. (3.31), m is the mass of the object thrown and × represents the vector product. Recall that the result of a vector product is a new vector perpendicular to the plane containing both multiplying vectors, and whose modulus is the product of the moduli of both vectors and the sine of the angle between them. It was assumed that in Figure 3.34c **V** is in a plane parallel to the platform, and since $\boldsymbol{\omega}$ is a vector perpendicular to the plane of rotation, $\theta = 90°$. The symbol **n** is a unit vector in the direction perpendicular to the plane containing $\boldsymbol{\omega}$ and **V**. Therefore, for the observer located on the platform, it happens as if on the object acts a force **F**c (the Coriolis force) that changes its trajectory.

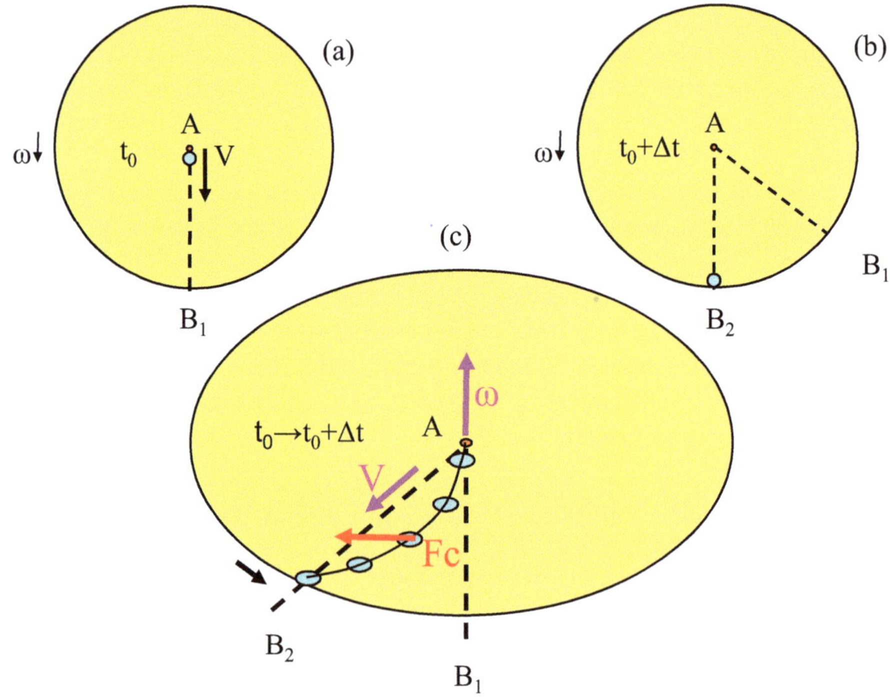

Fig. 3.34: (a) and (b) A person standing at the center of a platform rotating with angular velocity **ω** throws an object with velocity **V** towards B_1. An observer from outside the platform will see the object describing a straight trajectory from A to B_2. (c) A person standing on the platform at B_1 will see a curved trajectory from A to B_2. Then, for Newton's second law to still be valid for the observer on the platform, it is required to add a fictitious force **Fc**, the Coriolis force, given by Eq. (3.31) and perpendicular to the plane determined by **ω** and **V**.

3.10.3 Examples of the Coriolis Force

(1) If a long-range gun is shot from a point located on the equator, aiming at a target in the northern hemisphere, at the time the projectile leaves the rotating platform (earth) it is subjected to two speeds, one of them is the tangential eastwards speed of the Earth at the equator, and the other, due north, imposed by the gun.

The projectile will land to the East of the place the gun was aimed at. This deviation is because at the moment that the projectile is fired, the gun moved eastward with a speed greater than the target in the North. Conversely, if the gun were fired from the northern hemisphere pointing to a target located on the equator, the projectile would land west of the target. In this case, the target moves quickly eastward than the projectile at the time it was shot.

(2) If a projectile is fired in the direction of a meridian, from a point located at a certain south latitude, to a point in the northern hemisphere at the same latitude, the projectile will fall on the same meridian, as both points rotate at the same speed. An observer standing on the same meridian but at the equator, who observes the firing cannon (i.e. the observer is looking southward), will see that the projectile is deflected to his right following a curved path, and will land behind him.

(3) Suppose an object at rest is left to fall from the top of a very tall imaginary tower located on the equator (Fig. 3.35). Since the tangential velocity of the object, at the time it starts to move, is greater than the tangential speed at the base of the tower, the object will impact some distance east of the base of the tower. This easterly drift is, however, very small. It can be shown that if an object at rest is dropped from an altitude z, it will experience an easterly drift (x) (regardless of the hemisphere) given by (Goldstein, 1959)

$$x = \frac{\omega}{3}\sqrt{\frac{(2z)^3}{g}}\cos\varphi$$

where ω is the angular velocity of the Earth ($7.29 \times 10^{-5}\ s^{-1}$), φ is the latitude, and g is the acceleration due to gravity ($9.8\ m\ s^{-2}$). For example, if $z = 30$ m, x will be zero at the poles ($\varphi = 90°$), 2.55 mm at $\varphi = 45°$, and 3.60 mm at the equator ($\varphi = 0°$). The real experiment is, however, difficult to carry out because the small drift is likely to be perturbed by the wind, the viscosity of the air and other atmospheric agents.

If the object were restricted to fall within a narrow flexible tube, the easterly drift due to the Coriolis force would make the object push the east wall of the tube. As before, this force would be proportional to the object's mass, its velocity and the angular velocity of the earth.

(4) If a fluid were pumped through the flexible tube of the above example from the top to the base of the tower, the liquid would experience a similar easterly drift as the object, again making the fluid push the east wall of the tube.

(5) The Coriolis force on a circulating fluid may be easily visualized by means of a hose, full of water, fixed and hanging from the two ends and oscillated as a pendulum. If the water is motionless, the hose will oscillate as a swing. This means that the hose will oscillate in a plane which rotates about an axis passing through the fixed points of the hose. Once water flows, forces appear on the hose which twist it out of the plane. This effect will be depicted in Chapter 5 when describing a mass flowmeter. This is a good and simple experiment to introduce the functioning principle of the Coriolis mass flowmeter.

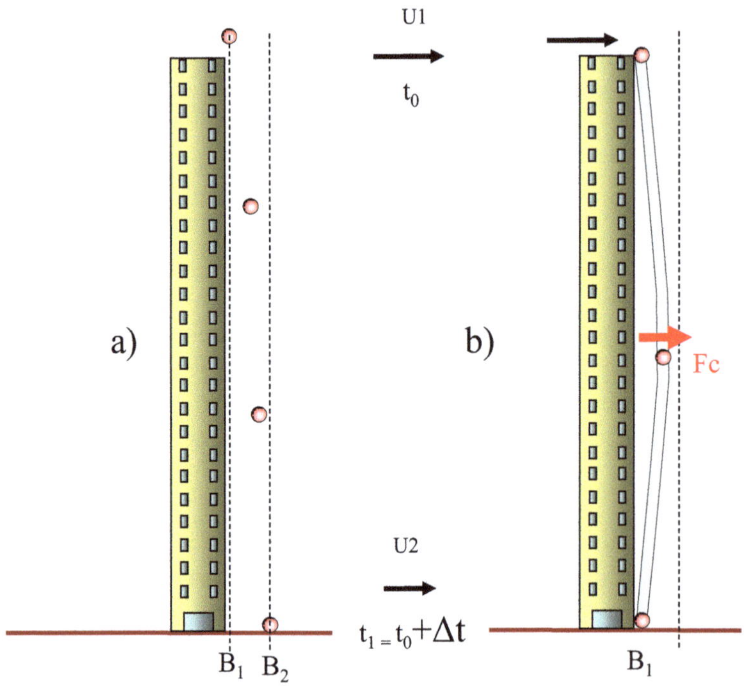

Fig. 3.35: (a) The tangential velocity U1 at the top of the imaginary very tall building is higher than U2 at its base, thus the object will impact at B_2, which is to the east of B_1. (b) If a flexible tube is placed on the wall of the building, the object falling within the tube would apply a force on the tube.

References

Bhattacharya, B. *Terfenol and Galfenols: Smart Magnetostrictive Metals for Intelligent Transduction.* http://www.iitk.ac.in/directions/dirnet7/P~BISHAKH~F~DIR7.pdf

Goldstein, H. (1959). *Classical Mechanics.* Reading, Massachusetts: Addison-Wesley Publishing Company, Inc.

Johnson, D. H., & Dudgeon, D. E. (1993). *Array Signal Processing: Concepts and Techniques.* Englewood Cliffs, NJ: Prentice Hall, Inc.

Kip, A. F. (1962). *Fundamentals of Electricity and Magnetism.* New York: McGraw-Hill Book Company, Inc.

Millman, J., & Halkias, C. C. (1967). *Electronic Devices and Circuits.* Tokyo, Japan:
McGraw-Hill Book Company.

NSC (National Semiconductor Corporation). (1980). *An Introduction to the Sampling Theorem.* Application Note 236.

NUSC (Naval Underwater Systems Center). (1990). *Acoustic Measurements for Sonar Transducer Test Personnel, Student Guide.* New London, CT. http://gbenthien.net/pdf/acoumeas.pdf

Piezo Systems, Inc. (2013). *Introduction to Piezo Transducers.* http://www.piezo.com/tech2intropiezotrans.html Last access: February 12, 2013.

Skilling, H. H. (1974). *Fundamentals of Electric Waves.* Huntington, New York: R. E. Krieger Publishing Company.

Smith, S. W. (2003). *Digital Signal Processing: A Practical Guide for Engineers and Scientists*, Newnes (an imprint of Elsevier).

http://www.electronics-tutorials.ws/electromagnetism/hall-effect.html

www.aurelienr.com/electronique/piezo/piezo.pdf

The Piezoelectric Effect Last access: February 12, 2013.

4 Sensors

4.1 Introduction

This chapter explains how a measurand can modify some property of a sensor in order to produce an electrical signal on which the measurand information is "impressed". In this way the information is passed to the electrical domain which permits managing it with all the tools available for electrical signal acquisition, transmission, processing and storing.

The description of some sensors will need information already presented in Chapter 3, so that several references to the previous chapter will be done to help those who skipped the reviewing chapter.

Before describing the sensors in detail some elementary electric circuits will be introduced in order to visualize how sensors are included into the circuits which are essential parts of the instruments. Although generic, these ideas are probably new for most readers, and because of this they are introduced early in the chapter to facilitate the understanding of the following material.

The concepts on electronics introduced in Section (4.2) could not only be considered as a simple summary for those with little experience in electric circuits, but they could also be of help when some simple laboratory measurements are required. Among these concepts are the dispositions of sensors in series or parallel according to their impedances, the Wheatstone bridge and the four wire technique, which are classical ways of connecting sensors. Also, a summary on operational amplifiers is presented to demystify these very useful devices for amplifying sensor signals. A simple to build amplifier will be described for students who could require amplifying small signals and do not have any experience in electronics.

The descriptions are first centered on how the elementary electrical devices such as impedances are modified by external parameters (measurands) in order to become a sensor. It would not be possible to understand how sensors work without a clear understanding on how an electrical signal "represents" a non-electrical measurand. We could call these elementary sensors proto sensors because they are the basis or origin of more sophisticated sensors. Grasping these simple devices prepares the readers for understanding any other sensor that they could find in the future.

Basically, any sensor with electrical output can be assimilated to one of these two cases:

1. An electrical generator, as is the case of transducers that transform some type of energy into electrical energy.
2. A variable impedance (resistance, capacitance or inductance).

The measurand, which is the variable that we want to know, is "transformed" into an electrical variable. In the transducers case the measurand produces some kind of energy that the transducer transforms into electrical energy. In the second case, the sensor is a variable impedance and the measurand just modifies this impedance.

In the transducer case the first electronic conditioning step is signal amplification to bring the signal to adequate levels for further treatment. In the second case, the sensor (variable impedance) must be part of an electric circuit to copy the measurand information into electrical information.

Therefore, the descriptions of the functioning principle of sensors to be carried out in this chapter are not merely explanations of sensors used in Environmental Sciences. They are also intended as a didactic tool to prepare the readers to understand the working principle of any sensor that they could meet over the course of their careers.

After grasping the simplest ideas on how sensors convert measurand information into electrical information, more complex sensors are presented which will be the first stage of the instruments described in the following chapters.

An introduction to oscillators is also presented here. It is essential to understand how they work because they are the base of clocks, alternating circuits, acoustic and electromagnetic waves generation and excitation sources for some sensors. The treatment of this topic has been postponed a bit since some knowledge on sensors and actuators is needed to make this subject more understandable.

4.2 Introduction to Electric and Electronic Circuits

In order to obtain an electrical signal that can be useful later as part of a measuring system, sensors must be part of an electric circuit. The most adequate circuit for each sensor depends on the sensor characteristics. If the sensor generates its own electrical energy, as it happens with transducers, the first electrical stage is usually a simple amplifier. If the sensor does not produce any electrical energy it must be included in an electric circuit whose electrical characteristics can be modified by the measurand. In this case the sensor's impedance is modified according to the phenomenon we want to measure. Then the circuits accompanying sensors have to be selected according to their electrical characteristics.

Some schematics of circuits are shown below with the purpose of introducing the reader to the subject. Because they are easier to understand, DC circuits using resistive sensors are used. In this case sensors change their **resistances** due to a change in the measurand. The concepts so introduced will be later extended to AC circuits where sensors change their **impedances** due to changes in the measurand.

Figure 4.1a shows a constant voltage battery connected in series with a resistive sensor and an ammeter which measures current. This kind of circuit is used when the sensor has a relatively high resistance compared with that of the measuring circuit (ammeter), thus almost all resistance in the circuit is due to the sensor. Therefore, the circuit results predominantly sensitive to sensor variations because the resistance of the measuring circuit can be disregarded (Section (3.7.6)). Changes in the measurand produce a considerable current modification in agreement with Ohm's law. The arrow crossing the resistor symbol indicates that it is a resistor that varies according to the external measurand. In this circuit the "exciting" source (which is fixed) is the voltage and the "output" variable signal is the current, on which the measurand information is "copied".

Another kind of circuit is used when the sensor has a relatively low resistance compared with that of the measuring circuit. The circuit consists of a constant current source (as the exciting source) connected in series with the sensor. The voltage on the sensor (output) is measured by a voltmeter placed in parallel with the sensor (Fig. 4.1b). The voltmeter is assumed to have a high input resistance and practically no current is derived through it. Thus the constant current passing through the variable sensors' resistance (which is modified by the measurand) produces a variable voltage. Therefore, in this case, the measurand information is "copied" on the voltage (Section (3.7.6)).

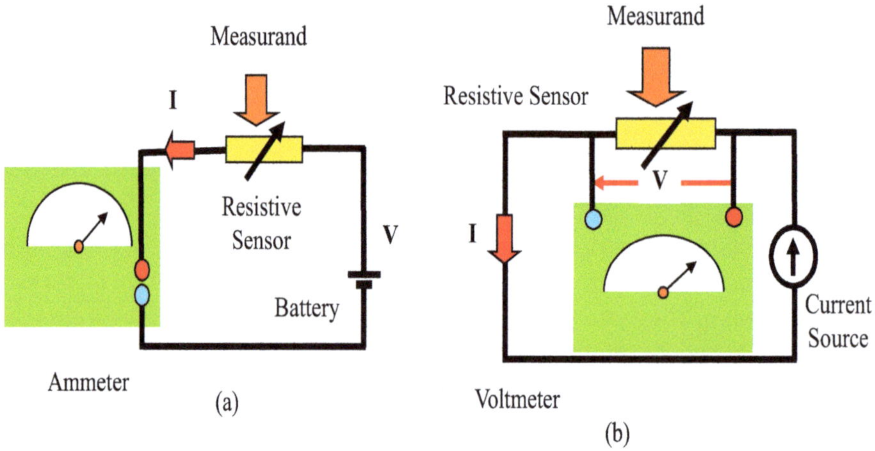

Fig. 4.1: (a) The sensor resistance is much larger than the ammeter resistance, thus almost all the measured current is due to the sensor variations. (b) The voltmeter resistance is much larger than the sensor resistance then the measured voltage is due to sensor variations.

4.2.1 Wheatstone Bridge

The Wheatstone bridge is a high sensitivity circuit often employed to process sensor signals analogically before entering the amplifying stage. It was originally a laboratory circuit composed of four resistors (some of them being the sensing elements), a voltage supply and a galvanometer used as a null-balance meter (Fig. 4.2a).

When the voltage drop between A and B is equal to zero, the galvanometer will indicate zero, and the bridge is said to be "balanced". The balance condition is achieved when the resistors are related by

$$\frac{R1}{R2} = \frac{R3}{R4} \tag{4.1}$$

The Wheatstone bridge allows the value of an unknown resistor to be determined when the other three are known and the bridge is led to the balance condition. With the purpose of obtaining the zero current condition through the galvanometer, a set of variable resistors with accurately known resistances should be available. For example (Fig. 4.2b), if *R*x is unknown, $R3/R4 = K$ is a known constant, and *R*v is a variable precise resistor which is adjusted to balance the bridge, then Rx is given by

$$Rx = K\,Rv \tag{4.2}$$

A modified version of the bridge is used in modern instruments where the galvanometer is replaced by the input of an instrumentation amplifier. The bridge is not led to balance but a small imbalance is admitted. Also, this bridge has been generalized to measure impedances. In this case the battery is replaced by an AC voltage of adequate known frequency. With this extended version of the bridge, capacitances and inductances can be measured. Both versions of the bridge will be seen later on as part of measuring instruments.

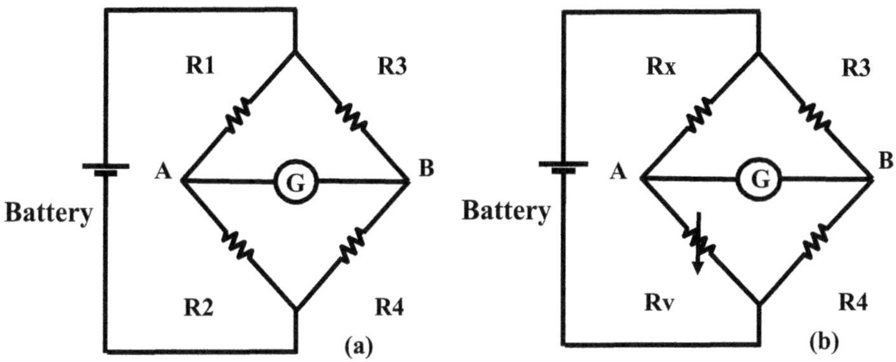

Fig. 4.2: (a) Wheatstone bridge, (b) The resistor Rv is used to measure Rx.

4.2.2 Four Wire Technique

Other electrical circuit frequently used with low-resistance resistive sensors is known as *four wire sensing*. Sometimes the sensor resistance is so low that it is comparable to the cable resistance. This can happen when long cables are needed because the sensor is far from the power supply (*V*s) and the measuring instruments. In these cases the voltage drops through the cables add to the sensor's voltage and could result a serious problem if the cable resistance vary with temperature, because cable variations become indistinguishable from sensor variations (Fig. 4.3).

The four wires sensing method separates the current supply circuit from the voltage measuring circuit (Fig. 4.4). At first glance it seems amazing that by adding more cables as *w*1 and *w*2, the influence of the cable drops decrease. It happens that the voltmeter is an instrument with high internal resistance and needs a very low current (*Im*). Then, the current in cables *w*1 and *w*2 are very low and so are the voltage drops in series with the voltmeter, thus the voltmeter measures practically only the voltage on the sensor.

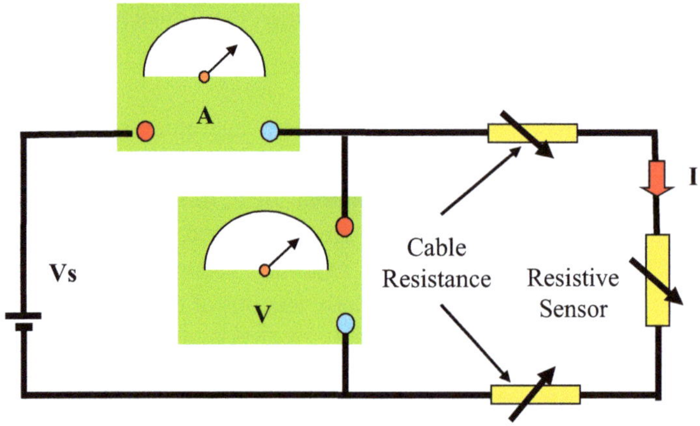

Fig. 4.3: An ammeter A and a voltmeter V are used to measure the sensor resistance. The cable resistance appears in series with the sensor, thus cable variations are indistinguishable from sensor variations, which are the signal we want to measure.

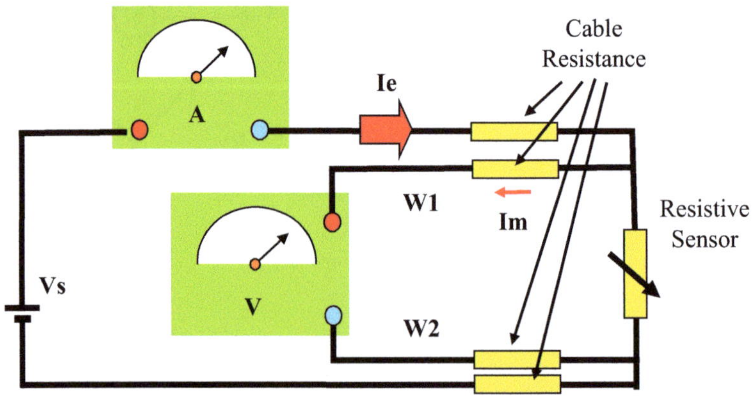

Fig. 4.4: The cables *w*1 and *w*2 separate the current supply (*Ie*) circuit from the voltmeter circuit. Because the voltmeter has a high internal resistance, the current *Im* is much lower than *Ie*, thus proportionally decreasing the voltage drops in series with the voltmeter. Therefore, the voltmeter measures approximately the resistive sensor voltage.

All the concepts explained above relative to circuits with resistances can be extended to cases where alternating voltages and impedances are used.

4.2.3 Operational Amplifier

Operational Amplifiers (OA) are devices very often used in electronics and they will be introduced conceptually to facilitate the understanding of signal amplification in instruments. In general, sensors require some degree of amplification and it would be interesting that researchers could have the ability to employ OA to adapt sensor's signals to data loggers or other measuring instruments such as testers and oscilloscopes. In order to adequately design with OA it is necessary to have some background on electronics. Nevertheless, a practical circuit that students could use in their projects will be introduced.

Two OA are depicted in Figure 4.5; the triangle represents the OA and the word GAIN indicates the amount of times that the input signal is amplified by the OA. The amplifier has two connections for power supply (+PS and –PS), two inputs (INV and NON-INV) and one output. Also shown are a voltage input signal *V* and an electrical potential adopted as reference (REF).

This is an OA which requires dual power supply, one positive (+PS) and one negative (–PS), with respect to the REF. The input signal might be the signal from a sensor and could be positive or negative with respect to REF. Because in both cases one input terminal is connected to the REF it is said that the OA has single-end input, in opposition to differential input OA which will be described below.

When the input signal to be amplified (*V*) is introduced between the REF and the NON-INV input, as in Figure 4.5 (a), the output will have an amplitude equal to that of the input signal times the GAIN, and will show the same polarity than the input. This means that if the input is increasing with time, the output will also be increasing.

When the input *V* is connected to INV the amplitude of the output will also be the input signal times the GAIN, but with the opposite polarity than the input, as shown in Figure 4.5 (b). In other words, if *V* is positive with respect to REF, the output will be negative.

The interesting capability of an OA is that the gain may be very large (10^6) so that very small signals can be greatly amplified. Frequently, most sensors have very low analog signals and OA are essential to amplify sensor signals before converting them into digital ones (see Section (3.6.7), Application example, data logger).

It should be taken into account that the amplifier also amplifies the noise. Then, certain precautions must be taken to avoid noise entering the amplifier.

The NON-INV input has very high input impedance; this means that it will practically take no current from the input signal *V*. This implies that the introduction of the OA will not perturb the signal source. If *V* is the output of a sensor with high internal impedance (such as a PH meter electrode) we would like to connect it to the

NON-INV input to disturb the sensor signal as little as possible. Remember that the input of the amplifier is in parallel with the sensor. In contrast, the INV input usually needs some current from *V* and should be avoided with sensors with high internal impedances.

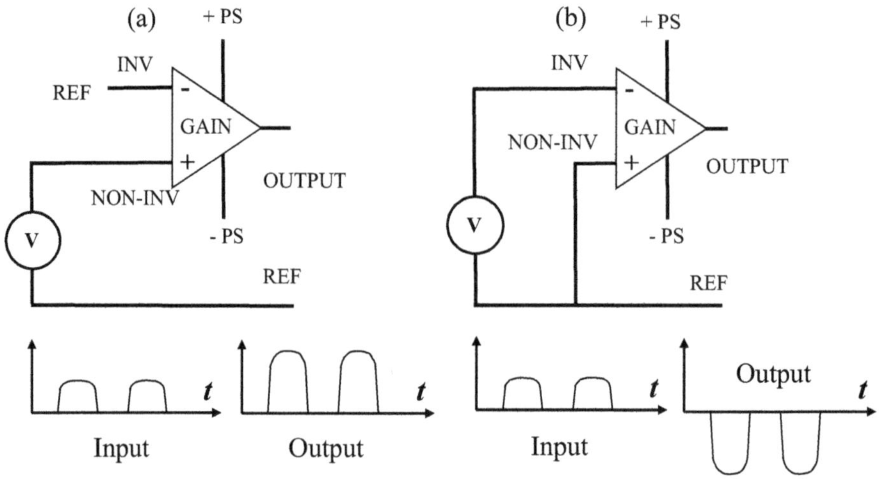

Fig. 4.5: The triangle with five connections indicates the OA; *V* is a voltage input signal, and REF an electrical potential adopted as a reference potential. In (a) the OA is connected in non-inverted mode and in (b) in inverted mode. In the first mode the output is an amplified version of the input with the same shape; in the second is also amplified but the waveform is inverted.

In sum, the OA is a device which can multiply the input voltage by the GAIN of the amplifier. The output waveform can have the same or opposite polarity than the input, depending if the input signal is connected to the NON-INV or INV input respectively. The OA requires a power supply for working (two in this case of a dual-power supply OA). The GAIN of the schematics of Figure 4.5 is fixed by external resistors not drawn.

There are other amplifiers in which the input signal is connected between two input terminals (Fig. 4.6); they are called *differential* OA. They can have two kinds of outputs. One, with only one output connector; the output signal is thus referred to a reference voltage as in Figure 4.5. The other with two output connectors; the signal is obtained between them. In the second case, where inputs and outputs have two terminals, OA are called *fully-differential*. In fully-differential OA the amplitude of the output voltage (between the two outputs terminals) is the signal at the input terminals times the gain of the amplifier.

For example, a differential input amplifier is usually employed to amplify the signal of a Wheatstone bridge. The input of the amplifier is connected replacing the galvanometer in Figure 4.2b.

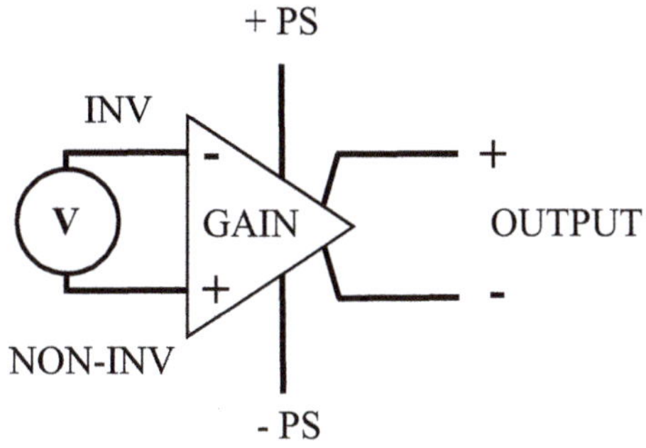

Fig. 4.6: Fully-differential Operational Amplifier.

Another application could be to replace the ammeter in Figure 4.1a by a low resistance (R) and a fully-differential amplifier (Fig. 4.7). Thus, by Ohms' Law, the input signal to the amplifier is the current times the value of the low resistance (voltage drop across R). The output voltage (V_{OUT}) of the OA is the input voltage times the gain (G) of the amplifier (Eq. (4.3)). Therefore,

$$V_{OUT} = I\,R\,G\,. \tag{4.3}$$

In this case, because R is low, the voltage drop across it will be also low, thus G has to be high to obtain a good output signal.

In a similar way, in a real instrument using the circuit of Figure 4.1b, the fully-differential amplifier could replace the voltmeter. In this case perhaps the gain could be low because the voltage on the sensor is large.

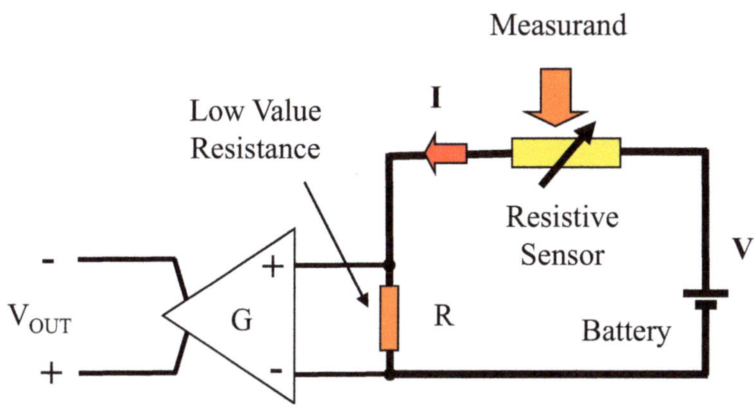

Fig. 4.7: The ammeter of Figure 4.1a is replaced by a low resistance and a differential amplifier.

An OA extensively used in instruments, which due to its good performance is still recommended for new designs, is the INA 111. This instrumentation amplifier will be presented with a double purpose: to illustrate the previous concepts, and to provide an easy to build amplifier for students which require amplifying small signals and do not have any experience in electronics.

The circuit of Figure 4.8 represents the INA 111 with some external components (http://www.ti.com/lit/ds/symlink/ina111.pdf.). Note that this is an ideal representation because the polarization resistances needed at the input of the amplifier are not shown. They will be added in Figure 4.9. This OA has differential input because it has two inputs (V_{IN} – and V_{IN} +) where the input signal (sensor signal) is connected.

The output voltage is obtained on the LOAD, between two terminals: the amplifier's output (V_{OUTPUT}) and the reference (REF); thus, because it has only one output connector it is not a fully differential amplifier. The load may be simply a resistor, the input of a data logger, a voltmeter, an oscilloscope, etc. In this case, the reference is connected to ground, which is the point where the positive and negative voltage supplies +PS and –PS are connected too. Two 0.1 µF capacitors (*C*) are connected in parallel to the power supplies just to reduce the electrical induced noise.

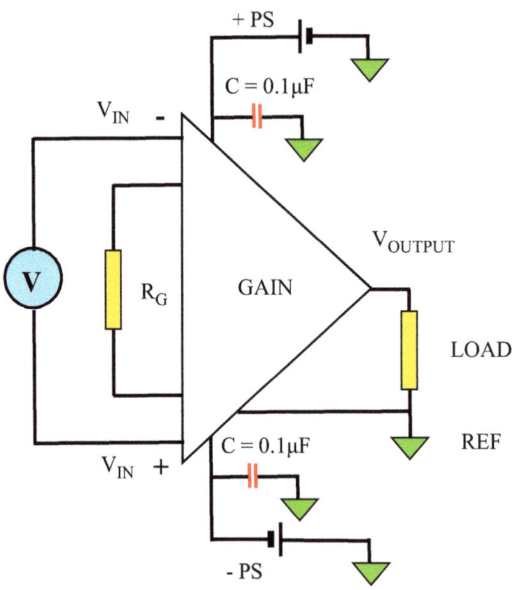

Fig. 4.8: Schematic general diagram of an INA 111.

The GAIN is user selectable; it is the factor by which it is desired to multiply the input voltage to obtain the required output voltage. In the INA 111, this factor is selected by placing a resistor R_G between two pins of the OA. The output voltage as a function of the input voltage between V_{IN} − and V_{IN} + and the gain (G) is shown in Eq. (4.4), where the gain has subsequently been replaced by a function of the resistor's value R_G expressed in ohms. This equation is provided by the OA manufacturer.

$$V_{OUTPUT} = GAIN \times (V_{IN+} - V_{IN-}) = \left(1 + \frac{50.000}{R_G}\right) \times (V_{IN+} - V_{IN-}) \tag{4.4}$$

Approximate values of R_G, using standard resistors are shown in Table 4.1 to obtain different gains (http://www.ti.com/lit/ds/symlink/ina111.pdf.).

Table 4.1: Gains and standard resistors R_G

Desired gain	**5**	**10**	**20**	**50**	**100**
R_G Standard resistor (kΩ)	12.4	5.62	2.61	1.02	0.511

The circuit of Figure 4.8 needs only a few more resistors for practical use. The value and position of these resistors depend on the signal source (sensor) characteristics. For the sake of clarity these resistors are drawn in Figure 4.9, where Figure 4.8 has been simplified (+PS, -PS, R_G and both C are not shown). The two circuits presented could cover many amplification applications. Circuit (a) has a signal source with a large internal resistor R_S such as crystal or ceramic transducers, and requires two large resistors for polarization, one at each input terminal; R_P = 1,000,000 Ω being an appropriate value for both resistors.

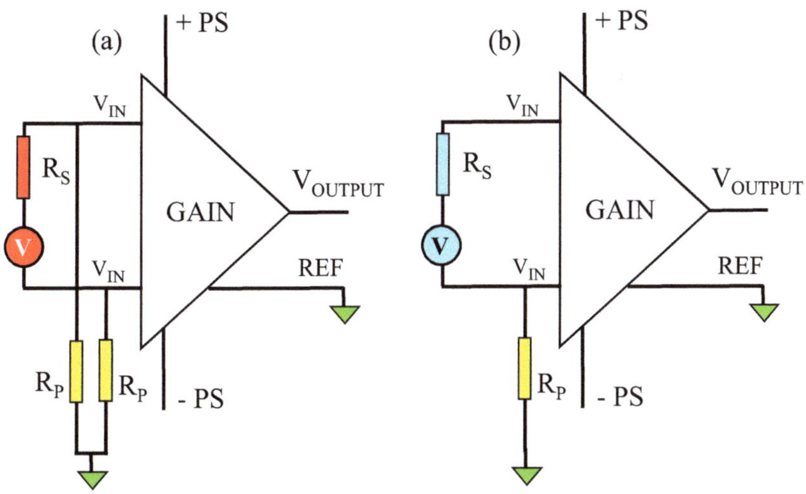

Fig. 4.9: Schematic general diagram for two application cases of an INA 111. In (a) R_S is large and two R_P are needed. In (b) R_S is small and only one R_P is required. For selecting R_P values see the text.

Circuit (b) is used with signal sources with small internal resistors R_S such as thermocouples; in this case only one $R_P = 10{,}000\ \Omega$ is adequate. Therefore, adding one or two resistors to the circuit of Figure 4.8 a practical amplifier can be built for low or high internal resistance signal sources respectively.

For applications where these circuits do not work appropriately some help could be found in http://www.ti.com/lit/ds/symlink/ina111.pdf.

4.3 Resistive Sensors

Resistive sensors are devices that modify their resistance due to changes in a given physical or chemical quantity (measurand) acting on them. As shown in Eq. (3.12) (repeated in Eq. (4.5)) the electrical resistance of a practical resistor depends on the conductivity of the material (σ), its length (l) and cross section (A) ($\rho = 1/\sigma$ is the resistivity of the material). Therefore, in order to have a **resistive sensor** some of these parameters have to vary in response to an external action, so that changes in resistance might be induced by the measurand,

$$R = \frac{\rho l}{A} \tag{4.5}$$

Some examples of resistive sensors are presented below. Sometimes, resistors are built of a material which has a constant resistivity, and mechanical strains induced by the measurand alter their geometry (length or cross section). In this way **R** varies due to a dimensional change of the resistor (Eq. (4.5)); this is the case of sensors for measuring mechanical stresses, called **strain gauges** (described below in Section (4.6)).

A different case is when the resistor's geometry (length and cross section) is kept constant, but the resistivity of the material that forms the resistor changes with a certain external parameter (measurand). For example, if the material forming the resistor were porous and capable of absorbing moisture, ρ in Eq. (4.5) could change because of the water entering the pores of the resistor's material. Thus, the resistor would change with the amount of water absorbed and a resistive **humidity sensor** could be obtained (described below in Section (4.9)).

4.3.1 Sensors Based on Potentiometers

So far we have been describing resistors that have two terminals. The potentiometer is a special case of a resistor which has three terminals and its resistance can be varied; it was originally known as rheostat. A schematic of a potentiometer is depicted in Figure 4.10. It consists of a resistive strip with two fixed terminal at the ends and a third terminal (called wiper) that can slide along the strip.

Let us define a number n, $0 \leq n \leq 1$, such that $n = 0$ when the sliding terminal touches the upper terminal and $n = 1$ when it touches the lower terminal; the wiper divides the resistor in two parts whose sum is R.

The circuit on Figure 4.10a is the potentiometer connection known as voltage divider. The battery voltage (V) is applied between the upper and lower ends of the resistor (R), then the current Ip in the circuit is constant $Ip = V/R$ and the voltage at the wiper, referred to the lower (negative) terminal of the resistor, is variable. Thus, by displacing the wiper it is possible to control de output voltage of the sliding terminal between V and zero.

The potentiometer on Figure 4.10b is used as a variable resistor; the battery is connected between the wiper and one of the fixed terminals. In this circuit voltage at the sliding terminal, referred to the lower (negative) terminal of the resistor is always constant and equal to the battery voltage. Instead, the current in the circuit is variable $I = V/[(1 - n) R]$; in practice, I is always finite because the battery cannot supply infinite current.

In many electrical devices the wiper is usually manually slipped by means of a knob or lever with the purpose of modifying the level of a voltage or current. For example, it can be used to set the volume of an audio power amplifier. Potentiometers are presented in this introduction because they can be part of electro-mechanical sensors. They can convert the motion of a mechanical device which displaces the wiper into an electrical signal. In other words, they may be a sensor, converting motion into electrical signal.

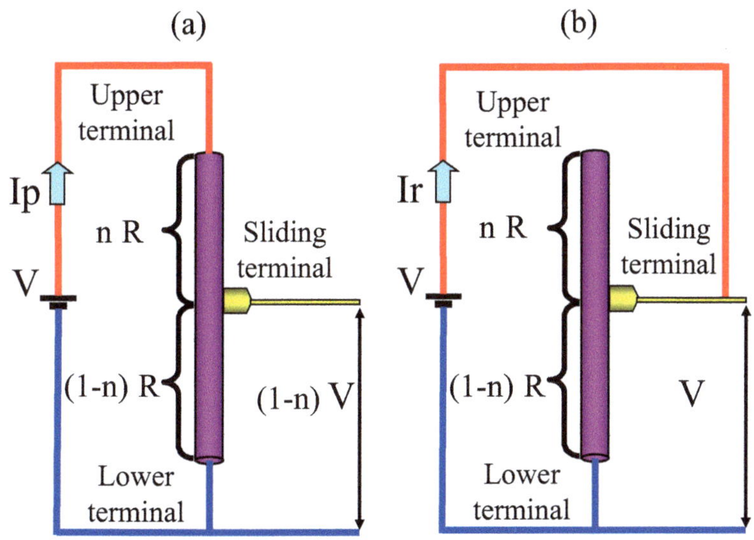

Fig. 4.10: (a) A potentiometer used as a voltage divider. (b) A potentiometer used as a variable resistor.

The concept of potentiometer as a voltage divider is used in well known devices as joysticks and touch-screen. Also they were used in tide gauges, anemometers, and instruments which need to account for mechanical displacements. At present, they are replaced by optoelectronic encoders in several applications.

4.4 Impedance Sensors

Devices that modify their electrical impedance due to changes in a measurand are called *impedance sensors*. They are of capacitive or inductive type, or a combination of both. The circuits used with impedances are similar to those discussed for resistive sensors (Figs. 4.1 to 4.10), but the exciting sources must have alternating waveforms. Most times they are sinusoidal voltages or currents with a frequency from kHz to MHz.

In some circuits the excitation is a voltage source whose frequency and amplitude are kept constant. The electrical output signal is produced due to changes in the impedance (capacitance or inductance). These changes are produced by the variation of the measurand that directly modifies some impedance characteristics, similar to the cases shown in Figure 4.1 for resistive sensors. Therefore, the current or the voltage is modified becoming the electrical signal output which represents the measurand.

In other circuits, the impedance is part of an oscillator whose resonant frequency is a function of the sensor's impedance. Thus, when the measurand modifies the impedance, the oscillator's frequency varies with the impedance changes. In this case the output signal where the measurand variation is "impressed" is the frequency. Perhaps these ideas will be better understood after reading Section (4.5).

In any case, to understand how impedance sensors work, we must study how the impedance could be changed by an external parameter such as the measurand. As shown for resistors, an impedance will also depends on its geometry (length and area) and on the material it is made of.

4.4.1 Capacitive Sensors

It was stated (Eq. (3.16)) (repeated in Eq. (4.6)) that the capacitance of a parallel plate capacitor is directly proportional to the facing surface of the plates (A) and to the permittivity of the dielectric material (ε) separating both plates, and inversely proportional to the distance between them (d),

$$C = \frac{A\varepsilon}{d} \tag{4.6}$$

The capacitance can thus vary due to geometric changes or changes in the permittivity of the dielectric material. In general, for capacitive sensors, the plate area is fixed in the manufacturing process, but the facing surface between the two plates may be changed as in Figure 4.11a. Also the distance between both plates can be modified as in Figure 4.11b. Some sensors have a fixed geometry but the dielectric material moves as in Figure 4.11c, changing the amount of dielectric between plates.

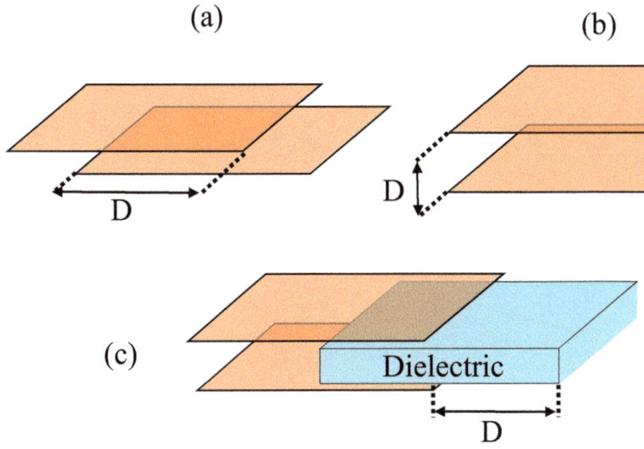

Fig. 4.11: A variable capacitor. (a) The facing surface varies. (b) The distance between plates varies. (c) The dielectric varies.

In Figure 4.11, *D* indicates a change of distance between the plates, or in the dielectric position with respect to the plates. When the sensor is used to measure displacements, the object whose position is being measured is coupled mechanically to that part of the capacitor that produces the change in the capacitance.

In other sensors all parts are fixed but the dielectric properties of the material change due to chemical or physical changes. For example, in some humidity sensors, moisture changes the permittivity of the dielectric material.

4.4.2 Inductive Sensors

A variable inductance (L) is basically a winding of N turns that has a movable core of magnetic material. Equation (3.18) (repeated as Eq (4.7)) indicates how to calculate the inductance for long coils.

$$L = \frac{\mu N^2 A}{l} \tag{4.7}$$

Theoretically, to modify the inductance, the geometry of the coil or the core magnetic permeability could be altered. In practice, the coil length and area cannot be easily modified, and then to have a variable inductance the permeability should be changed. But because the magnetic permeability is a property of the core material, once constructed it cannot be changed. Thus, the easiest way to change the inductance is by displacing the core a given distance *D* into or out of the coil, as shown in Figure 4.12. This device could be used to measure displacements. In general, the coil remains fixed and the core is connected to the piece which transmits the measured displacement.

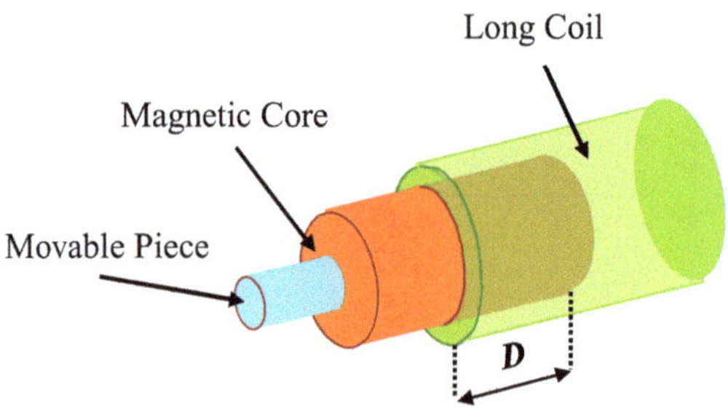

Fig. 4.12: A movable piece attached to a magnetic core displaces it inside a coil, changing the coil inductance as *D* varies.

Another inductive sensor is the proximity sensor (Fig. 4.13). It is composed of a coil and a magnetic core. When the coil is excited with alternating voltage an electromagnetic field is induced in the core. Any conductive piece placed close to the core will modify this magnetic field and the current through the coil will change. Then measuring the current in the coil, the presence of a conductive piece can be detected. This operating principle is used to detect the proximity of conducting materials. It is very used in industries, appliances, cars and instruments where motion has to be measured.

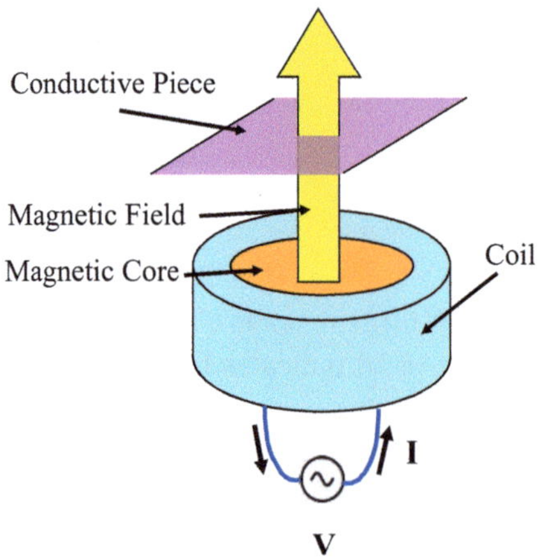

Fig. 4.13: A coil with a magnetic core is excited by an alternating voltage source thus generating a magnetic field. A conductive piece passing through the field takes energy from it "reflecting" this energy change in the coil current. Thus by measuring the coil current the conductive piece can be detected.

4.4.3 Inductive Actuators

Although they are not sensors, it seems adequate to describe this device here due to its inductive nature and because this concept is needed to explain how an oscillator works. This kind of actuator is based on the electromagnetic force, and variations of the elementary device described below are used in many applications.

As was introduced in Section (3.7.12), a coil through which a current is flowing behaves as a magnet, and it is called an electromagnet. Electromagnets are very useful as actuator devices. As explained in Section (2.1), an actuator is a device that converts an electrical signal into a mechanical motion.

The same as with transformers, practical electromagnets convey the electromagnetic field through a closed magnetic core because it greatly concentrates the effect of magnetic fields. Some cores increase the magnetic force by thousands of times with respect to the coils in air. A simplified electromagnet is depicted in Figure 4.14 where the coil and the core are shown. In practical electromagnets there is also a hollow ferromagnetic cylinder covering the coil (not shown in Figure 4.14 for the sake of clarity) which improves the efficiency of the electromagnet.

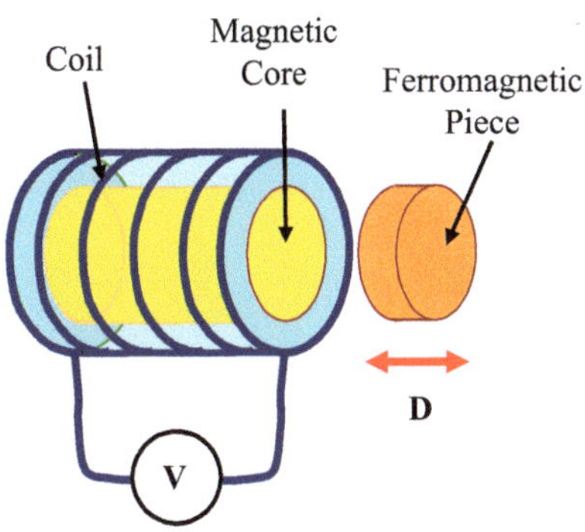

Fig. 4.14: An electromagnet is comprised of a coil wound on a magnetic core and a voltage source (*V*) which can be DC, AC or pulsed. The electromagnetic force moves the ferromagnetic piece a distance *D*. According to the application, the ferromagnetic piece may represent a part of a mechanical actuator or several tons of steel; the theory behind both is the same.

When the electromagnet is supplied with a DC voltage it works just as a magnet, attracting ferromagnetic materials. These electromagnets are very used in equipment we use everyday, as a door lock. They permit a mechanical device to be operated from an electronic circuit. Special electromagnets are able to hold ferromagnetic pieces of several tons and they are used to move big pieces in steel industries.

When the electromagnet is supplied with a pulsed voltage, a transitory force is produced that pushes or attracts a piece; this is the kind of application we will see later in tuning forks. If an AC voltage is applied, a periodical force is generated. This is the case of a doorbell: people pushing the button at the entrance close a circuit which, by means of AC current, makes the electromagnet vibrate a hammer over the bell.

4.4.4 Transformer Sensors

The most widespread transformer sensor is the Linear Variable Differential Transformer known as LVDT, which is used to measure small displacements.

Figure 4.15 shows the components of an LVDT schematically. The transformer consists of one primary (P) winding and two secondary windings (S1 and S2), P being centered between S1 and S2. The coils are wound on a hollow cylinder made of polymer. The transformer has a movable core (C) made of magnetic material which displaces axially inside the coil's hollow. The object whose displacement is to be measured is attached to the core which slides smoothly inside the coil (no physical contact between both is required).

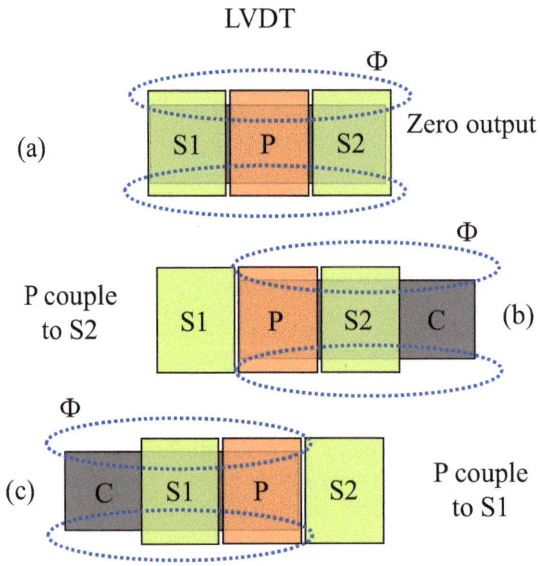

Fig. 4.15: A Linear Variable Differential Transformer (LVDT). (a) The primary coil P generates a flux Φ which is coupled equally to both secondary coils (S1 and S2) through the magnetic movable core, thus the output is null. (b) The displacement of the core to the right increases the coupling with S2, decreasing it with S1. The differential voltage (algebraic sum of both coil outputs) reflects this change. (c) The shift to the left produces the opposite effect and the differential voltage changes sign.

An alternating voltage of constant amplitude is connected to the transformer's primary (P) and the voltage **difference** between the secondary coils is taken as the output signal. For practical use, this difference is generally amplified and conditioned by an electronic circuit. When the core is at the middle point of the transformer, it magnetically couples the primary magnetic flux (F) to both secondary coils in a similar way (Fig. 4.15a). Then, both secondary voltages are similar giving a zero voltage difference. As explained for transformers, the core confines and guides the magnetic field. Then when the core is shifted to the right or to the left (Figs. 4.15b and c), it guides the magnetic field towards S2 or S1, respectively. Therefore, the output has the maximum values at (b) and (c). According to the phase between the input and output signals one of the maximums is considered negative while the other positive.

LVDT are typically used to measure displacements. They are manufactured to work in ranges from ±1 mm to ±100 mm having an accuracy of 0.1 μm at the lower end of the range.

4.5 Oscillators

All timers and clocks used nowadays have some kind of crystal oscillator, and some sensors based on piezoelectric transducers are part of an oscillator circuit. Therefore, a conceptual approach to these devices follows.

4.5.1 Introduction to Oscillators

In electronics engineering an oscillator is a circuit that produces a periodic signal output, often a sine wave. In order to introduce this subject to people without an electronics background, the tuning fork oscillator example has been selected (Fig. 4.14). It is an old circuit, simple to understand.

As a first step towards the fork oscillator let us consider a mechanical pendulum clock; the pendulum has its own oscillating frequency which depends mainly on the pendulum length. The pendulum oscillation may be initiated by separating the pendulum mass from its lowest potential energy position, giving it an initial swing. Because of friction the amplitude of the oscillation will decrease, and after a while the pendulum will stop if enough energy to overcome friction is not supplied to it. This energy must be supplied by a gentle push on the pendulum, in the proper direction and time. It means that for the pendulum to continue oscillating, the energy has to be supplied in a way synchronized with the pendulum movement.

The tuning fork, likewise the pendulum, has it own oscillating frequency (which is a function of its length and mass) and, if it is desired to keep it oscillating, some external energy will also be needed to compensate frictional looses which tend to decrease the oscillation amplitude. This energy has to be supplied also in a synchronized mode to keep the fork moving.

Figure 4.16 illustrates a two prongs tuning fork with the minimum components required to keep the fork oscillating: an excitation coil, a pick up coil and an amplifier.

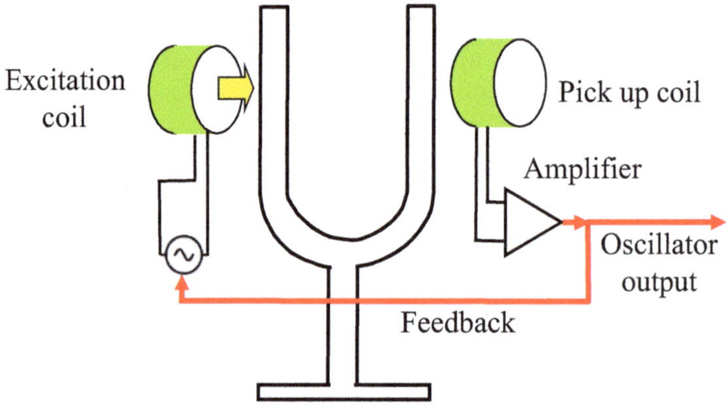

Fig. 4.16: When excited by a stroke a tuning fork oscillates at its own oscillating frequency. To keep it oscillating the pick up coil takes information about the oscillation as an inductive sensor can do. This information is used to compensate the energy lost by friction through the excitation coil which works as an electromagnet.

A coil, through which a current flows, works as an electromagnet which attracts or repels ferromagnetic materials. Then if the fork is made of a ferromagnetic material a coil excited by a current could be used to excite the fork mechanically. Therefore, the excitation coil has the task of transforming electrical energy into mechanical energy. In other words it works as an actuator, supplying the tuning fork with the necessary push to keep it vibrating. The push should have enough energy to compensate the energy lost by friction and has to be applied in due course.

The pick up coil captures the vibration of the right prong as in the case of the proximity sensors (Fig. 4.13), producing an electric signal of the same frequency as that of the fork oscillation. The oscillation waveform of the prongs is used to synchronize the delivery (through the excitation coil) of the energy needed by the tuning fork to remain vibrating. Through the feedback loop, the pick up coil together with the amplifier synchronize the needed push to the left leg at the exact moment and adequate direction. Thus, the fork keeps oscillating and the electrical signal at the oscillator output may be used for timing purposes. Assuming that the fork has an oscillating frequency of 1000 Hz, a clock with a period of 1 ms can be built. The energy to keep the system working comes from an electric power supply (not shown to keep the figure legible) that feeds the amplifier.

One problem of these oscillators, used in the first electronic timing devices, is that due to temperature the length of the fork can change, thus modifying the oscillating frequency. Also, because of the size of the fork the oscillating frequencies are in the audio frequency range (low frequency).

A simplified and more general schematic of an oscillator circuit is shown in Figure 4.17. The power supply (PS) provides the electrical energy to keep the system oscillating. The triangle denotes the amplifier and the box is a mechanical and electrical filter composed of the two coils and the fork. The filter transforms electrical energy into mechanical energy and vice-versa, and works at the tuning fork oscillating frequency (which is a function of its mechanical characteristics). Thus, if a mechanical filter more stable in frequency and of a smaller size (higher frequency) were available, we would have a clock with a better resolution, more precise and accurate.

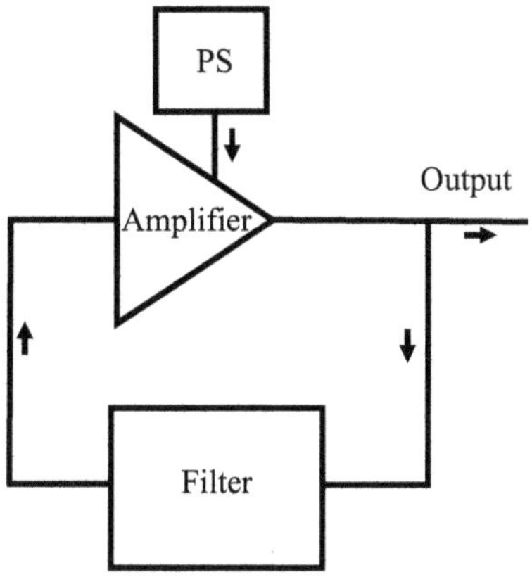

Fig. 4.17: Schematic of a tuning-fork or a piezoelectric oscillator circuit with an amplifier and a power supply.

4.5.2 Piezoelectric Oscillators

We will call piezoelectric oscillator any oscillator that uses the piezoelectric effect, based either on quartz crystal or on piezoelectric ceramics, because both have the same operation principle. Piezoelectric oscillators follow the same concept described for the tuning fork and can be represented also by the schematic of Figure 4.17. There is a mechanical vibration as well which is sustained by a synchronized supply of electrical energy. The vibrating element is a piezoelectric crystal as those depicted in Figure 3.30. Recall that piezoelectric crystals have two electrically conductive plates (electrodes) and they transform electrical energy into mechanical energy and vice-versa; as the mechanical and electrical filter described above does. As stated in

Section (3.8.1), the oscillating frequency of piezoelectric crystals depends on their mechanical characteristics, mainly the thickness of the crystal which defines the resonance frequency. Therefore, it could be admitted intuitively that the piezoelectric crystal could replace the tuning fork with some advantages.

Conceptually the process could be described as follows. When the crystal is exposed to a mechanical shock, a mechanical vibration is initiated at the resonant frequency of the crystal and an alternating voltage of the same frequency than the vibration will appear on its electrodes. This signal is amplified and feed backed to the electrodes of the piezoelectric crystal. Thus the oscillation is sustained because the energy delivered to the crystal compensates the losses.

In fact, piezoelectric crystals replace the tuning fork with many advantages. Piezoelectric crystals need little external energy to oscillate; their oscillating frequency is very stable (does not change with temperature); they are small and of low cost. Because they are small and the oscillating frequency depends on their size, very high frequency oscillators can be constructed. Very stable oscillators for several megahertz can be manufactured at a very low cost.

4.5.3 Piezoelectric Oscillator Applications

The piezoelectric crystal properties define the stability of the oscillator frequency. Significant research efforts have been done to build crystals with certain desired properties. Some quartz crystals are very stable and they are used as the time base for clocks in almost all time keeping devices.

Conversely, some piezoelectric crystals have been manufactured in such a way that their resonant frequencies be sensitive to some desired parameter such as temperature or force. Thus, when these crystals are employed as part of an oscillator, the oscillator output frequency will contain information on those external parameters. This is the working principle of several sensors in which the measurand modifies the resonant frequency of an oscillator.

4.6 Strain Gauges (or Force Sensors)

They are resistive sensors in which the dimensions of the resistor change under the action of a force. To understand the working principle of a strain gauge suppose that the resistor shown in Figure 3.18 is fixed at one end and a force is applied to the other, stretching it; then, the cross sectional area A will decrease and the length l will increase. If the deformation is small enough such that the material is within its range of elasticity, when the force is removed the material will return to its initial dimensions. The resistance of such a resistor will increase when stretched (Eq. (3.12)) and will come back to its initial value when the force is released. Unfortunately for short resistors as that shown in Figure 3.18, the change in resistance is very low and very difficult to measure.

This property by which resistors change their values when subjected to a force is exploited to build sensors that allow a mechanical stress to be transformed into an electrical signal. A particular resistor designed to maximize the change in resistance when stretched is shown schematically in Fig. 4.18. A long wire of small diameter is bonded on an elastic material such that most of its length is located along the same direction, which will be called the sensing direction. When this device is stretched in the sensing direction, changes in the cross sectional area and length of the wire will increase its resistance.

When the ends of the wires are connected to a battery by means of two electrical cables (Fig. 4.18), and a variable force is applied in the sensing direction, changes in resistance produce a variable current. Then, a sensor that converts force signals into current signals is obtained. It is worth noting that this is just a demonstrative circuit, usually a Wheatstone bridge is used to detect the resistor changes (Fig. 4.2a).

The wires are very fragile and difficult to handle, so they are placed on an elastic backing material used to support them. This support is made of a dielectric material (usually plastic). Backing and wire form a device known as **strain gauge**.

When it is desired to measure the bending of a beam of a bridge or the compression of a column, a strain gauge is rigidly bonded to the specimen whose strain is being measured. The sensing direction of the wires is placed in the direction of the expected compression or extension. The backing material provides a good electrical insulation between the wires and the specimen.

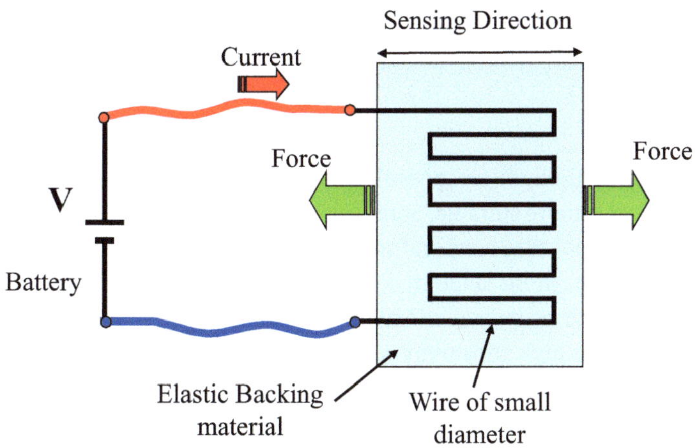

Fig. 4.18: A strain gauge. A long wire of small diameter is disposed such that most of its length is located in the horizontal direction. It is bonded on an elastic support which, when stretched horizontally, changes the length and diameter of the wire, changing its resistance.

Today strain gauges are made in different shapes. Some have two perpendicular sensing directions. One is used to measure the desired axial deformation and the other is placed in a perpendicular direction not affected by the deformation. The second is used as part of a measuring Wheatstone bridge to compensate for changes in the resistance due to temperature, such that the Wheatstone bridge does not result unbalanced because of temperature changes. Other strain gauges have several radial sensing directions to measure in different directions simultaneously. Strain gauges are often used to study the structure of buildings and the dynamic behavior of cars, planes and machinery, and also in many instruments for environmental and hydraulic applications. Because of the ability of strain gauges to measure deformation, they are incorporated as part of some sensors such as pressure sensors.

Devices used to measure loads are called **load cells.** They consist of a deformable body made of a metal piece on which strain gauges are bonded. When the applied load deforms the body of the load cell, a change in the strain gauge resistance is measured by an electronic circuit which gives an output voltage. A calibration process relates the applied force to the output voltage and the transference of the load cell is established.

So far, strain gauges based in metal wires have been described, but there are also strain gauges based on semiconductor materials. Some semiconductors exhibit a piezoresistance effect which is defined as the change in electrical resistance when a stress is applied. These materials are thus used to manufacture strain gauges. The main advantage of these gauges is that they are several times more sensitive than gauges made of metals, but they are also more affected by temperature.

4.7 Sensors and Instruments to Measure Temperature

4.7.1 Introduction

There are a large number of methods for measuring temperature. Each one has a set of applications for which it is more suitable. Some questions that should be answered when selecting a method to measure temperature are: What is the temperature range of use? Should the measurement be made in contact with or at a distance from the object whose temperature is to be measured? Which is the desired response time? Is long-term stability needed? Is high sensitivity required? What level of complexity of electronic circuits is accepted? Which is the size of the sample to be measured? What is the necessary accuracy, and which is the affordable cost? For some applications several methods could satisfy the measuring requirements, but for others only one may be adequate.

The characteristics of some commonly used sensors for measuring temperature will be described first. They are resistance temperature detectors, thermistors, thermocouples and I.C. (integrated circuit) sensors. All of them need some kind of electronic circuitry in order to present the results adequately. Sometimes, standard electronics easily accessible to researchers can be used with these sensors, such as data loggers which incorporate electronic conditioning circuits for most popular temperature sensors. The main characteristics of these sensors are presented to help users select the most suited for their applications.

Towards the end of this Section (4.7), two methods known as infrared thermometry and infrared thermography will be presented. They use particular optical and electronic components, and have to be purchased as complete instruments. They cannot be used with standard electronics as in the previous cases.

4.7.2 Resistance Temperature Detectors (RTD)

RTD are resistive sensors used to measure temperature. They are based on the conducting properties of metals, which have electrons moving freely throughout the metal lattice. When temperature increases, the atoms in the metal lattice (ion cores) vibrate and the conduction electrons tend to collide more frequently with the stationary lattice, which hinders the movement of the electrons. This atomic behavior is macroscopically perceived as a rise in the resistivity of the material. For some metals the resistivity as a function of temperature is a well-known and highly repeatable relation. For platinum, for example, this relation is approximately linear over a great temperature range (-150 to 600 °C approximately). Even when some non linearity is identified in this relation, for smaller partitions of the range (about 100°C) the relation is quite linear. Thus, for many applications linear approximations or simple analytical expressions are used to relate temperature and resistivity easily.

A RTD is basically a resistor made of a metal wire whose resistance varies with temperature. As shown in Eq. (3.12), the resistance is proportional to the resistivity. For a RTD the length and area of the resistor are kept constant and the resistivity increases with temperature. Platinum and nickel are often employed to manufacture RTD. Platinum RTD are very repeatable but expensive. Nickel's are not so repeatable but are less expensive.

An approximation of the resistance value (R) as a function of temperature (T) for a RTD can be calculated as

$$R = R_0 \left(1 + \alpha_1 T + \alpha_2 T^2\right) \tag{4.8}$$

where R_0 is the resistance at a reference temperature (usually 0 °C) and α_1 and α_2 are the fractional changes in resistance per degree. They are constant characteristics of the material used to manufacture the RTD; e.g. for platinum $\alpha_1 = 3.84 \times 10^{-3}$ 1/°C and $\alpha_2 = 5.83 \times 10^{-6}$ 1/°C^2 (Allocca and Stuart, 1983). RTD made of platinum are usually called Platinum Resistance Thermometers (PRT), and are commercially available under names such as PT100, PT500 or PT1000, the number following the PT letters being the resistance of the device at 0 °C. The practical range of use of the PT100 is from -200 to 650 °C. Because of the low resistance of PT100, the resistance of the cables connecting the device to the electronic circuit could introduce errors; therefore, a four wire technique should be used (Fig. 4.4).

When it is desired to know the temperature of a given environment, RTD have to be placed in contact with such environment, and some time is required for the sensor to reach the equilibrium temperature. In general, RTD have a slow response time, for example, an industrial PT100 has a response time of 3 s in liquid and 15 s in air. Then they are not suitable to measure fast changes in temperature.

The sensitivity of a PT100 sensor is low, for a 1 °C change in temperature its resistance will vary 0.384 Ω. A large current through the sensor could produce a good voltage signal but it would also cause heating of the sensor. At the same time, heating introduces error because the RTD would report a higher temperature than that of the environment whose temperature is to be measured; this behavior is known as self heating, and errors due to it are larger when measuring in gases, due to the lower heat dissipation from the sensor.

Generally currents on the order of 1 mA are used to avoid self heating, and then signals of 384 μV /°C are obtained. Because of these low signal levels it is necessary to avoid noise entering the circuits to have and adequate signal to noise ratio. In this regard, it is necessary to avoid long cables, and to place cables away from devices that may emit electrical noise such as motors. The use of shielded cables is recommended.

In summary, RTD have large temperature ranges, high repeatability but low sensitivity and slow response time.

4.7.3 Thermistors

Thermistors are a kind of resistive temperature sensors made of semiconductor materials. This gives a greater sensitivity to them but their resistances vary nonlinearly with temperature.

There are two types of technologies; one of them produces devices with positive temperature coefficient (PTC) - resistance increases with temperature as in the RTD case. They are frequently used to limit current and protect electrical devices.

The other technology produces thermistors with negative temperature coefficients (NTC); these being the most used as temperature sensor for environmental applications. The transfer function for NTC thermistors has the form:

$$R(T) = R(T_0)\exp\left[\beta\left(\frac{1}{T} - \frac{1}{T_0}\right)\right] \tag{4.9}$$

where R is the value of the thermistor's electrical resistance (in Ω) at the absolute temperature T (K); $R(T_0)$ is the resistance at a reference temperature T_0 (generally 298.15 K or 25 °C); and β is a constant value, characteristic of the thermistor's manufacturing process. Manufacturers do not specify each of the thermistor characteristics, but they provide average values of β, $R(T_0)$ and T_0 and their respective dispersions. Then, if users want to have the best approximation to the transfer function, they must measure R_1 and R_2 at two different known temperatures T_1 and T_2 to obtain β for their particular sensor. Besides, generally, thermistor transferences vary with time and the sensor requires periodical calibrations to keep it accurate.

For β = 3000 K, $R(T_0)$ = 1000 Ω and T_0 = 25 °C the transfer function of an NTC thermistor is plotted in Figure 4.19. It is clearly more sensitive at low temperatures. Then, due to this nonlinearity, if sensor's users want a better knowledge of the transference, they should measure several calibration points along the curve and fit the function by a least squares method.

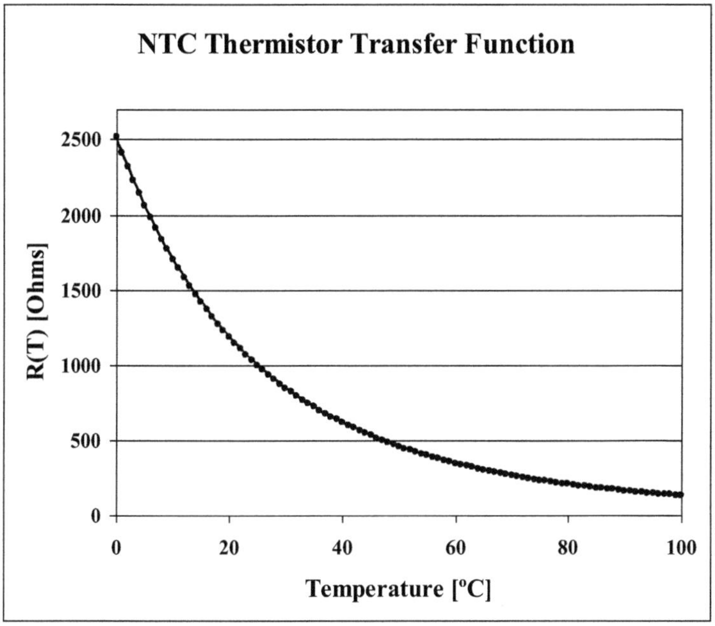

Fig. 4.19: NTC thermistor transfer function.

Many circuits have been developed with the purpose of compensating the nonlinearity of thermistor transferences, but a simple way to obtain the temperature as a function of resistance which does not need special circuits is to use some approximations. Two frequently used approximations are (Hewlett Packard, 1980; http://www.princeton.edu)

$$\frac{1}{T} = A + B(\ln R) + C(\ln R)^3; \qquad T = \frac{B}{\ln R - A} - C \tag{4.10}$$

where T is the absolute temperature (in K), R the resistance of the thermistor (in Ω, at temperature T) and A, B and C are constants to be found by selecting three (R_i, T_i)

data points and solving the three resultant simultaneous equations. The first of these approximations is known as the Steinhart-Hart equation. For a temperature range smaller than 100 °C the equation approximates the real value with errors less than ±0.02 °C.

The second approximation is easier to compute, but to produce similar errors to those from the Steinhart-Hart equation it should be used in a narrower temperature span (for example to measure the temperature of water in the sea, where the span is generally of a few Celsius degrees).

It is easy to find thermistors with high resistances; hence the resistance of the cables connecting them to the electronic circuit is not as important as in the RTD case.

Because thermistors are made of semiconductor materials their maximum measuring temperature is limited to about 200 °C; prolonged exposures of them close to the maximum operating limits will cause the thermistor to drift out of calibration. Manufacturing processes allow thermistors to be made with a small thermal mass; thus response times smaller than those of RTD are feasible.

Figure 4.20 shows at the center a metal encapsulated thermistor and on the right side a ceramic bed thermistor in which the sensing element is only the spherical tip, the metal wires being covered by an insulating tube (a pen is included for comparison purposes).

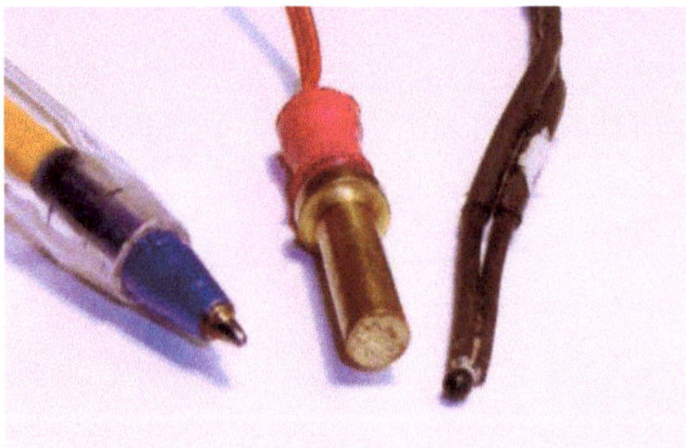

Fig. 4.20: Two types of thermistors (the scale is given by the pen).

4.7.4 Thermocouples

4.7.4.1 Thermoelectric Effect

It is a reversible effect by which thermal energy is converted into electrical energy or vice-versa. The first phenomenon is known as the Seebeck effect, and the second as the Peltier effect. More specifically, temperature differences are converted into voltage (or current) and backwards. Both effects have practical use in measuring temperature or changing the temperature of objects.

4.7.4.2 Thermocouple Description

Thermocouples are temperature sensors based on the thermoelectric effect. They are composed of two wires of different metals (such as constantan and copper) joined at two points (junctions). When one of the junctions is heated, a current flows through the wires due to a thermal emf (ε). Figure 4.21 shows a device which converts a temperature difference into a current. In this example the current is measured by an ammeter whose input impedance is almost zero. The current is proportional to the temperature difference between both junctions ($T_1 - T_2$).

If the two copper wires are connected to a voltmeter with a great resistance a weak current will flow, but a thermally induced electromotive force ε will be measured by the voltmeter. This emf (ε) is proportional to the temperature difference between both junctions ($T_1 - T_2$). The proportionality constant (k) (Eq. 4.11) depends on the type of metals the thermocouple is made of.

$$\varepsilon = k\,(T_2 - T_1) \tag{4.11}$$

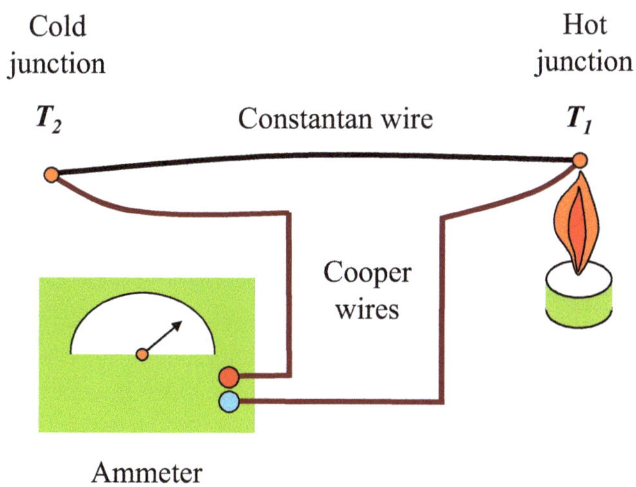

Fig. 4.21: A constantan-copper thermocouple subjected to a temperature difference between both junctions.

Usually the thermally induced emf generated by the Seebeck effect is very small. The most used metals to construct thermocouples are copper and constantan. At room temperature they produce an emf of 40 μV per degree centigrade of temperature difference between the two junctions.

Because the voltage measured is proportional to the *difference* in temperature, if the temperature T_1 is desired, it would be necessary to know T_2. One way to force a known temperature T_2 is to place the cold junction in an ice bath, then T_2 = 0°C becomes the reference temperature. Because the use of an ice bath is not a practical solution, an electronic compensating circuit known as *electronic ice point reference* is used to measure the temperature T_1.

Thermocouples have a wide temperature range, for example, those made with wires of platinum and platinum-rhodium can measure over 1500 °C. Thermocouples are not linear in all the measuring range, but because they are very well known there are standard tables that allow the temperature to be known by measuring the thermocouple output voltage. Also, because they have small masses they can respond fast to changes in temperature.

4.7.5 I.C. Sensor

The I.C. letters stand for integrated circuit. These are sensors based on certain properties of semiconductor junctions, and for this reason they are also known as semiconductor sensors. A semiconductor forward-biased PN junction has a linear voltage (*V*pn) to temperature (*T*) relationship; for a silicon diode this relationship is given by

$$\frac{\Delta V\text{pn}}{\Delta T} \approx -2.3 \frac{\text{mV}}{^\circ\text{C}} \tag{4.12}$$

This relationship is exploited to manufacture temperature sensors. Because these sensors are made of semiconductor materials the measuring temperature range is limited to -55 to 150 °C. The response time is slow due to their thermal masses. Today, these kinds of sensors have reached such a degree of development that they can be easily used, even by people with little electronic skill.

There are basically two types of I.C. sensors: with analog output and with digital output. Sensors with analog output require a minimum amount of external components to become a thermometer (National Semiconductors, 2000). In some cases only a power supply and a resistor is enough to have an analog voltage proportional to temperature. This voltage can be read on a simple voltmeter or recorded by a data logger.

Some specifications for analog output I.C. sensors are: accuracy of ± 0.25 °C at room temperature, and of ± 0.75 °C over the full range (-55 to 150 °C); they have a self-heating less than 0.1°C in still air and are suitable for remote applications.

I.C. sensors with digital output can be read directly from microprocessor standard digital buses such as I^2C (Texas Instruments, 2009). Then, a user with some skill in programming could use this technology to assemble a network of temperature sensors. Some I.C. allow several devices to be networked in parallel in just one bus (few cables) No external components are required and accuracies of 0.5°C in the range from 0 to +65 °C, or of 1.0°C in the range from −40 to +125 °C, are available. Figure 4.22 shows an analog sensor on the right and a modern digital one at the center.

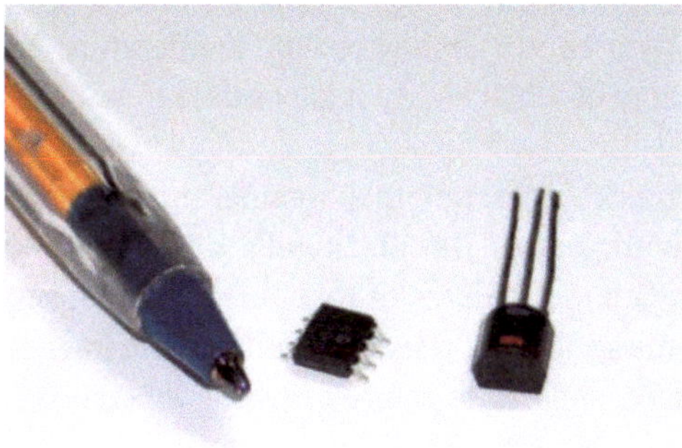

Fig. 4.22: (Right) An analog I.C. sensor. (Center) A digital I.C. sensor (the scale is given by the pen).

It should be noted that manufacturers specify the accuracy that the user could expect using the factory calibration. This is the accuracy that we can suppose from a temperature measuring instrument when the sensor has to be changed for another coming from the factory.

But because in some sensors the resolution is 0.0625°C, the accuracy could be increased several times if the sensor is previously calibrated by the user. In this case, if the sensor has to be changed, the new one has to be recalibrated again by the user in order to keep the improved accuracy. If not recalibrated, the accuracy will be that specified by the manufacturer.

Digital I.C. sensors are a good option when measurements of slow temperature changes at several points in a reduced volume are required such as in the case of a silo for grain storage or a greenhouse.

4.7.6 Infrared Thermometers (IRT)

Infrared thermometers, also known as infrared pyrometers, measure the surface temperature of an object at a certain distance from it. IRT are useful in applications where contact measurements are impracticable as in the case of melting metals or dangerous areas. They measure the electromagnetic radiation emitted by the object due to the object's temperature. Some ideas about electromagnetic radiation will be introduced to facilitate the understanding of this measuring method.

A blackbody is an idealized object that absorbs all radiation arriving at it and emits a specific spectrum of energy. The intensity of the radiated energy at any particular wavelength increases as a function of temperature. The total power emitted by a black body per unit area of its surface is proportional to the fourth power of its absolute temperature (Stefan's law).

$$E \propto T^4 \tag{4.13}$$

where E is the power per unit area (W/m^2) and T is the absolute temperature (K). Therefore, by measuring the emitted energy of a blackbody it is possible to know its temperature.

Most materials do not behave as blackbodies but their temperature may be estimated by comparing their radiation to that of the blackbody. In this respect correction factors are used to relate the radiation curves of real objects to that of a blackbody. These relations give origin to the concept of *emissivity* which is the fraction of energy being emitted by a given surface, relative to that emitted by a blackbody surface, when both surfaces are at the same temperature.

A blackbody is a perfect emitter of heat energy, it emits all the energy that it absorbs so it is said that it has an emissivity of 1. On the contrary, a perfect thermal mirror will reflect all the energy that reaches it and is thus said to have an emissivity of 0. Most objects are neither blackbodies nor thermal mirrors; so they have emissivity values between 0 and 1. In order to obtain accurate temperature measurements from objects that are not blackbodies it is required to know the emissivity of the object being measured.

It could be difficult to know the emissivity of some surfaces whose temperature is to be measured. This hinders the exact temperature to be known by an IRT. Moreover, when there are several objects in the field of view of the IRT and each one has a different emissivity it is not correct to assign one emissivity as representative of them all.

Also, because emissivity is a property of the surface of an object, it could change with time. For example, a new polished metal surface will look shiny and will have a certain emissivity, but when it becomes corroded it will look dark and the emissivity will change. Then the same object will have different emissivities as time passes by.

Most of the emitted radiation is concentrated in a wavelength band called the infrared spectrum. Infrared energy is electromagnetic energy in a range of frequencies below the visible light; all objects radiate infrared energy not perceived by the human eye. The human eye cannot detect electromagnetic wavelengths longer than 0.7 μm (red) (http://www.x26.com/articles.html#). Some IRT are designed to be sensitive to a wavelength band around 8 to 14 μm. Very hot objects such as molten steel emit visible light and their colors are related to the temperature of the object. Visible light is a very narrow portion of the entire electromagnetic spectrum (≈ 0.4-0.7 μm).

There are different models of IRT but a generic one is pictured in Figure 4.23 to give an approximate idea on how these devices work.

Radiation from the emitting surface passes through a filter which limits the radiation wavelength range to the infrared band and focuses the energy onto a mirror. The mirror concentrates the energy on a detector which could consist of a blackened disc that heats up due to the received energy. The temperature of the disc can be measured by some temperature sensor such as a micro thermocouple or a small thermistor. As explained above, these sensors have not a very fast response; then, when it is desired to detect the temperature of fast moving objects, the detector should be a photomultiplier tube.

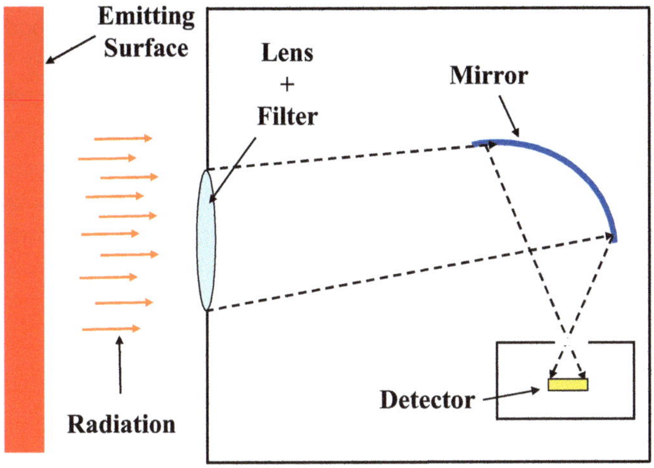

Fig. 4.23: Schematic of a generic IRT.

A photomultiplier is a device designed for the detection of photons. It exploits the secondary emission of electrons in a vacuum tube. When photons strike the tube's cathode, electrons are emitted and attracted to a first positively charged electrode (anode). When electrons collide with the first electrode (or anode), more electrons are released and attracted to a second positively charged electrode (or anode). This process is repeated several times until there is an easily detectable current flow through the last anode.

IRT are not intended to measure temperature exactly, but only approximately. Fortunately, for most applications it is only necessary to compare the temperature of different objects, not to know their absolute values.

For example, if it is desired to know the temperature of a high voltage line contacts, all contacts would probably have a similar unknown emissivity. Then, the absolute temperature of the contacts will be measured with error because the actual emissivity is ignored. But if one of the contacts is hot due to a defective connection, the increase in its temperature relative to the others will be easily detected.

IRT have a wide temperature range of use, for example, some commercial equipments have ranges from -20 to 1000 °C for wavelengths from 8 to 14 μm, and others from 550 to 4000 °C for wavelengths from 0.85 to 1.1 μm. The first are used in ceramics, rubber, paper, food, asphalt and plastics industries; and the second in steel, glass, induction heating and forging industries. Instruments with a response time between 10 ms to 1.5 s are available.

IRT are a promising tool in environmental sciences applications where it is required to measure a considerable area from a certain distance. It has been found that low-cost consumer-quality IRT are useful for non-contact measurement of plant canopy temperature (Mahana & Yeaterb, 2008). They can also be used for studies on land-atmosphere interaction, surface energy balance, human an animal body temperature measurements, etc.

Because the energy radiated from an object whose temperature is being measured is transmitted through the air and received by the thermometer, the results may be affected by smoke and dust. Thermometers receive radiation caused by the ambient temperature in the room together with the radiated energy from the object being measured; to avoid measuring undesired background temperatures, the infrared energy collected by the thermometer should be originated by the particular area whose temperature is measured.

4.7.7 Infrared Thermography

Infrared thermometers measure the *average temperature of an area*. Instead infrared thermography uses infrared cameras which detect energy emitted from individual points of an object, convert them into temperature data, and *display an image of temperature distribution*. Thermography converts thus infrared radiation into a visual image of the temperature distribution similar to that shown in Figure 4.24.

Fig. 4.24: (a) Photography of a drying leave. (b) The same leave as seen in a thermographic image.

Different temperatures on a surface are represented by different colors and temperature comparisons over a large area are possible. This permits hot spots to be found quickly. A temperature image consists of a matrix of pixels and the number of pixels depends on the number of detectors. Some thermal cameras have for example 320 × 240 pixels which are colored up pixel by pixel by the onboard software according to the temperature measured by each detector.

This technology let thermal anomalies of rotating equipments, such as big engines to be detected by periodical inspection under the same running conditions. It also permits safe inspection of live electrical equipment difficult to access. Food temperature can be measured in a sanitary fashion.

In environmental sciences, thermography allows semiautomatic analysis of large areas of canopy with effective replication of measurements (Jones et al., 2002). It has also been demonstrated the system's ability to survey populations of several wildlife species. It permits different species in the same habitat to be detected, and it is expected that with the aid of computer-assisted analysis, infrared thermography may become a useful wildlife population survey tool (Garner et al., 1995).

4.7.8 Comparison of Temperature Sensors and Instruments

In order to select the most adequate temperature sensor for a given application, their general characteristics are shown in Table 4.2.

Table 4.2: Characteristics of temperature sensors

Temperature sensor	Advantages	Disadvantages
RTD	Most repeatable Accurate Linear Wide temperature range	Small sensitivity Low resistance Slow response Self heating
Thermistor	High sensitivity Fast response High resistance	Non linear Limited temperature range Fragile Self heating
Thermocouple	Wide temperature range Robust	Non linear Small sensitivity Temperature reference needed
I.C. sensor	Linear High sensitivity Inexpensive Analog and digital output	Limited temperature range Slow response Poor factory calibration
Infra Red Thermometer	Wide temperature range Measure at distance	Not accurate Periodical calibration required
Infra Red Thermography	Same as IR thermometer + Image of temperature Distribution	Same as IR thermometer + Cost

4.8 Pressure Sensors

4.8.1 Introduction

Pressure sensors give an electrical signal output proportional to the applied pressure. Figure 4.25 shows a device to convert an applied pressure into a voltage signal. This is a schematic of an idealized device which meets certain characteristics common to various technologies used for pressure sensors. This schematic allows a simple explanation on how these sensors work.

The device comprises a circular membrane placed on the top of a rigid cylinder. The applied pressure bends the membrane made of an elastic material such as a thin metal. While the material is in the elastic range, the membrane will return to its initial conditions when pressure is released. The material, shape and size of the membrane are selected to have a deformation proportional to the applied pressure.

One way to account for membrane deformation is by means of a *strain gauge* (Section (4.6)).Then, bonding a strain gauge to the elastic membrane, connecting it to a Wheatstone bridge (Fig. 4.2) and the bridge to an amplifier, a voltage output proportional to the applied pressure is obtained.

In summary, pressure deforms the membrane and the deformation is measured by a strain gauge which converts deformation into an electrical signal. Therefore, by determining the relationship between the applied pressure and the measured voltage the transfer curve of the sensor can be established.

It should be pointed out that the membrane is free to deform and that the only restoring force is due to the elastic force of the material. No force is applied to the back of the membrane. A reference port is depicted in Figure 4.25 which vents the reference chamber located inside the cylinder to the atmospheric pressure, so that air is not compressed by the displacement of the membrane.

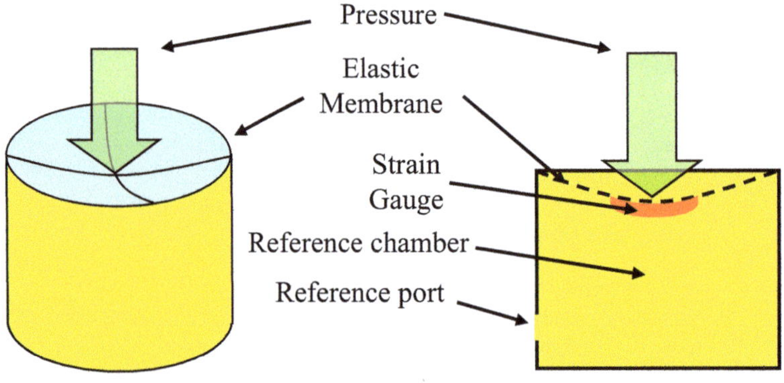

Fig. 4.25: Schematic of a device for converting an applied pressure into a voltage signal.

4.8.2 Types of Sensors

4.8.2.1 Differential Sensors

One way to know the difference between two pressures is to measure both with two independent sensors such as that previously described. Each one will give an output voltage proportional to the pressure to which it is subjected. Each membrane is referred to the atmospheric pressure through the reference port. The voltage difference that results from subtracting both outputs is proportional to the pressure difference.

A more accurate method for measuring pressure differences than that explained above follows. The membrane of the sensor of Figure 4.25 can receive pressure from the top and the bottom. Then, if the upper side of the membrane is exposed to one of the pressures to be measured and the reference port to the other, the membrane will deform according to the pressure difference. Thus, the strain gauge with the associate circuit will give an output voltage directly proportional to the differential pressure.

Manufacturers of pressure sensors utilize a particular classification that users should know to correctly select sensors for diverse applications. Sensors that have two pressure ports accessible to users (as in this case) are called "*differential sensors*". The pressure port related to the upper side of the membrane is called the measuring port and that related to the back of the membrane, the reference port.

4.8.2.2 Vented Gauge Sensor

When measuring fluid levels with a pressure sensor, the pressure that receives the measuring port is that of the fluid column plus the atmospheric pressure. Therefore, if the reference port is left open to air, it will sense the atmospheric pressure and the membrane will measure the difference, compensating for atmospheric pressure changes. This is a particular case of differential pressure sensors where the reference port is vented to the local atmospheric pressure. This kind of sensors is known as "*vented gauge sensors.*" This is the case depicted in Figure 4.25.

In most industrial processes it is quite simple to measure fluid levels with a vented gauge sensor because, in general, the pressure exerted by the fluid column is connected to the measuring port by means of a tube and it is not difficult to expose the reference port to atmospheric pressure, but it could be particularly complex in other circumstances and precautions should be taken.

For example, some manufacturers offer submersible vented pressure sensors to measure water level in lakes, rivers and harbors. In order to expose the reference port to the atmospheric pressure, they utilize a vent tube into the same cable used for the electric power supply and output signals. This vent tube would allow atmospheric pressure variations to be compensated, but in practice any obstruction restricting air movement along the vent tube could prevent the atmospheric pressure from reaching the sensing membrane and would introduce errors in the measurements. Sometimes humidity condensation is the cause of the vent tube obstruction.

Because humidity in the tube can cause inaccurate readings, a desiccant must be installed and periodically replaced to avoid condensation. This routine is not always possible and sometimes unpractical, as it could be the case of unattended instruments. According to the particular application, it should be careful evaluated when it is convenient to use this kind of sensors with compensating vented tube.

4.8.2.3 Sealed Gauge Sensors

These sensors have the reference port hermetically sealed and the reference pressure chamber is at a certain pressure within the atmospheric pressure range. This pressure is defined by the manufacturer at the moment of sealing the reference port. Sealed gauge sensors are useful to measure dynamic pressures and allow positive and negative pressures to be measured with respect to the sealed atmospheric pressure.

When sealed gauge sensors are used to measure water levels, the varying atmospheric pressure is superimposed to the water column pressure and may cause measuring fluctuation on the order of ±0.1 m. Therefore, to evaluate the exact level it is needed to measure and subtract the atmospheric pressure.

4.8.2.4 Absolute Pressure Sensor

If a vacuum is created into the reference pressure chamber and then the port is sealed, an *absolute pressure sensor* is obtained. In this case the external pressure on the measuring port is relative to vacuum.

4.8.3 Other Pressure Sensor Technologies

For most pressure sensors, as those described above, the measuring pressure produces a mechanical stress on the sensor which changes some electrical parameter in it. In the previous case the changing parameter was the resistance of a strain gage bonded to a membrane. In other sensors the mechanical stress may change the properties of a semiconductor or the frequency of a resonant circuit.

4.8.3.1 Solid State Pressure Sensors

In solid state pressure sensors the elastic membrane and the variable resistances are integrated in a small wafer of silicon where the sensing element and some additional circuit components are mounted on a ceramic substrate. They are known as IC (Integrated Circuit) pressure transducers and some manufacturers package them in a dual-in-line configuration (two parallel rows of pins, similar to that of IC), then sensors may be mounted on printed circuit boards. Often, these sensors require only power supply to provide a voltage output proportional to pressure.

Silicon micromachining techniques are used to implant piezoresistive strain gages into a Wheatstone bridge configuration on board the chip. Because massive production techniques are used, these sensors are of low cost. Some sensors include temperature compensation but, in general, IC pressure sensors are more sensitive to temperature changes than those from other technologies. Frequently they are used where high accuracy is not required and low cost is considered.

4.8.3.2 Quartz Crystal Pressure Sensors

When high accuracy measurements are needed, special quartz crystal resonator sensors are used. As explained above (Section (4.5.2)), crystal oscillators may have frequency outputs related to a measurand. It is only required that the measurand change the resonant frequency of the crystal. In the pressure sensor case, by means of certain special designs, crystal frequency is a function of the stress exerted on the crystal. The resonant frequency output is sustained and detected with electronic circuits similar to those used in precision clocks, as was previously described (Section (4.5.3)).

Different types of quartz crystal stress-sensitive resonators have been developed (http://www.paroscientific.com/qtechnology.htm). One of them consists of two similar beams with two mounting pads, on which stress is applied (Fig. 4.26a). The beams are electrically excited by a pair of electrodes. Frequency is a function of the applied stress, increasing with tension such as it happens with a guitar string.

Until now we have a stress sensor, but we want to measure pressure; then, by means of some mechanical device, pressure has to be transformed into stress and the stress applied to the crystal (Fig. 4.26b). In this schematic the quartz crystal sensor is fastened at the bottom to a fixed plate and aneroid bellows transform the input pressure into a deformation, such that a force is generated and applied to the other end of the sensor.

This kind of sensors can achieve 0.01 % accuracy, 0.0001 % resolution, and have high reliability and stability; their output frequency makes them less susceptible to interference. Also, they are easy to interface to counters, microprocessors and computers. This kind of technology is more expensive than those previously described.

One of the manufacturers of this kind of sensors claims that a sensor installed at a depth of 6,000 meters was able to detect an earthquake-generated wave (tsunami) (Yilmaz et al., 2004). The real signals were resolved to 1 mm of water (1 part in 6 million) and clearly show the characteristics of the tsunami which was only several centimeters at the deployed depth of thousands of meters. This pressure transducers have accuracy better than 100 parts per million of full scale pressure and maintain this accuracy for a long time

Long-term stability tests indicate that the median drift rate of three units, tested during a fifteen-year test period, is 7 parts per million per year, which is an excellent figure.

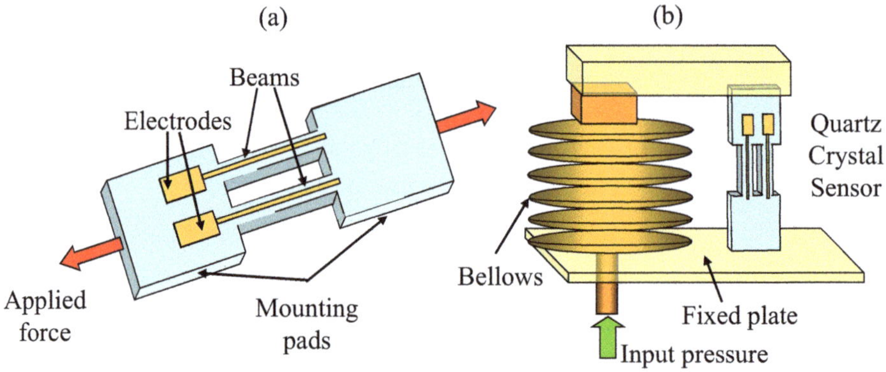

Fig. 4.26: (a) Two similar beams with two mounting pads on which stress is applied; beams are electrically excited by a pair of electrodes. (b) Aneroid bellows transform pressure into a deformation, such that a force is applied to the sensor.

4.8.4 Application of Pressure Sensors to Measure Water Level

When applying pressure sensors to water level measurements it is convenient to evaluate some constraints. A real case example will be introduced with the aim of weighing up different features of the problem.

It was needed to simultaneously measure water level at several points of the drainage channels of an urban area to know how they are filled during rain events. The channels run below the streets of a city and they are about 3 m high and 7 m wide.

Initially, submersible pressure sensors of the type *vented gauge sensor* were considered for this application due to their ability to compensate for atmospheric pressure changes, but some problems arose that influenced to change the first thought.

Since the vent tube of the vented gauge sensor is included in the same cable used for the electric wires, it is convenient to buy cables with the suitable length required for the installation such that not electrical and pressure connections have to be done in the field.

Because access to the channels was difficult and required logistic support from the police force and the fire-brigade, at the time the sensors had to be bought there was no information about how to run the cables from the drainage channels to the surface. Therefore, it was required to have the flexibility to define the length of the cables during the installation of the instrument in the field, which is easier done with cables that only have electrical wires.

The environmental humidity at the instrument installation places was quite high and condensation in the cable vent tube could happen. The use of silica gel at the end of the tube requires a periodical service that was decided to be avoided because there was no possibility of a frequent visit to the instruments.

The above reasons and the lack of experience using vented tubes, added to the client's demands to avoid collecting erroneous data, biased the decision towards the use of *sealed gauge sensors*. They entail the use of extra instruments to measure atmospheric pressure but they looked less prone to failures or errors. A fact that contributed to this choice was that all the installations were in a reduced area where the atmospheric pressure could be considered the same. Then only one atmospheric pressure recorder was needed to compensate all sealed sensors.

It has to be underlined that it is not intended to say that this is the best possible solution to the problem presented in this example; other valuables solutions could be chosen as well. Sometimes, the solutions are selected according to the user's previous experiences and circumstantial constraints. Therefore, sensors with sealed pressure reference ports were selected for this application to avoid humidity access, and a simultaneous atmospheric pressure sensor used to compensate atmospheric variations.

Figure 4.27 shows the installation of a pressure sensor in a drainage channel as described before. Sensors are not installed on the floor of the channel because water carries sediment which might obstruct the sensor pressure port. Also, the sensor was placed inside a steel tube mounted on the wall to protect it from sharp and heavy objects transported by the flow. Sensors were made of stainless steel, then, to avoid galvanic corrosion they were isolated from the tube and plastic bolts were used to fix them.

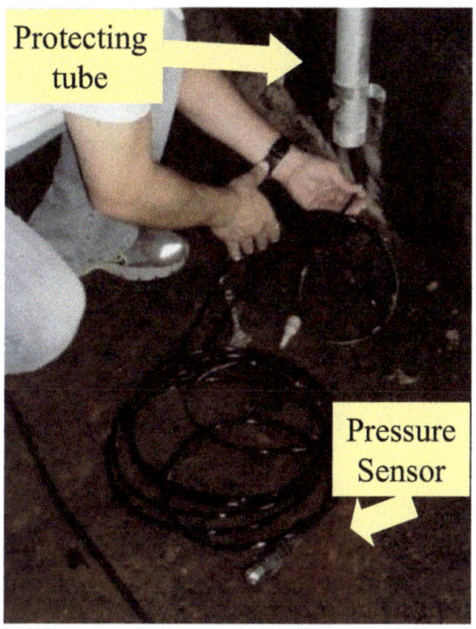

Fig. 4.27: Installation of a pressure sensor in a drainage channel.

Figure 4.28 shows the water level measured in the aforementioned drainage; when the drainage is dry, as in Figure 4.27, the atmospheric pressure is the only pressure recorded. When it rains in the area, the channel begins to fill, the water surface arrives to the sensor level and it begins to record the atmospheric pressure plus the column of water over it.

Because sensors do not measure from the floor of the channel, to know the real water column the atmospheric pressure column has to be subtracted and the installation height added. Obviously, changes in the water level below the sensor's level were not measured, but they are not data of interest in this study.

As it happen with most sensors, pressure sensors can suffer some drift on their transference as time passes. Therefore, it is recommended to perform some simple tests before spending time and money in field work. Some tests on five pressure sensors are presented on Section (11.2), with the purpose of underlining the idea that sensors are neither perfect nor invariant, and they need to be controlled periodically.

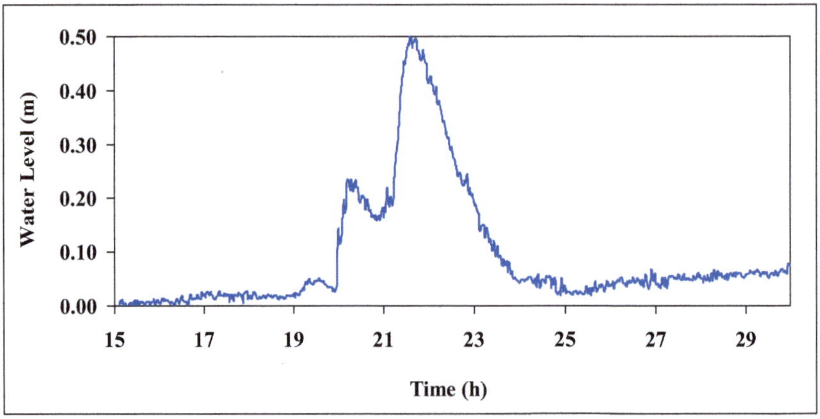

Fig. 4.28: Water level measured in the drainage channel of Figure 4.27.

4.9 Humidity Sensors

This is another example of a resistive sensor. In this case the conductivity of the material of which the resistor is made varies with the measurand. These sensors consist of two electrodes encapsulated in a block of a porous material that in the early designs was gypsum.

Contrary to what happens with strain gauges, for humidity sensors the geometry of the resistor is fixed because the size and separation of electrodes and the size of the porous block remain constant. What changes is the conductivity of the material. Initially, the block is dry and the electrical conductivity of the sensor is low because gypsum pores are full of dry air, which is a poor electrical conductor. When the sensor is buried in wet soil, humidity will be absorbed by the porous material and air replaced by water (more conductive than dry air), thus the total conductivity of the block will increase.

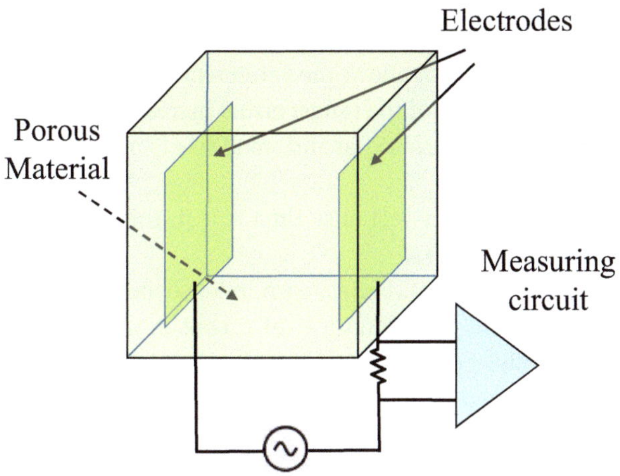

Fig. 4.29: Humidity sensor. Two electrodes are embedded in a porous material. When water replaces air in the porous material the electrical conductivity of the material changes. A voltage source, a resistor and an OA give an output voltage proportional to conductivity.

The transference of electrical conductivity to humidity may be obtained by a calibration process. A circuit as shown in Figure 4.29 (with two resistors in series as in Figs. 3.19a and 4.7) may be used to measure the resistance between electrodes; then, the conductivity is calculated and the humidity inferred. Most times an AC voltage is used to decrease electrode corrosion and to improve signal to noise ratio.

This kind of sensor has some drawbacks: changes in temperature and in soil water salinity modify the sensor transference; the time response of the sensor is long because it takes several hours for the block to reach soil humidity. The size of the sensor is not suited for using it in potted plants and the life of the sensors was short in early designs. At present, electrodes are made of materials which suffer low corrosion such as stainless steel, and porous materials such as fiberglass are used to give a longer life to sensors.

The same schematic of Figure 4.29 can be used to explain capacitive humidity sensors. In this case, the electrodes are the plates of a capacitor and a dielectric material is placed between the plates. The permittivity of the dielectric material (Eq. (4.6)) varies with the humidity.

Capacitance sensors are small compared with the above conductive sensors, they are flat and sizes are below 5 mm by 5 mm. The capacitance of the sensor with relative humidity (RH) of 30 % is about 150 pF and a sensitivity of 0.25 pF / % RH is available. The sensor has to be excited with an AC voltage with frequencies in the range 1 to 100 kHz (http://www.ist-usadivision.com/sensors/humidity/). In general, humidity sensors measure with a relative good accuracy up to a point slightly away from both ends of the relative humidity range where the accuracy reduces.

Other capacitive air or gas humidity sensor allows measurements in the range 0 – 100% RH with errors ± 2% in the central part of the range; errors increases at the low and high 10 % extremes of the range. They are also flat and small size (3 mm by 3 mm)

(http://www.sensirion.com/en/home/

Because capacitive sensors are small, the response time is fast, they can reach 63% of a step humidity change in less than 10 s.

Sensors to measure grain humidity (wheat, barley, corn, rice, soybean, etc.) are also capacitive. In this case the capacitor is cylindrical of a considerable volume (about half a liter). The grain is the dielectric material. They have a higher accuracy (0.5%) but they measure only in a reduced range (5 to 40% RH) which is the range of interest for cereals.

4.10 Conductivity Sensors for Fluids

Electrical conductivity of liquids allows the amount of dissolved salts to be known, a parameter of interest in oceanography, hydrology, industries, etc. Instruments to measure the electrical conductivity of liquids are composed of a sensor and an electronic circuit. The first is introduced into the liquid, whereas the second excites the sensor, process the signal and presents the results to the users. Results may be shown in a display or recorded in a non volatile memory for further analysis.

Because the conductivity of some liquids is very sensitive to temperature (e.g. for water the sensitivity coefficient is about 2 % per °C), conductivity measurements are ordinarily referred to a standard temperature of 25 °C. In order to refer the conductivity measured at the actual temperature to the standard temperature, the liquid temperature has to be measured close to the conductivity sensor. For a number of solutions, temperature coefficients are not constant and some kind of correction table should be stored in the memory of the conductivity meter to provide a means for automatic corrections.

There are two prevailing technologies for measuring conductivity; they are based on two quite different kinds of sensors: **conductivity cell and inductive probe**. Each one has their advantages and limitations.

4.10.1 Conductivity Cell

This kind of sensor uses two electrodes which are in direct contact with the liquid. Electrodes are made of different materials according to the liquid in which they will be used (stainless steel, platinum, graphite, platinum coated with platinum black, titanium, gold-plated nickel, etc). The simplest conductivity cell consists of two parallel plates (electrodes) separated by a fixed distance. This cell is shown in Figure 4.30; this figure is helpful to introduce the concept of "cell constant" and to show how conductivity units are derived.

In order to measure the conductivity of the liquid, the cell electrodes are excited with an alternating voltage, and the current through the liquid is measured (Fig. 4.30). From Eq. (3.13), if V and I are known, the conductance G may be calculated. This conductance corresponds approximately to the conductance of the volume of liquid contained between both electrodes.

The conductance is a parameter that depends on the geometry of the electrodes and the conductivity of the liquid (σ) (Eq. (3.12b)) (shown again in Eq. (4.14)).

$$G = \frac{A\sigma}{l} \tag{4.14}$$

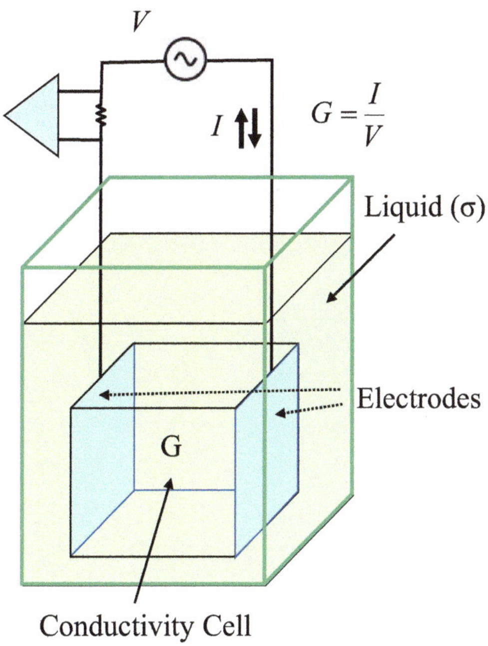

Fig. 4.30: Conductivity cell. The size and separation of the electrodes are fixed, but the conductivity is variable.

Let us assume that both electrode plates have an area $A = 1\ \text{cm}^2$ and are separated by $l = 1$ cm; keeping in mind that the unit of G is the siemens (S), the conductivity of the liquid may be calculated from Eq. (4.15), resulting the unit frequently used for conductivity, S/cm. The unit adopted by the International System of Units is S/m, but μS/cm = 10^{-6} S/cm = 10^{-4} S/m is also used for many purposes.

$$\sigma = \frac{Gl}{A} = GK\ \frac{\text{S}}{\text{cm}} \tag{4.15}$$

The cell constant $K = l/A$ (cm^{-1}) represents the dimensions and shape of the electrodes. It has been adopted in conductivity literature that for a cell having two electrodes with the above dimensions ($A = 1\ \text{cm}^2$ and $l = 1$ cm) $K = 1\ \text{cm}^{-1}$.

Conductivity meters should cover a broad conductivity range; pure water has conductivity below 0.5 μS/cm, drinking water, about 50 μS/cm, and sea water about 50 mS/cm. Therefore, to keep measuring errors due to electronic circuits as low as possible it is necessary to maintain V and I within a certain range (not very small values so as to keep them greater than the background noise). To achieve this purpose it is required to change the cell constant (cells of different dimensions) according to the expected range of the conductivity of the liquid inside the cell. Therefore, it seems reasonable to have different cells for each conductivity meter. For example, for low conductivity liquids, the current would result small, then the cell used to measure them should have a cell constant also low, i.e. large A and small l. This assures to have reasonably large current which can be easily measured with small error. Table 4.3 presents three cell constants and their recommended measuring range.

Some meters have a fixed AC frequency to excite the cell, but others may allow the user to adjust the frequency to optimize signal to noise ratio avoiding noisy frequency ranges; in the last case frequency range is usually between 20 Hz and 20 kHz.

Table 4.3: Cell constants and measuring ranges

Cell constant	**Recommended conductivity range (μS/cm)**
0.1	0.5 to 400
1	10 to 2000
10	1000 to 200,000

Sometimes, with the purpose of expressing the ability of liquids to conduct electricity, the resistivity of the solution is used instead of its conductivity. The unit of resistivity is Ω m. As stated in Section (3.7.5), resistivity is the reciprocal of conductivity, so values of μS/cm correspond to values of MΩ cm.

Because electrodes are in direct contact with the liquid, measurements may result affected by different kinds of phenomena that could alter electrodes, such as fouling, electrochemical plating and polarization. In these cases electrodes must be cleaned to renew the active surface of the cell. Abrasives or sharp objects should not be used to clean electrodes because they could be permanently damaged, but water with liquid detergent could help. A piece of cotton can be used with great caution to avoid modifying the gap between electrodes because it would change the cell constant. For most cells the cotton may be soaked in acetone or a chlorine solution. Before using conductivity meters, it is recommended that they be calibrated to a standard solution with conductivity close to the solution to be measured.

Conductivity cells are used in laboratory conditions, and frequent electrode cleaning is thus feasible in this environment, but could result impractical for autonomous instruments deployed in the field. Therefore, for field applications other type of sensors may be more adequate.

4.10.2 Inductive Probe

The working principle of this type of sensor is based on concepts discussed in Section (3.7.17). The sensor is composed of two toroids, each one with its own coil wound around the core (Fig. 4.31). Both toroids are submerged in the liquid whose conductivity is being measured, thus the liquid fills the central hole of the toroids. One of them, which will be called the primary, is excited with an AC voltage so that the primary current generates a variable flux Φ that is confined to the primary core. This flux produces an AC electrical potential around the core which creates an electrical current in the liquid. The conductive liquid acts like a transformer whose secondary has only one turn.

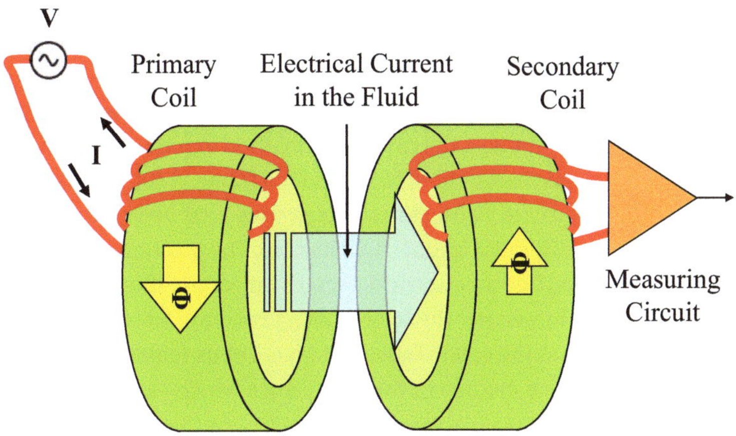

Fig. 4.31: Inductive probe. The primary toroid generates an electric current in the fluid proportional to fluid conductivity and the secondary measure a voltage proportional to the fluid current.

The current in the liquid acts as the primary of a second transformer and generates a flux Φ in the second toroid. This flux, which is confined to the second core, induces a voltage in the secondary coil. This voltage is the input of a measuring circuit. The higher the conductivity of the liquid the higher is the resulting current in the liquid and the secondary voltage. As usual, a transference curve that relates output voltage to conductivity is obtained by a calibration process.

Actually, when manufactured, the set of toroids together with their coils are encapsulated in materials (e.g., synthetic resin) not affected by the liquid properties (i.e. corrosivity), then they are not in direct contact with liquids, and no metal/solution contact exists. Inductive probes require low maintenance and suffer minimal effects due to fouling, thus resulting more adequate for long time deployment.

4.11 Accelerometers

Sensors that measure acceleration are called accelerometers. The most familiar way to make an accelerometer is with a mass and a spring mounted on a frame. The frame is attached to the body whose acceleration is measured. The displacement of the mass relative to its supporting frame is used to measure acceleration. In this sensor it is supposed that this relative mass displacement equals the frame displacement induced by the accelerated frame, as measured from outside the accelerometer. This assumption is true for a certain range of frequencies of the accelerating forces.

A net force *F* acting on a mass *m*, subject to an acceleration *a*, deforms the spring whose displacement *x* is proportional to the force through the spring constant *k* (Hooke's law) (Weinberg, 1999),

$$F = m\,a \quad \text{and} \quad F = k\,x; \qquad x = \frac{m}{k}a \tag{4.16}$$

Because *m* and *k* are constant, measuring the displacement *x* allows the acceleration to be known.

There are different electrical ways to measure the relative motion of the mass and the frame, the simplest to understand is the capacitive method explained below.

Figure 4.32 shows the seismic mass, the springs and the frame. A central electrode attached to the mass, together with two external electrodes attached to the frame, forms two variable capacitors of similar initial capacitances. The position of the mass, relative to the frame is converted into an electrical signal by means of these capacitors. When the frame is accelerated upwards the lower spring is compressed, the central electrode moves downwards and the capacitances change. Let us remember that this change in capacitance produces a change in the impedance which can be measured by an electronic circuit.

Because of the upward acceleration, the distance between the capacitor's plates are modified (Fig. 4.32). The central electrode moves downwards, thus decreasing the distance between the plates of the lower capacitor and increasing its capacitance (Eq. (3.16)). The opposite happens with the upper capacitor.

A simplified electric circuit is shown on the right side of the Figure 4.33. An alternating voltage V excites a bridge with two capacitors (C) and two resistors (R). The voltage drop between the mid points of both the resistive and the capacitive branches enters the amplifier (A). The bridge is initially balanced and the signal entering the amplifier is thus zero. When the seismic mass is displaced, the capacitances change due to the change in the distance between the plates. The bridge becomes unbalanced and a voltage difference between the mid points of both branches appears; this difference is amplified. The dashed rectangle relates the mechanical and electrical representation of the capacitors.

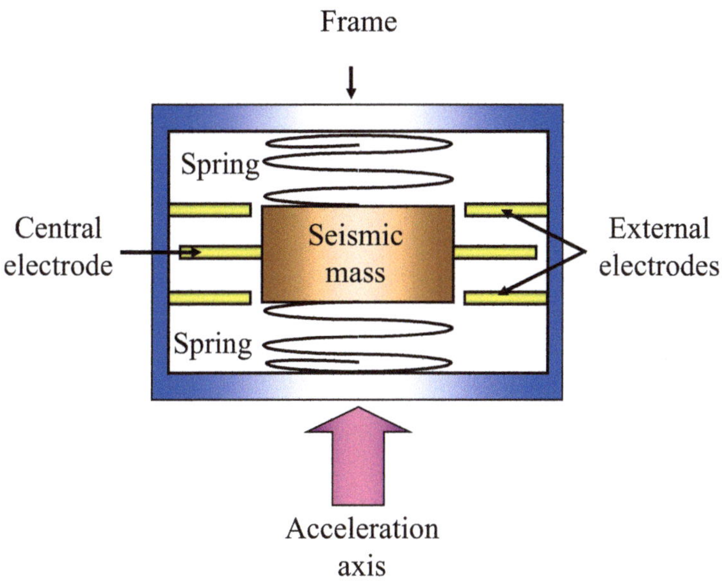

Fig. 4.32: Accelerometer: the mechanical set is composed of a mass, springs and a frame. The electrical sensor that measures the relative motion between the frame and the mass is a capacitor whose plates are attached to the frame and the mass.

Summarizing: acceleration produces a relative motion of the mass and the capacitor plates which is directly related to it. The change in the plate distances produces an impedance variation which is determined by a bridge, providing a voltage output directly related to the acceleration.

During the design of an accelerometer the two parameters under control are the spring stiffness (k) and the mass (m). The resonant frequency of accelerometers and the useful frequency bandwidth are established by these two parameters.

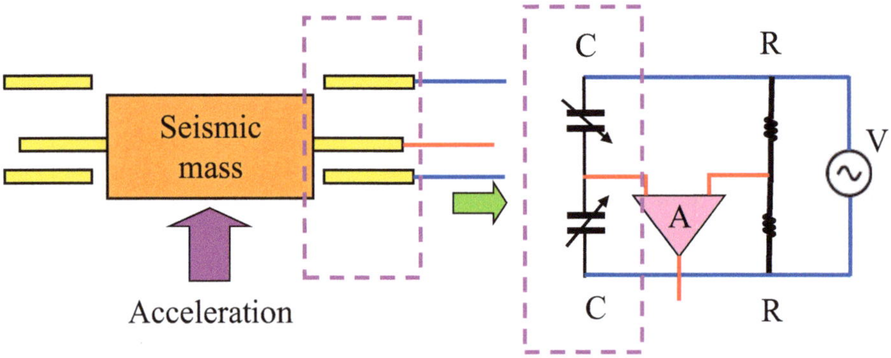

Fig. 4.33: An accelerometer is composed of a mechanical part depicted on the left, and an electrical one depicted on the right. Dashed box shows the capacitors in both schematics.

The type of accelerometer previously described belongs to the kind used for vehicle guidance, navigation and speed measurement and attitude determinations. Accelerometers for other applications require different mass and spring designs. An accelerometer similar to one used in wave measuring buoys is schematically depicted in Figure 4.34. It has a flexible beam supported at only one end (or cantilever) which accomplishes both the roles of mass and spring. The distance between the cantilever support and the fixed electrodes is constant, but when the whole system moves, the cantilever electrode shifts with respect to the fixed electrodes. As before, the capacitances between the moving and fixed electrodes change.

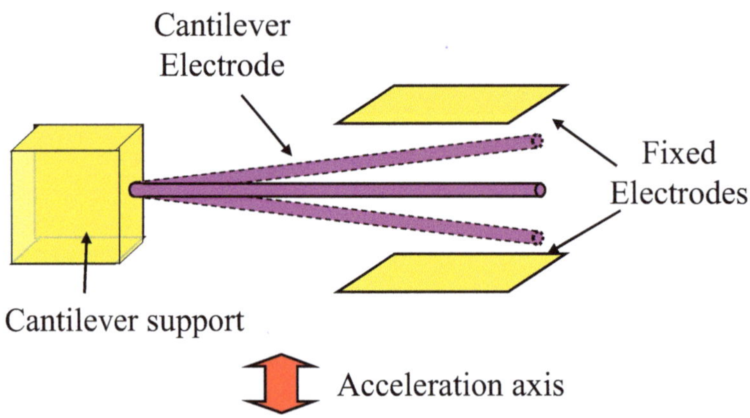

Fig. 4.34: Another kind of accelerometer. The cantilever plays simultaneously the role of mass and spring. In air, the cantilever and the fixed electrodes form two capacitors. In conductive fluid they form a potentiometer.

The diameter of the cantilever is about 0.1 mm and the gaps between the cantilever and the fixed electrodes are about 2 mm. The device is housed in a box full of liquid. The liquid plays two roles: one mechanical, damping the beam movement and another electrical, as the dielectric material of a capacitor.

Electrical wires are connected to each electrode and a circuit similar to that shown in Figure 4.33 could be employed to have a voltage proportional to acceleration. This kind of accelerometer produces a transference curve similar to that shown in Figures 2.12a and b (Section (2.4.3)).

If a conducting liquid were used, the two variable capacitors would be transformed into a potentiometer as in Figure 4.10a (or an impedance with resistive and capacitive components); the fixed electrodes being the ends of the potentiometer and the cantilever the center. The circuit of Figure 4.33 would change to a bridge with four resistors (or impedances), where the two variable capacitors are replaced by the potentiometer.

4.11.1 MEMS Accelerometers

In the last two decades Micro-Electro-Mechanical Systems (MEMS) have evolved fast resulting in many micro-machined devices. MEMS have strongly influenced the development of small sensors. One of these sensors is the surface-micro-machined monolithic accelerometer. This device includes both the signal conditioning circuitry and the sensor, manufactured together on a single monolithic chip. Conceptually they do not differ from the previously described capacitive accelerometers. The great change can be found in its small size and low cost of production.

These devices also comprise a seismic mass, springs and variable capacitors. Figure 4.35 represents the mechanical part of a MEM accelerometer where the seismic mass and central electrodes have the appearance of a fish backbone. In the figure, the central electrodes are called movable fingers and the external electrodes are called fixed fingers. For one of the commercially available MEM accelerometers (Weinberg, 1999) the overlapped finger's length is 125 μm, the finger's thick is 2 μm and the gaps between the central electrode and any of the external electrodes is 1.3 μm.

From an electrical point of view, electrodes play the same role than in Figure 4.33, but because each capacitor is very small, many capacitors are required to reach some minimum capacitance required to have an adequate sensitivity, then more than 60 small capacitors are summed up (Samuels, 1996).

When the seismic mass with the central electrodes are perfectly centered, both sides of the differential capacitor have similar capacitance, and the output of the sensor is adopted as zero output. However, if the seismic mass is displaced because the device is accelerating, the variable capacitor becomes unbalanced and an output signal related to the acceleration is obtained.

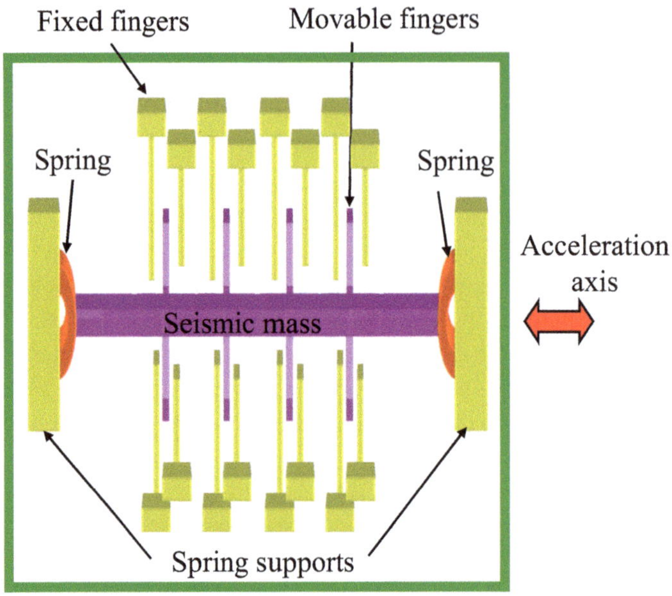

Fig. 4.35: Micro-electro-mechanical accelerometer.

The electronic circuit used to get a voltage proportional to acceleration is more complex than that shown for the previous accelerometers, but because it is integrated in the chip, users do not have to care about it. Nowadays, three axes accelerometers are commercially available at low cost.

4.11.2 Piezoelectric Accelerometers

Some accelerometers use the piezoelectric effect, explained in Section (3.8), to generate a voltage proportional to the acceleration. Figure 4.36 depicts this sensor which comprises a pre-loaded spring, a seismic mass, a piezoelectric crystal, a base and a frame. The crystal structure is initially stressed by the spring force and acceleration forces make the mass move, which produces changes on the stress applied to the crystal. A stress change on the crystal generates a voltage on the crystal electrodes as explained in Section (3.8). Then a voltage output on the crystal electrodes, proportional to acceleration, is obtained within some acceleration's frequency range.

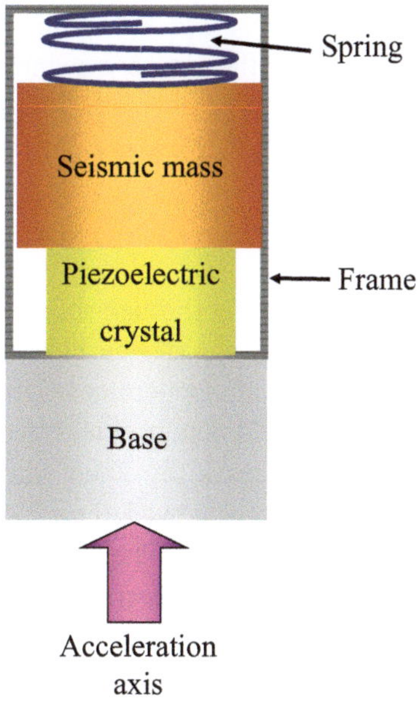

Fig. 4.36: Piezoelectric accelerometer. The preloaded spring stresses the piezoelectric crystal; when the acceleration moves the seismic mass, the stress changes and the crystal generates a voltage.

4.11.3 Accelerometer Applications

Accelerometers are mounted on objects, animals or vehicles to know their dynamic behavior. They produce a signal output $a(t)$ proportional to the acceleration that they experience. The transfer function of an accelerometer relates the output voltage to the input acceleration; its sensitivity is expressed in V/g or Vs^2/m; where g (the acceleration due to gravity) ≈ 9.8 m/s^2. The velocity v and the distance x can be calculated from the acceleration as follows:

$$a(t) = \frac{dv}{dt}; \quad v = \int a(t)\,dt; \quad v(t) = \frac{dx}{dt}; \quad x = \int v(t)\,dt \tag{4.17}$$

When a three orthogonal axes accelerometer is used, it is possible to describe the motion of a moving body in space. Accelerometers are widely used in cars to switch the air bags, in airplanes and ships to measure their dynamic response along three orthogonal axes. In ocean sciences they are used as part of the navigation systems of remote operated vehicles. Accelerometers can be used for determining the position of moving objects floating on the sea surface and then, indirectly, the surface motion can be estimated.

In particular, due to their small size, MEM accelerometers have a wide range of applications to know motion, for example in robots or to study the dynamic behavior of racing cars. Nowadays, there is an increasing use of these sensors to understand the dynamic behavior of animals and human beings, for example the arms of a swimmer or the motion of a runner.

4.12 Geophones

Geophones are devices used to measure small vibrations passing through soils. The device depicted in Figure 4.37 represents a geophone. It comprises a seismic mass (m) suspended by springs from a frame in a way similar to that of an accelerometer. Here the seismic mass is a coil of wire that moves in a magnetic field. The magnet is fixed to the frame (Oome, 2008), which is in contact with the ground or the object whose motion is to be measured. When the geophone's frame is vertically moved, the springs are stretched and compressed and there is a relative motion between the frame and the mass. This motion is detected by the coil, which generates an electrical signal output. If there is no motion of the coil relative to the magnet, there is no electrical signal.

These geophones work in the same way as inductive microphones, where a magnet is surrounded by a coil (Fig. 4.37). When a vertical velocity is imposed on the geophone the mass moves and a voltage is generated, according to Faraday's law (Eq. (3.23)).

Faraday's law relates the induced emf (ε) on the coil with the rate of change of flux through the coil. In the device depicted in Figure 4.37 the spatial distribution of the flux varies, therefore, when the coil moves the flux in the coil changes, thus giving a voltage at the coil terminals. Equation (4.18) shows that the output voltage V is proportional to the velocity of the frame, S being the sensitivity (in V s/m).

$$V = S\,\frac{dx}{dt} \tag{4.18}$$

This sensor in which the relative velocity between the mass and the frame is converted into a voltage by means of a coil and a magnet is called an electromagnetic geophone. In this sensor, it is supposed that the mass velocity equals the frame velocity, which is true for a certain range of frequencies.

A generic transference is plotted in Figure 4.38 where the high pass characteristic of geophones is observed. In this figure the amplitude of the signal is expressed in dB relative to the maximum.

Geophones are insensitive to low frequencies and some recent research indicates that the use of MEM accelerometers may have advantages over geophones to measure in the low frequency range (Hons, 2008). One advantage of geophones over accelerometers is that they do not require electrical energy to give a signal output because they are transducers that transforms mechanical energy into electrical energy.

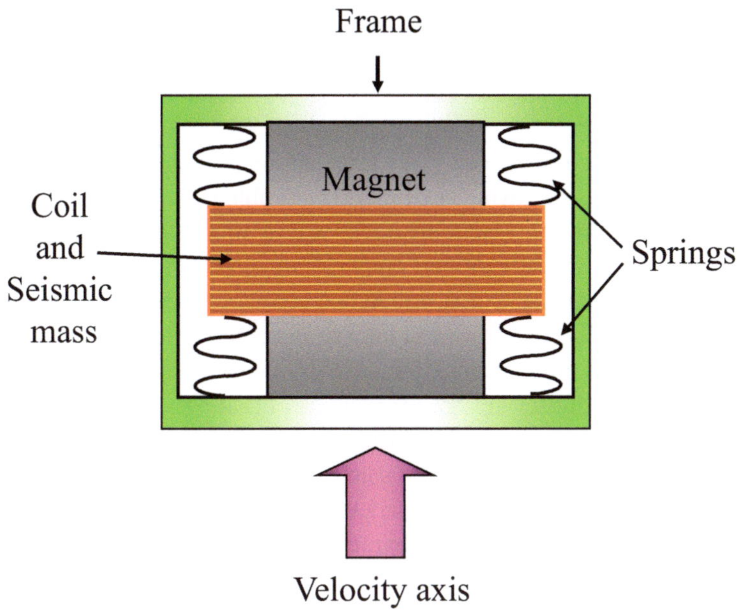

Fig. 4.37: Geophone. A coil is suspended from a frame by springs. When the frame moves, a relative motion appears between the magnet and the coil generating a voltage at its output.

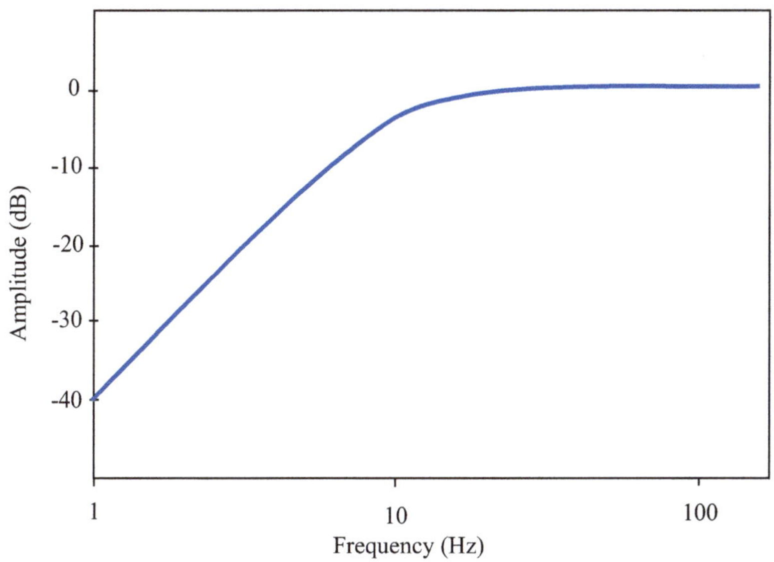

Fig. 4.38: Generic geophone transference indicating its low sensitivity at low frequencies.

4.13 Acoustic Transducers

4.13.1 Introduction

Many devices based on electromagnetic wave propagation have been successfully developed with the purpose of studying environmental problems in the atmosphere. However the attenuation of the electromagnetic waves limits their underwater operation and hinders their underground application also. For these reasons, acoustic waves are employed instead of electromagnetic waves. Such are the cases of underwater communication between divers or underwater control of remote operated vehicles. Also, acoustic wave propagation is frequently employed to study the seabed. This is why acoustic transducers are so important in oceanography, hydraulics, geology and geophysics. This key role of the acoustic transducers in environmental research suggests developing this subject with some detail.

This topic dedicated to acoustic transducers will include both acoustic sensors and acoustic generators. They have several characteristics in common, and in many cases these transducers work in both ways; sending (generators) and receiving (sensors) acoustic signals.

According to their working principle, three groups of acoustic transducers should be considered; one based on Faraday's law, another based on the piezoelectric effect and a third one based on the magnetostrictive effect. The first two groups are the most widespread and used.

Speakers and microphones used in audio to play and record voice and music are acoustic transducers designed to work in air. A magnet and a coil of wire are employed to transfer mechanical energy into electrical energy and vice versa. They are based on Faraday's law and the Lorentz force, and are known as inductive transducers

Most of the attention of this topic will be paid on piezoelectric transducers which are used in many applications related to environmental studies. Some sound generators based in the magnetostrictive effect have particular features and they will also be briefly described.

4.13.2 Inductive Transducers

Figure 4.39 depicts a **generic bidirectional inductive transducer**. It can work either as a sensor or a generator, as a microphone or as a loudspeaker. It has an element whose function is to transform wave pressure into motion or motion into pressure waves; it is a flexible component which is called *diaphragm* in microphones and *cone* in loudspeakers. The edge of this element is fixed to the frame but pressure waves can move it back and forth at the center, where a coil of wire is wound. A magnet is attached to the frame such that the coil is located in its magnetic field. Two electrical wires are connected to the ends of the coil.

4.13.2.1 Working as Sensor

When the transducer of Figure 4.39 works as a sensor or microphone, the mechanical energy from the pressure wave moves the diaphragm and is converted into electrical energy by the coil placed in the magnetic field. Thus, the input energy to the transducer is the sound and the output signal is a voltage between the two electrical wires.

This device is also known as constant-velocity microphone or moving coil microphone, and is based on Eq. (3.23) (Faraday's law) which is applied to the microphone example and shown in Eq. (4.19),

$$V_0 = Blu \tag{4.19}$$

where V_0 is the output voltage (emf ε), B the magnetic field, l the conductor length of the coil and u the coil velocity.

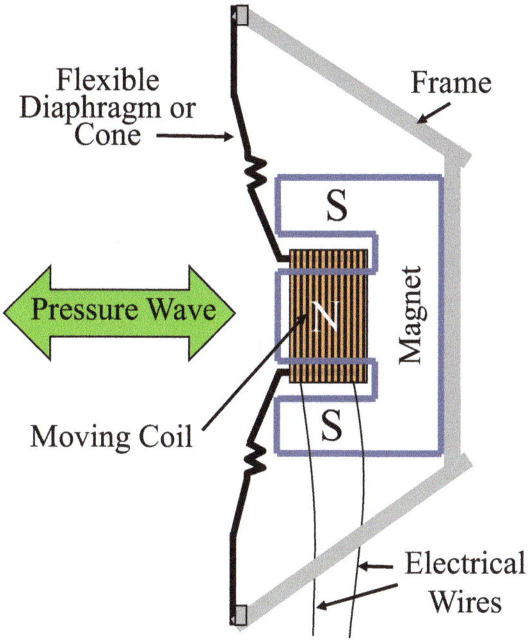

Fig. 4.39: A generic bidirectional inductive transducer

4.13.2.2 Working as Generator

In an inductive generator an electrical energy is transformed into a pressure wave. In Figure 4.39 an alternating voltage is connected to the coil by the two electrical wires and current flows through the coil placed in a magnetic field. As a result a magnetic force appears on the coil (Lorentz force) as shown in Eq. (3.28). This force displaces the center of the cone, thus generating a pressure wave. This is the way a loudspeaker works; if the current fluctuates at a given frequency, say 1000 Hz, a person will listen a sound of the same frequency. In general, the frequency range of these devices is between 20 to 20.000 Hz, so the frequency band of the voice is well reproduced.

4.13.3 Piezoelectric Transducers

Piezoelectric transducers are based on the piezoelectric effect (Section (3.8)) and are widely used as part of many instruments employed in environmental sciences. They are used to transform electric energy into mechanical energy and vice versa. Usually they work in air and water, but they have to be specifically designed for the environment where they will be used due to the different properties of media. Air is largely compressible and sound propagates in it at about 342 m/s, whereas water is very little compressible and sound travels in it at about 1480 m/s. Then the transference in each environment is different. It is said that the transducer "sees" different mechanical impedances in air than in water. Also, electrical parts in underwater sensors must be sealed from the medium, and the sensor's housing has to be designed so as to support greater pressures than in air.

Figure 4.40 represents a piezoelectric disc with two electrodes; **P** is its polar axis and *V* the applied voltage.

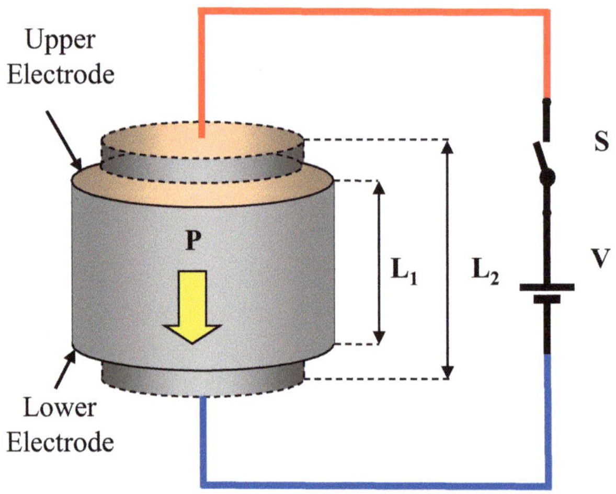

Fig. 4.40: Piezoelectric disc with electrodes. Voltage on the electrodes deforms the ceramic. Stress on the ceramic generates voltage between electrodes.

When the switch (S) is closed, the piezoelectric disc changes shape from the solid line to the dashed line. As it was introduced in Section (3.8), a DC voltage applied to the bar will deform it constantly, whereas an AC electrical signal will make the disk vibrate at the frequency of the electrical signal. Conversely, a mechanical vibration applied to the material generates an electric alternating voltage of the same frequency than the vibration.

The electrical energy deforms the piezoelectric disc which, in turn, produces a mechanical energy proportional to the change in its length $\Delta L = L_1 - L_2$. Therefore, with the purpose of increasing the converted energy the length variation must be maximized. With this purpose, several discs are usually stacked, alternating the direction of their polar axes to reduce the number of electrodes; electrical connections and electrodes (E), are shown in Figure 4.41.

Therefore, the active vibrating element of some transducers consists of a stack of individual piezoelectric crystals. They can have shapes different to a disc but they are stacked so as to increase the energy transferred. A frequent piezoelectric transducer used in the sea is made of a stack of piezoelectric rings. They have the advantage that a bolt can be threaded through the hole to fix the stack.

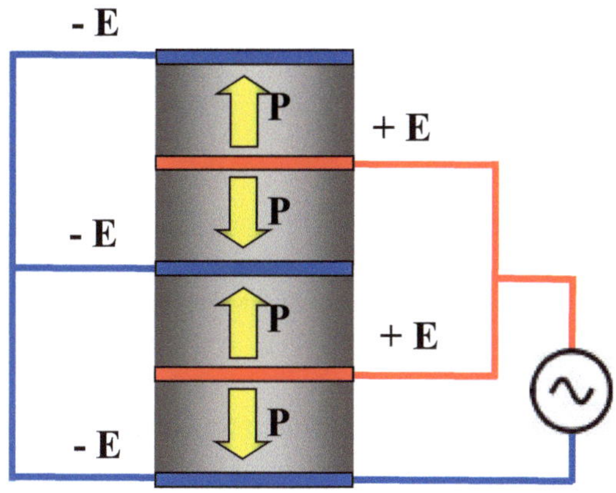

Fig. 4.41: Piezoelectric ceramic discs are stacked to increase the transducer power. Their polar axe directions are alternated to reduce the number of electrodes. The symbols –E and +E should be considered as instantaneous values; half a cycle later the polarities are exactly the opposite.

4.13.4 Piezoelectric Sound Generators

A generic sound generator for use in the sea is depicted in Figure 4.42. It comprises three parts: the tail mass, the head mass and the active elements. Because these generators work underwater most of their parts are incorporated in a waterproof housing (not shown). The active elements are piezoelectric rings threaded by a stud bolt. The bolt is screwed at both ends into the tail mass and the head mass compressing the rings. Then, when the piezoelectric elements are excited by an alternating voltage (V) they vibrate. Because the tail mass is big it has a great inertia and remains at rest. The head mass which is much lighter than the tail mass is pushed away from the tail mass producing a pressure wave in the water. This pressure wave will have the shape of the voltage source waveform (sinusoidal, pulsed, etc.).

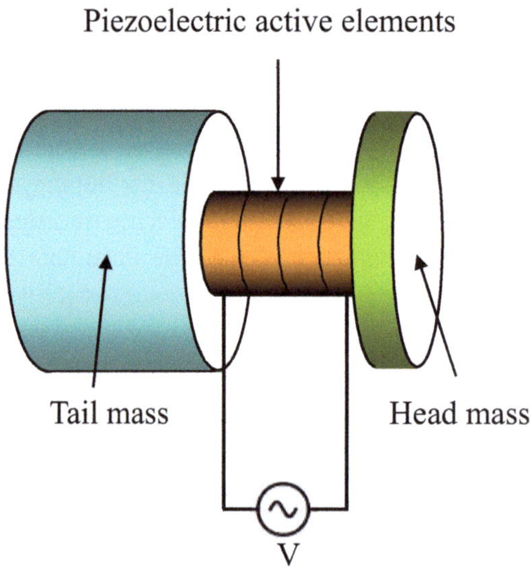

Fig. 4.42: The sound generator is composed of a tail mass, a head mass and the piezoelectric active elements. These elements move the head with respect to the tail generating a pressure wave.

Table 4.4 shows the technical characteristics of two quite different transducers for use in air. Typical applications of piezoelectric transducers in environmental sciences for use in air are: tide gauges, ultrasonic anemometers, acoustic radars (for measuring the atmospheric boundary layer and the wind speed) and geophones. Transducers for water are used in acoustic current meters, echo sounders, side scan sonars, water level meters, ocean and shallow water survey, commercial fishing equipment, instruments for measuring sediment concentration, underwater voice and data communications, submersible positioning, obstacle avoidance and hydrophones, among others.

Sound transducers for higher frequencies (in the order of megahertz) have the same principle but a different embodiment (Fig. 4.43). The same active element is generally used to emit and receive sound. The backing material is usually a high density material that absorbs part of the energy radiated backwards. The acoustic properties of this material define the spatial resolution (beamwidth) and the output signal amplitude or the receiving sensitivity.

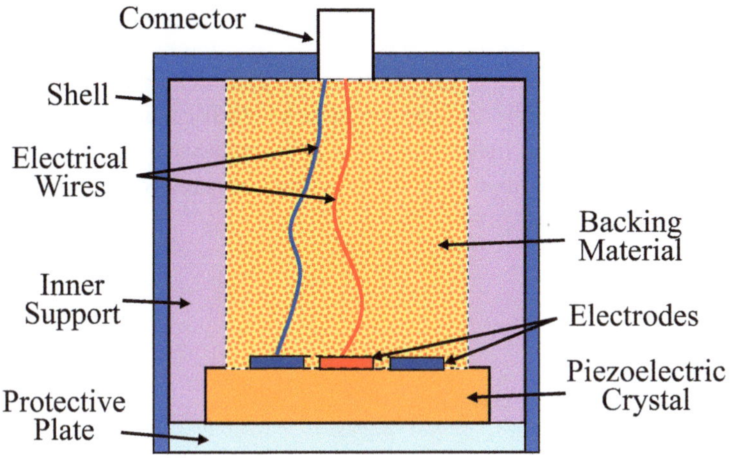

Fig. 4.43: Sound transducer for high frequency. Usually low power is needed and a single piezoelectric element may be used. The backing material contributes to define the transducer beamwidth.

In those cases where it is not required a great amount of energy, a single piezoelectric element may be used (stack of elements is not necessary). As a sound generator, the piezoelectric crystal is energized by a voltage source through the connector, the electrical wires and the electrodes. Signals follow the inverse path when the device works as a receiver.

Table 4.4: Piezoelectric transducers in air

Operating Frequency: 300 kHz	Operating Frequency: 30 kHz
Typical Sensing Range: 5 cm to 50 cm	Typical Sensing Range: 80 cm to 25 m
Beamwidth: 10° ± 2°	Beamwidth: 12° ± 2°
Diameter = 12 mm	Diameter = 106 mm
Length = 10 mm	Length = 141 mm

The protective plate protects the transducer from the environment. For underwater transducers it has also the purpose of serving as an acoustic transformer between the high acoustic impedance of the active element and the water.

4.13.5 Hydrophones

A hydrophone (Figure 4.44) is a sensor that works submerged in a liquid, generally water. It receives sound energy (pressure) and produces an electrical output. It works like an underwater microphone. A hydrophone is a listening passive device usually composed of several piezoelectric rings threaded by a stud bolt screwed into both end heads (Naval Underwater Systems Center, 1990).

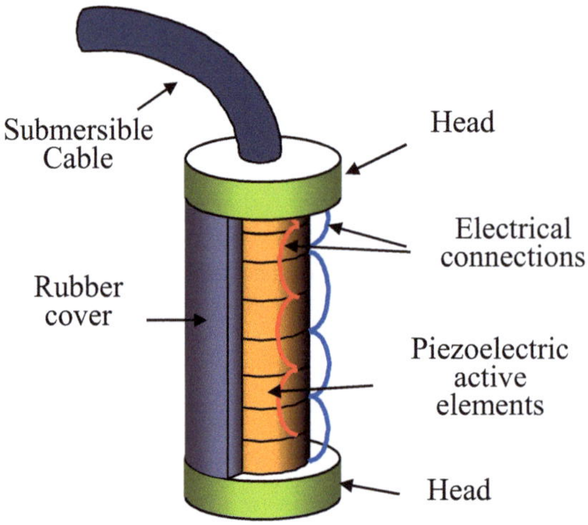

Fig. 4.44: Hydrophone. A stack of piezoelectric rings connected to sum up the voltage generated by each one. The electrical output is conducted through the submersible cable to the amplifier. The set is sealed by a rubber cover.

Ring elements have metallic electrodes which are connected alternating the direction of the polar axis (to reduce the number of electrodes) as illustrated in Figure 4.41. A cover of rubber, which is shown cut in the figure, protects the positive and negative electrodes, the internal connections and the ceramic rings from water. The wires from the electrodes are connected to a submersible cable which leads the signal to an amplifier. The schematic presented here is one of the several embodiments that a hydrophone may have.

4.13.6 Magnetostrictive Transducers

Magnetostrictive transducers use the magnetostrictive effect (Section (3.9)) of a material to convert magnetic energy into mechanical energy and vice-versa. A kind of magnetostrictive transducer for low power applications may be built arranging in parallel a large number of plates made of a magnetostrictive material. A coil of wire is placed around the plate's stack (Fig. 4.45). When an alternating current (*I*) is forced through the coil, a magnetic field causes the magnetostrictive material to change its length, resulting in a mechanical vibration.

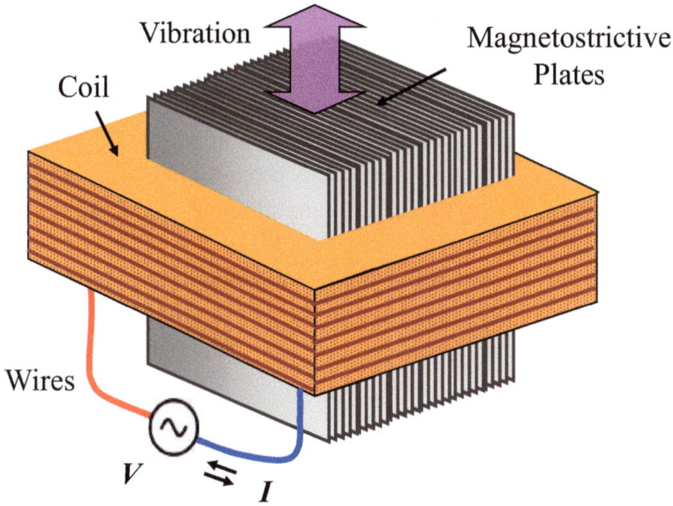

Fig. 4.45: A magnetostrictive transducer formed by plates of magnetostrictive material and a coil. Voltage power supply generates the magnetic field which makes the plates vibrate and change their length as shown by the arrows.

The optimal operation frequency in magnetostrictive transducers depends on the length of the transducer; higher frequencies require shorter lengths. Physical size limitations restrict the operating frequencies much below those reached with piezoelectric sensors (http://www.ctgclean.com).

4.14 Rotation Sensors

In many measuring systems it is required to know the angle of rotation of a shaft, as in the case of an anemometer's direction sensor (vane) or the angular speed as in the cups and propeller of wind speed sensors.

A simple way to measure both parameters is to couple a certain kind of piece to the shaft. This piece is just used to amplify the mechanical information, facilitating the mounting of the required measuring sensors. The shape and material of the piece will depend on the operating principle selected for the sensor. There are several classic operating principles based on optical and electrical effects.

4.14.1 Rotation Speed

We have described in Section (3.7.13) (Fig. 3.22) the way in which electrical energy can be generated from mechanical energy. A magnet was coupled to a rotating shaft and coils placed in the varying magnetic field to generate an electromotive force (voltage). A pulse is generated when one of the magnetic poles passes close to the coil. Then, counting the amount of pulses produced by the coils in a certain time, the number of turns of the shaft may be estimated.

Also, it was stated that the amplitude of the generated voltage is proportional to the time rate of change of the magnetic flux, which is proportional to the rotation speed of the magnet. The average voltage output is thus proportional to the average rotation speed of the shaft. Then coupling a voltage generator to the shaft is another way used to know its rotation speed.

A different electrical device for measuring the above mentioned magnitude is depicted in Figure 4.46, where a voltage is applied to the coil terminals producing a magnetic field around the coil.

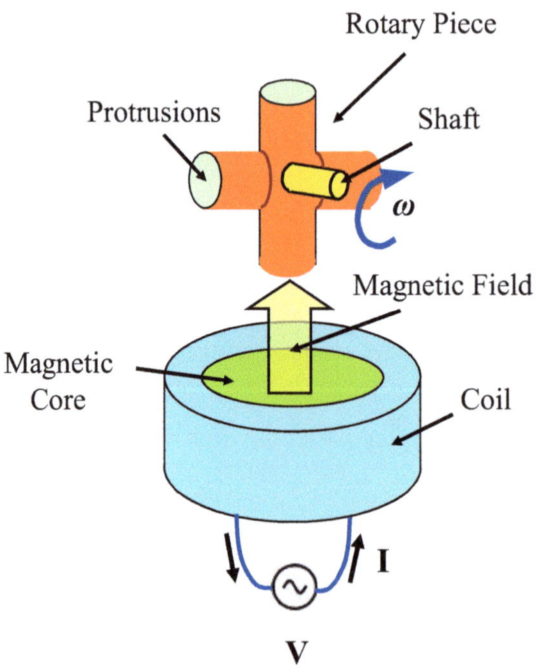

Fig. 4.46: The voltage applied to the coil produces a magnetic field that is disturbed by the rotating protrusions. The alteration of the magnetic field produces current changes in the coil. Counting the current changes permits protrusions passes and shaft rotation to be determined.

As explained above for inductive sensors (Section (4.4.2)), an electrical conductive piece close to the coil would disturb the produced field, modifying the current flowing through the coil. A conductive rotary piece with protrusions is then coupled to the shaft. The rotation of the shaft will produce a pulse-shaped disturbance of the magnetic field so that a pulsed perturbation of the current will be generated in the coil. This perturbation is amplified and analogically processed by an electronic circuit to generate a clear pulsed signal which in turn is counted over a known period of time. The amount of pulses counted is proportional to the average speed of the shaft.

A different operating principle consists in coupling to the shaft a disc with small magnets attached to it. A variable magnetic field in the vicinity of the disc is thus generated when the shaft rotates (Fig. 4.47). This changing field can be detected by a magnetic field detector which could be a Hall effect switch (Section 3.7.18)), a reed switch (Section (3.7.19)) or a coil; as explained in Section (3.7.12), if a magnet moves in the vicinity of a coil a voltage will be produced at the coil terminals

When the magnets pass close to the magnetic field detector, the detector will perceive changes in the magnetic field and an electronic circuit associated with it will produce pulses. Again, counting pulses over a fixed period of time gives the average rotating speed.

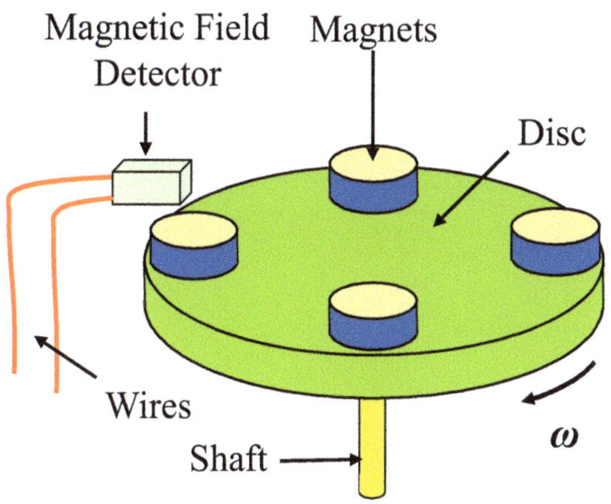

Fig. 4.47: A disc is attached to the shaft and magnets are placed on the disc. The magnetic field detector may by a coil, a Hall effect switch or a reed switch.

In the case of the coil, some mechanical energy from the shaft is taken by the coil to produce the electrical signal. It can be thought that the induced current circulating by the coil becomes an electromagnet that attracts the passing magnet tending to stop the disc. Then it is said that the shaft suffers a "mechanical charge" that opposes its rotation and it appears a braking torque on the shaft. This torque is of electrical origin and must be differentiated from the friction torque. In those methods where a sensing coil is used, a change in the coil current is produced by the change in the magnetic field; this means that an electrical energy is generated by means of the rotating shaft through the measuring system, and also some electrical braking torque appears. No electrical energy can be created without some mechanical work.

Coils are thus not the most adequate device for detecting rotation when the mechanical energy available is low because they tend to stop the rotation of the shaft. This effect is observed as a threshold in the transference of the instrument, which is a certain inability of the device to rotate when the velocity is low.

Another way of measuring the speed of a rotating shaft that does not have the above problem is by using a slotted disk to interrupt the optical path between an emitter and a receiver (Fig. 4.48). The emitter may be a LED and the receiver a photo detector. In the example of Figure 4.48 the disc lets the light beam pass twice per turn, generating two narrow pulses in one complete rotation. These devices are known as encoders, and today's technology permits them to be manufactured so as to produce thousands of pulses per turn by adding a large number of slots.

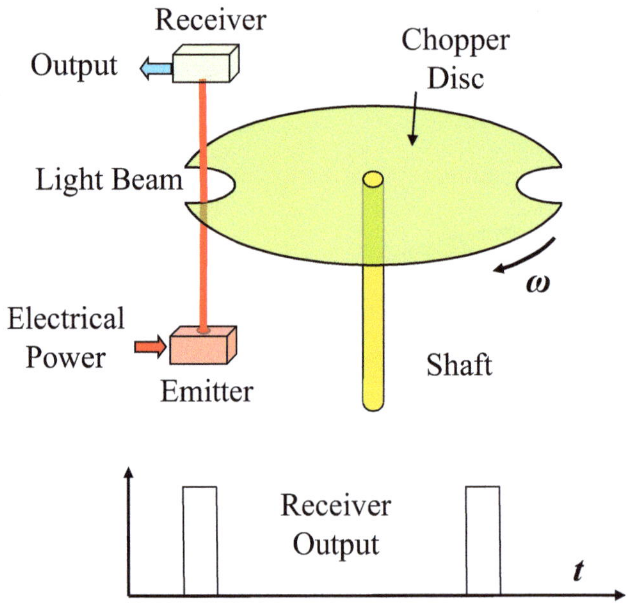

Fig. 4.48: A chopper disc attached to the shaft interrupts a light beam. The amount of pulses received in a know period is counted and it is proportional to the shaft speed.

It is worth noting that there is also a threshold in the transference of the instrument in the optical case, but it is only due to the friction and inertia of the mechanical parts.

4.14.2 Rotation Angle

The rotation angle of a shaft may be known by an optical system. It consists in attaching a coded disc to the shaft. The disc should have transparent and opaque zones which let light pass, or stop it. The shape and distribution of these zones are appropriately selected so that the detection of light through the disc permits the disc position to be determined. Figure 4.49 is the top view of one of such discs. Each of the three small squares represents a pair of optical transmitter and receiver as seen from above.

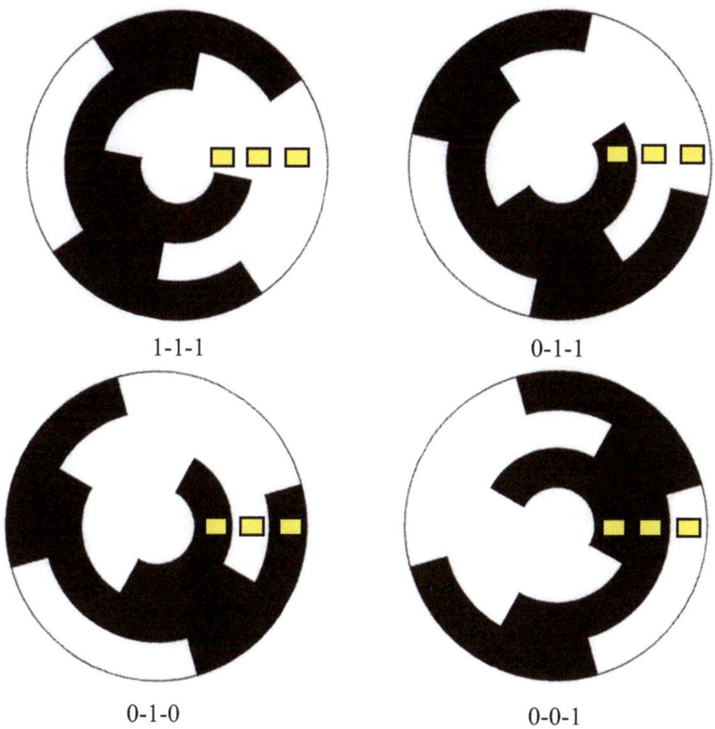

Fig. 4.49: A three bit coded disc. The squares are the light beams perpendicular to the disc plane. The binary code for each position is illustrated below each disc. For example the bottom left disc shows 0-1-0 which indicates that the first and the last beam are in dark zones, then, they are interrupted and the output at the receiver is "0". The central beam passes by the disk and is detected by the receiver giving as a result a digit "1".

Each light beam is perpendicular to the disc plane, and the optical pairs are disposed as was shown in Figure 4.48. In this example, three circular bands have been coded with transparent and opaque zones. The central zone is all transparent, so it does not contribute to the angular detection. Each receiver has only two possible states: the disc let the light pass, and its output is high, or the disc obstructs the light beam and its output is low. Because of this restricted dual state capability it is an inherently digital system. The detection of light through the disc is illustrated below the discs, binary-coded for each position. Each optical pair corresponds to a bit of the digital word and with three bits it is possible to determine $2^3 = 8$ positions, then the resolution of this example is 45°.

The angular resolution of this rotation sensor in increased by accommodating more circular bands on the discs and placing more optical pairs, which requires larger discs. At present, 12 bit angular resolution ($360°/2^{12} = 0.088°$) is achieved with discs whose diameters are about 75 mm. Usually, sensor digital outputs are acquired by microprocessors and presented on a display or stored in memory. This kind of sensor is very reliable, has long life and requires very low torque to rotate the disc, but the optical emitters consume non-negligible amounts of electrical power.

A more simple rotation angle detector may be implemented with a circular potentiometer (Fig. 4.50). Both end terminals of the potentiometer must be connected to an electrical power supply, and the shaft whose angular displacement has to be measured, must be coupled to the sliding terminal. The position of this terminal over the potentiometer resistance defines the output voltage. Thus, voltage is directly related to the angular position. The transference curve of this sensor is liner and the constant transference is expressed in units of voltage per degree. For example, for 0° the output will be 0 V, for 180° will be half the power supply voltage ($V/2$) and for 360° the output will be V.

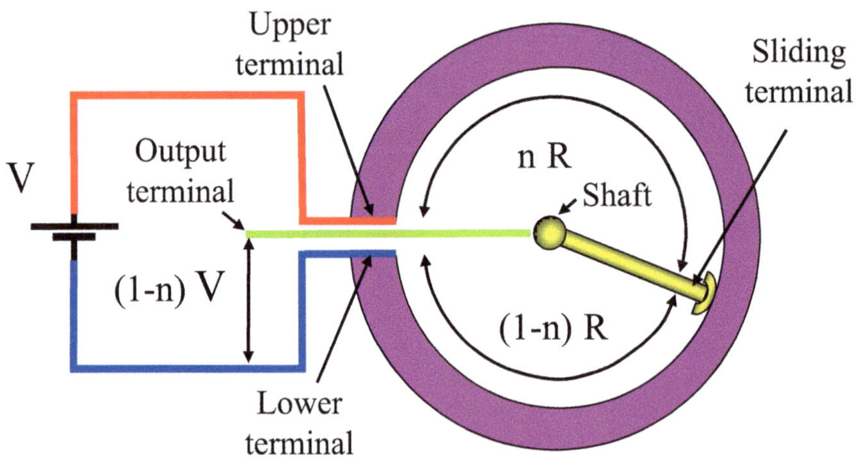

Fig. 4.50: A circular potentiometer with the sliding terminal attached to the shaft and the end terminals to the voltage supply. Voltage at the sliding terminal is proportional to the shaft angle.

In general, this kind of potentiometer has a small gap between the two end terminals which precludes measuring over the entire 360°. Furthermore, because there is a friction between the sliding terminal and the body of the potentiometer, the shaft needs to produce a certain torque to move the sliding terminal. For example, if the shaft is coupled to a vane to measure wind direction, low winds could not have enough energy to displace the sliding terminal. Fortunately, in general, low winds are of little interest. Also, for continuous moving shafts, the friction could wear the potentiometer adding noise to the output voltage.

Sometimes, it is desired to average the output of the angular sensor which is vibrating about the real direction when, for example, it is desired to obtain an average wind direction or to decrease random noise. Using potentiometers, the averages have to be performed taking some precautions, because when the sliding terminal is jumping between the two end terminals, say between 0° and 360°, a simple average will give as a result 180° which is almost the opposite to the real direction.

4.15 Concluding Remarks

As was noted in Section (1.7), the sensors we are interested in are those with electrical outputs. Therefore, to begin the description of these sensors it was needed to introduce the principles on which they are based on. Thus, it was unavoidable an introduction to electrical and electronic circuits to familiarize instrument users with general concepts needed to understand their working principles.

Subsequently, it was necessary to study the most elementary sensors which are based on changes produced by measurands on resistances, capacitances and inductances. Once these concepts were explained, other more realistic sensors based on these principles were studied: sensors to measure temperature, force, pressure, humidity, fluid conductivity, acceleration, sound, vibration, distance, velocity, rotation, position, etc.

Also, to help understanding automatic measurement systems and oscillator principles, it was needed to introduce some simple inductive actuators. Whenever time must be measured, or some kind of periodic waveform must be generated, oscillators are necessary. Then a conceptual approach to them was developed.

The purpose of the sensor description made here was to introduce readers to the essential nature of sensors with electrical output. It is expected that understanding these basic concepts will help readers to understand any other new sensor that they would find in the future.

Even when the sensors described are used in many fields of science and technology, it must be emphasized the role they have as parts of instruments used in Environmental Sciences. In the following paragraphs there is a summary relating these sensors with those instruments in which they could be used. Finally, in Table 4.5 there is a synopsis of sensor specifications and their applications.

Temperature sensors whose characteristics were outlined in Table 4.1 are extensively used in Environmental Sciences to record the temperature of soil, air and water. Also more sophisticated instruments such as infrared thermometers and infrared thermography are used to assess plant canopy temperature; studies on land-atmosphere interaction, surface energy balance; human an animal body temperature measurements, etc. They are also used as part of radiometers and of groundwater flowmeters.

Pressure sensors are also of massive use; their specifications and application depend on the technologies used for manufacturing them. In our field of interest, they are used to measure groundwater and river levels, wind waves, tsunamis, atmospheric pressure, pressure of drilling systems (geology), pressure on the wings of birds and even insects (http://environmentalresearchweb.org/cws/article/news/50265).

Humidity sensors are used in instruments to record humidity in soils, air, woods and grains (wheat, barley, corn, soil bean, etc.); they are also applied in irrigation studies and research of evaporation processes and soil – atmosphere interaction.

Electrical conductivity of liquids is measured to know the amount of dissolved salts, to detect the trajectory of tracers in groundwater, to detect the presence of contaminants, to evaluate aquifer risk under exploitation, etc.

Accelerometers are used in studying the dynamics of animals and human beings; they are deployed in ocean buoys to measure waves, also in robots to study hard to reach places such as deep oceanic waters or the surface of the moon. They are also used to measure speed and displacement in underwater vehicles for oceanographic research.

Geophones are devices used to measure small earth vibrations. They are used to study different layers of soils and to estimate their components, and are required in studies to detect the presence of minerals, water or oil.

Acoustic transducers are present in countless applications either as acoustic receivers or acoustic generators. They are used to measure water level, waves, the speed of particles in air or water, the speed of sound, wind speed, thickness of tissues, ocean currents, and water velocity in rivers and in the laboratory. They are also used in underwater vehicles positioning systems, flowmeters, underwater communication equipments, etc.

Rotation Speed and Angle sensors are used to measure heave, pitch and roll of sea buoys or vehicles, in instruments such as anemometers or groundwater velocimeters.

Strain gauges are used, for example, to measure dynamic forces exerted by living beings and wind force on trees or crops (Cleugh et al., 1998).

Table 4.5: Synopsis of sensors

Measurand	Approximate full scale error	Field of applications in Environmental Sciences
Pressure (semiconductor)	±1% - ±10%	Low accuracy and low cost applications.
Pressure (strain gauges)	±0.1% - ±1%	Water table, tides, waves, atmospheric pressure, pressure of drilling systems.
Pressure (quartz crystal)	0.0001% - 0.01%	Precision water levels, tsunamis.
Humidity (conductive)	5% - 100%	Soil humidity.
Humidity (capacitive – full range)	±2% - ±5%	Air, gas.
Humidity (capacitive – limited range)	± 0.5%	Grains.
Electrical conductivity of fluids	±1% - ±5%	Dissolved salts, contaminants, and aquifer risk evaluation.
Acceleration	≈ 5%	Dynamics of living beings, waves, underwater vehicles.
Soil vibration	±2.5%	Geology, detection of minerals, water and oil.
Acoustic waves	Depends on the application	Water level, waves, thickness of tissues, underwater communications, air and water velocity profilers.
Rotation speed and angle	Depends on the application	Measure of heave, pitch and roll. Anemometers, winches.
Strain gauges	≈ 1%	Forces exerted by living beings, wind force on trees or crops.

References

Allocca, J. A., & Stuart, A. (1983). *Transducers Theory and Applications.* Virginia: Reston Publishing Company, Inc.

Cleugh, H. A., Miller, J. M., & Böhm, M. (1998). Direct mechanical effects of wind on crops. *Agroforestry Systems*, *41*(1), 85-112.

Garner, D. L., H. B. Underwood, H. B., & Porter, W. F. (1995). Use of modern infrared thermography for wildlife population surveys. *Environmental Management*, 19(2), 233-238.

Hewlett Packard. (1980). *Practical Temperature Measurements*. Application Note 290, 5952-8801, USA.

Hons, M. S. (2008). *Seismic sensing: Comparison of geophones and accelerometers using laboratory and field data*. University of Calgary, Master of Science Thesis, Department of Geosciences, Calgary, Alberta, Canada.

Jones H. G., Stoll, M., Santos, T., de Sousa, C., Chaves, M. M., & Grant, O. M. (2002). Use of infrared thermography for monitoring stomatal closure in the field: application to grapevine. *Journal of Experimental Botany*, *53*(378), 2249-2260.

Mahan, J. R., & Yeater, K. M. (2008). Agricultural applications of a low-cost infrared thermometer. *Computers and Electronics in Agriculture*, *64*(2), 262–267.

National Semiconductors. (2000). *LM35 Precision Centigrade Temperature Sensors*. Data Sheet.

Naval Underwater Systems Center. (1990). *Acoustic measurements for sonar transducer test personnel, Student guide*. New London, Connecticut, USA. Available at: http://gbenthien.net/pdf/acoumeas.pdf

Oome, T. (2008). *Modeling of an Electromagnetic Geophone with Passive Magnetic Spring*. Eindhoven University of Technology, Department of Electrical Engineering, Electro Mechanics and Power Electronics group. The Netherlands. Available at: http://w3.ele.tue.nl/nl/

Samuels, H. (1996). Single- and Dual-Axis Micromachined Accelerometers. *Analog Dialogue 30*(4). Analog Devices.

Texas Instruments. (2009). *High-Accuracy, Low-Power, Digital Temperature Sensor with MBus™/Two-Wire Serial Interface in SOT563*. TMP112 Data Sheet.

Weinberg, H. (1999). Dual Axis, Low g Fully Integrated Accelerometers. *Analog Dialogue 33*(1). Analog Devices.

Yilmaz, M., Migliacio, P., & Bernard, E. (2004). Broadband Vibrating Quartz Pressure Sensors for Tsunameter and Other Oceanographic Applications. *Oceans 2004*. Kobe, Japan. Available at: http://www.paroscientific.com/pdf/realtimetsunami.pdf

http://www.princeton.edu/~cavalab/tutorials/public/Thermocouples.pdf

http://www.x26.com/articles.html#

http://www.paroscientific.com/qtechnology.htm

http://www.ctgclean.com/technology-library/articles/magnetostrictive-versus-piezoelectric-transducers-for-power-ultrasonic-applications/

5 Flow Measurements

5.1 Introduction

This chapter will include current meters used in environmental sciences and industrial flowmeters because they share many common characteristics and it was considered satisfactory to present them together. Orifice plate, Venturi, Pitot, Annubar and positive-displacement flowmeters are not included because they are traditional flowmeters that may be found in many books (Miller, 1996).

In general, the operating principle is the same whether the instrument is used in industry or in science and they have constructive differences only, as it is the measuring range. Usually industrial speeds are higher than current velocities in natural environments. Another difference is that in industrial applications, most of the times flows are measured along only one direction. Instead, in environmental sciences the velocity of the current is required along two or three orthogonal directions.

In industrial instruments, designers do not have to care about the amount of energy required to power them because they are connected to the main alternating voltage power supply so they can employ large signals to obtain a good signal to noise ratio. This is necessary to get high reliable measurements because industrial environments are usually noisy, from an electrical point of view.

Industrial flowmeters may be used to evaluate the amount of fluid product that an industry is transferring to another through a pipe, and the billing depends on the flowmeter measurements. Then these flowmeters have to be very reliable and stable, normally, more than those current meters used for research applications. Typically, each of the industries has its own flowmeter on the same pipe and both meters should measure the same quantity with errors below 0.5 %.

Environmental scientific research instruments are usually powered by batteries, so power availability constrains their design, giving as a result low exciting voltages and currents, which could generate weak signals. Fortunately, electrical noise in nature is generally much less than in industries, and a good signal to noise relation can be achieved even with low power supply.

The overall consequence of the above-mentioned differences is that an instrument designed for industrial use cannot be used in field research and vice-versa, but from a didactic point of view it is convenient to treat them together; starting with the description of the operating principles, noting the differences in the way they are constructed and showing similarities in the specifications. This strategy is applicable to instruments based on electromagnetic, acoustic and thermal working principles. There are, however, some technologies that have been developed for measuring only inside pipes, and so far they have not been used for measuring in open channels; one of them is the Coriolis mass flowmeter.

When trying to measure flow in industry, it is important to note that industrial processes very often require measuring the **mass flow rate**. The interest in knowing the mass transfer is because the revenue and some actions in automatic control are associated to the mass. For example, the amount of energy of fuel is associated to its mass. Also, in automatic chemical processes where different quantities of products have to be mixed, the equations that govern the mix are generally expressed in amounts of mass. Some flow meters such as those based on the Coriolis force and thermal transportation measure **mass flow rate** directly.

Other flowmeters, such as those based on acoustics, vortex, electromagnetism, etc., measure the speed of the fluid, which associated to the pipe cross sectional area gives the **volumetric flow rate**. Volumetric flowmeters can obtain the **mass flow rate** as an indirect measure by introducing the density of the fluid.

5.2 Vortex Flowmeter

These flowmeters base their operation in the existence of vortices in the fluid where flow is to be measured. A vortex is an eddy or swirl of fluid. The flutter of a flag, the whistle of the wind in the branches of trees and eddies at the stern of moving boats are common examples of vortices.

If an object is moved in a static fluid, vortices are formed within the fluid. If a fluid moves inside a pipe, vortices can be generated by placing an object with a flat front as shown in Figure 5.1. This object is a non-streamlined body called a "shedder bar" and the phenomenon is called vortex shedding.

In vortex flowmeters it is required to generate vortices by a shedder bar. The flowmeter manufacturer adopts different shapes for the bar, which is often installed vertically in the pipe (it is shown horizontally in Figure 5.1 to make the picture clearer).

Under turbulent conditions fluid cannot follow the shape of the vortex generator and it is separated from the bar forming two wakes. Vortices are areas where fluid speed is higher and pressure is lower than in the surrounding fluid. They rotate clockwise in one wake and counterclockwise in the other. Vortices are formed one at a time in alternating wakes, resulting in an oscillating pressure difference (Kármán vortex street). Usually piezoelectric sensors are placed downstream to detect this fluctuating pressure gradient.

According to Eq. (5.1), as fluid velocity (V) increases, so does vortex frequency (f),

$$f = St\frac{V}{b}\ ;\ V = \frac{f_b}{St} \tag{5.1}$$

where b is the shedder bar width and St is a dimensionless number called the Strouhal number. A calibration process allows St to be found for a certain shedder bar shape. Fluid velocity can then be calculated by measuring the vortex frequency.

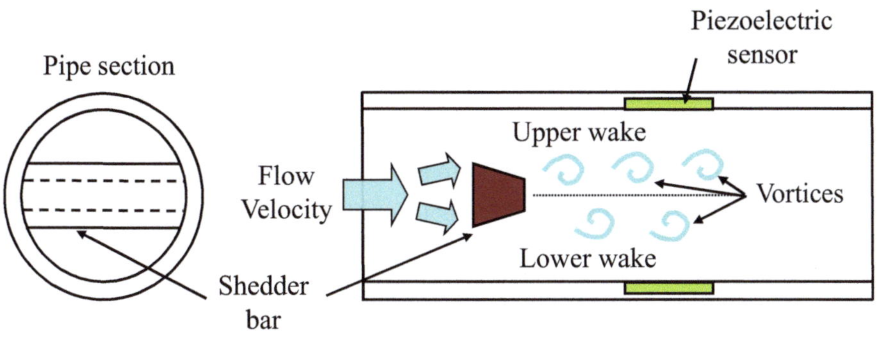

Fig. 5.1: Sectional view of a Vortex flowmeter. The shedder bar generates alternating vortices which are detected by means of piezoelectric sensors.

Since the Strouhal number may vary for low Reynolds numbers, flowmeters should be used in ranges where *St* is constant. Figure 5.2 shows *St* as a function of the Reynolds number (Re). The linear zone in this picture depends on the vortex generator shape. Today's flowmeters, which have solid state memory and intelligent processors, are able to correct the transfer function, extending their useful range down.

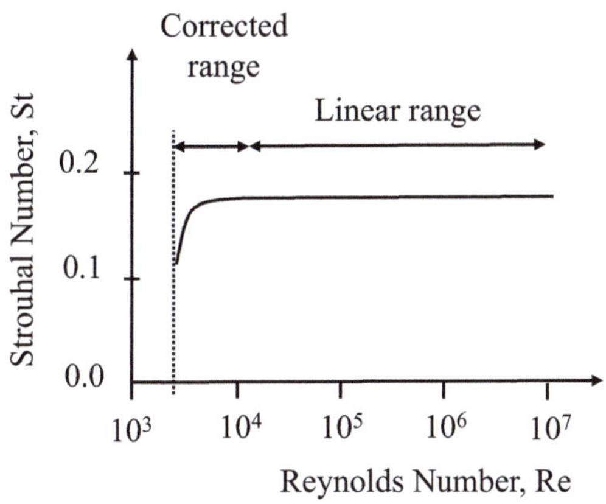

Fig. 5.2: Strouhal number as a function of Reynolds number. Nowadays instruments are able to correct the transfer function in the non linear lower range.

Some manufacturers also measure temperature (T) and pressure (P). Then using equations stored in the flowmeter memory they are able to calculate the fluid density and viscosity, the Reynolds number (to select the correction factor for low-velocity fluids) and the volumetric and mass flow rates.

In addition to the piezoelectric sensors, there are other ways of detecting vortex frequency which are based on capacitive, thermal and acoustic sensors.

5.2.1 Vortex Flowmeter Characteristics

Vortex flowmeters can be used with gases, liquids and steam. Some do not allow measuring bidirectional flows. They are sensitive to distortions of the speed profile, and then stable flow conditions upstream are required. The presence of the shedder bar results in pressure losses that are about half of that produced by an orifice plate flow meter. Excessive wear or corrosion in the shedder bar can change the calibration constant.

Vortex flowmeters can be used in dynamic ranges of 1:10 for liquids and 1:20 for gases, and can measure up to about 10 m/s in liquids, and 100 m/s in gases (see Section (2.2.1) for dynamic range definition). In general, viscous fluids cannot be measured due to their low Reynolds number.

5.3 Electromagnetic Flowmeters

5.3.1 Principle of Operation

The principle of operation of an electromagnetic flowmeter is based on Faraday's law, and is shown schematically in Figure 5.3. A current I flows through a coil generating a magnetic field **B** in a fluid moving with a velocity **V**. Let's assume that **B** is constant, so that $\partial \mathbf{B}/\partial t = 0$ in Eq. (3.26). It remains only the term related to the movement of the conductor in space, and for this simple example the motional electromotive force (emf) (ε) is as in Eq. (3.25) (slightly modified in Eq. (5.2)).

$$\varepsilon = VBL \tag{5.2}$$

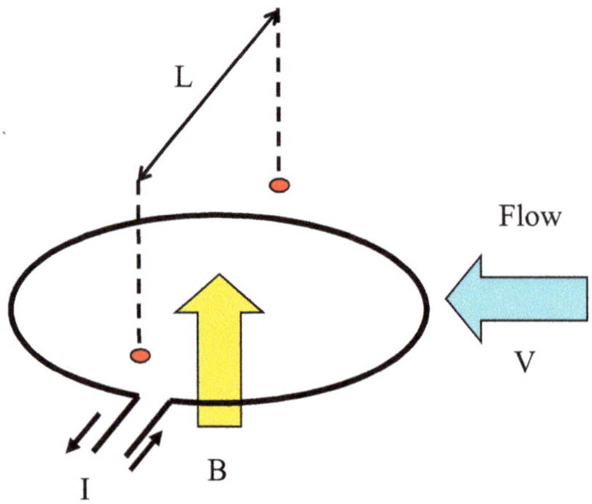

Fig. 5.3: Principle of operation of an electromagnetic flow meter. A current *I* flows through a coil generating a magnetic field **B** in a fluid moving with velocity **V**. Two electrodes separated a distance *L* measure the emf (ε).

In Figure 5.3 the rod of Figure 3.23 is replaced by a fluid passing through the constant **B**, and then according to Faraday's law an emf will be induced on the moving fluid. In order to measure the emf two electrodes are placed separated by a distance L in a direction perpendicular to the plane formed by **B** and **V**. The voltage (v) collected by electrodes is proportional to the emf and then to V.

Therefore, generating a magnetic field and measuring the voltage on two electrodes the velocity of a fluid could be known. This is the working principle of an electromagnetic flowmeter.

5.3.2 Current Meters for Environmental Applications

Figure 5.4 shows a practical scheme of an electromagnetic current meter (EMCM) that measures two components of the currents on the horizontal plane. This kind of instrument is used to know the velocity of water in rivers, lakes and seas. It has two main parts, the coil and the sensing head. The coil is formed by several turns of copper wire wound on a ferromagnetic material which produces the field **B** in the water.

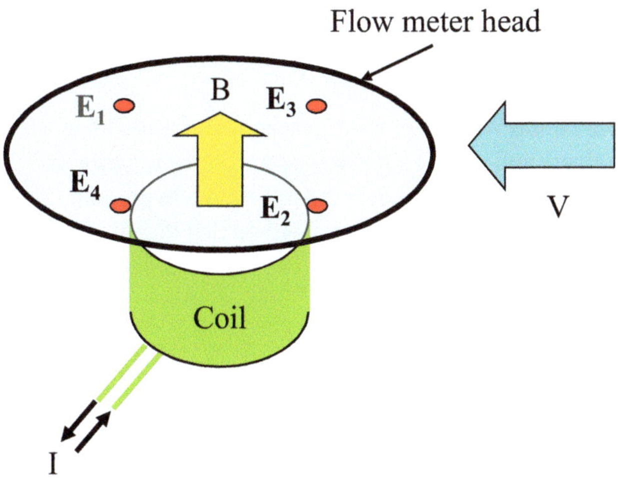

Fig. 5.4: EMCM used to measure the velocity of water on the horizontal plane. The coil generates the magnetic field **B** on the sensing head. The sensing head has two pairs of electrodes E_1 - E_2 and E_3 - E_4 arranged in perpendicular directions. They measure two components of the flow velocity based on voltage differences along their respective axes.

The flowmeter head has four electrodes. The pair E_1 and E_2 is in a direction perpendicular to that of the pair E_3 and E_4. Each pair measures a voltage proportional to the emf generated by the flow in its own direction. According to Eq. (5.2) the emfs are proportional to the flow velocity, so two components of the velocity along perpendicular axes are measured based on the voltage differences. Usually EMCM have a magnetic compass or a flux-gate compass to refer the velocity to the geographical coordinates giving as a result the *north-south and the east-west components of* **V**.

EMCM are available with different sensor shapes. The scheme presented in Figure 5.4 corresponds to the shape known as "discus", which is particularly suitable for laminar flow measurements. There are also "spherical" and "annular" shapes for other flow applications. The spatial resolution of the discus shape is approximately equal to the sensor's diameter, whereas for spherical shape it may be 3 times its diameter. Diameters of commercially available EMCM range from a few centimeters to about 20 cm.

Velocities up to 7.5 m/s are measurable with accuracies about ±1% of the reading. EMCM usually refer their measures to the north–south direction, for this reason manufacturers must specify the compass error or heading error, which may be about ± 1°. Some EMCM include a pressure sensor used to measure waves. Wave data combined with current measurements provides wave directional information.

Normally, autonomous current meters have low power consumption and are powered by low voltage cells. Because instruments internally perform some kind of mathematical filtering, each measurement is usually the average of several samples. In most cases the sampling rate and the amount of filtering is user selectable. Another specification that users must be aware of is the depth rating, which usually exceeds the requirements of most common applications.

There are some portable EMCM which are used to take manual readings while wading in shallow water of rivers and streams. In these applications the user should verify that the current meter is able to measure over the full conductivity range, from fresh water to saline water. Sometimes EMCM are mounted on Sea Remote Operated Vehicles to know their speed.

5.3.3 Industrial Electromagnetic Flowmeter

Electromagnetic flowmeters (EMFM) for industrial applications have a different design. Generally they are mounted on pipes; Figure 5.5 presents a typical scheme. In this case, the flow has the pipe direction and only one pair of electrodes is enough to know the flow speed. Because industrial environments have a high level of electrical noise, a strong **B** has to be produced in order to get a signal that may be recovered from noise. Therefore, two coils adding their magnetic field are employed.

The two coils are interconnected so as to generate an approximately uniform **B** across the pipe. Two electrodes are placed inside the pipe in a direction perpendicular to the plane formed by **B** and **V**. They are electrically isolated from the pipe wall and measure the emf induced on the moving fluid so only one of them can be observed in Figure 5.5. As before, the voltage measured by the electrodes is proportional to the flow speed.

Electromagnetic flowmeters are the most used in processes involving minerals, food, water, sewage waste and paper pulp. The majority of the aqueous solutions, whether they are clean or dirty, can be measured without any problems. Some important advantages over other flowmeters is that they are easy to maintain, do not produce pressure losses in pipes (because there are no parts protruding inside the tube), and allow flow measurements in both directions.

EMFM are built in diameters ranging from 3 mm to 2.5 m. The tube used for the construction of the meter must be internally coated with non-conductive material and both the tube and the coating must be non-magnetic. Ceramic and rubber internally covering the tubes allows temperature and chemical resistant instruments to be constructed. Electrodes that are not in direct contact with the fluid have been developed for some applications.

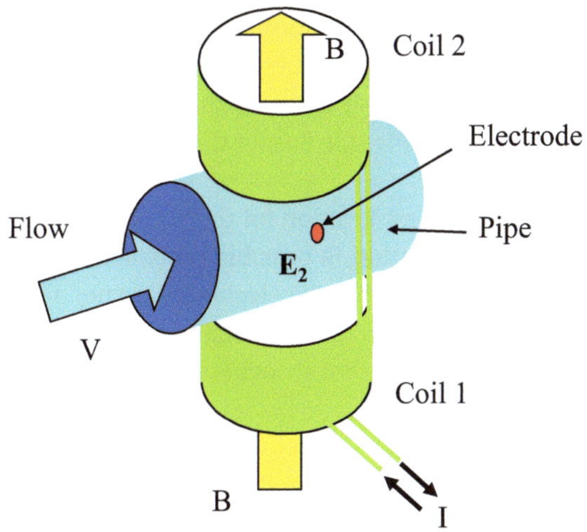

Fig. 5.5: EMFM used to measure the speed of water in pipes. Coils generate a uniform **B** inside the pipe. One pair of electrodes E_1 - E_2 is placed perpendicularly to the pipe direction. It measures a voltage difference proportional to the fluid velocity. Only E_2 can be seen in the figure.

The contribution of the fluid velocity to the generation of the emf is greater near the surface of the electrodes and decreases with distance from the electrodes to the center of the tube. This makes flowmeters sensitive to distortion of the speed profile. Some EMFM use four electrodes to be less sensitive to this problem.

Most EMFM measure with errors less than 0.5% of the measured value for speeds between 1 and 12 m/s, but for speeds less than 1 m/s errors increase significantly, as shown in the Table 5.1, which corresponds to an industrial meter.

Table 5.1: Error for an EMFM

Velocity (m/s)	Error (% of measured value)
0.05	3
0.1	2
0.2	1
0.25	0.5
1	0.3
12	0.3

Then, one limitation of this kind of flowmeters is that errors are important for very low velocities. Some EMFM specially designed for low speed flows are able to measure velocities about 0.1 m/s with errors of ± 1%.

It is important that the conductivity of the fluid be uniform. Otherwise the output signal will be noisy. Also, when measuring liquid which has entrained air, bubbles will cause the EMFM to measure in excess because it assumes that all the fluid is liquid. If the trapped air has the size of the electrode, the output signal could also be noisy.

EMFM must always be full of liquid so the best location for them to be placed is in vertical upward flow lines. Horizontal installation is possible provided that the EMFM is placed at a low point of the pipe line and that the electrodes are not at the top of the pipe (Fig. 5.6). According to manufacturer's recommendation, to keep their accuracy, EMFM require a length of five diameters (Φ) of straight pipe upstream and two diameters downstream.

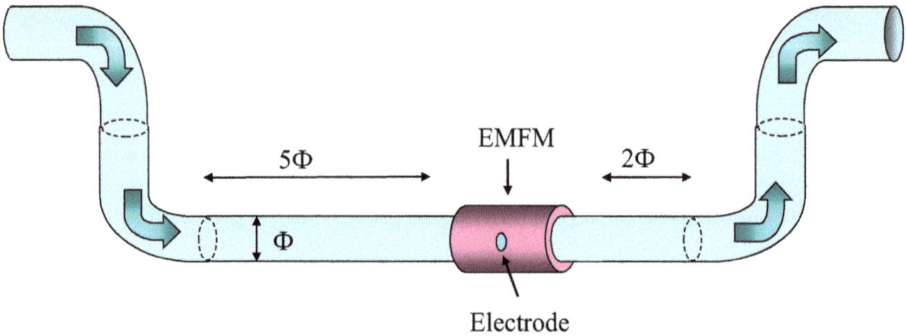

Fig. 5.6: When it is not possible to install the flow meter in upward sections of the line it could be installed in horizontal pipes provided that there exists a length of five diameters (Φ) of straight pipe upstream and two diameters downstream the flow meter. Installation of electrodes at the top or the bottom of the pipe should be avoided in these cases.

EMFM are also designed as probes to be inserted into pipes through taps. In these designs voltages at the electrodes reflects the velocity at the probe tip and not the average fluid velocity across the pipe. Also, this kind of EMFM produces pressure losses.

5.3.4 Common Characteristics of EMCM and EMFM

Until now, to keep the explanations simple, the magnetic field induced in the fluid has been supposed constant, as being produced either by a magnet or by an electromagnet supplied with a DC current. Even when this is theoretically correct, commercial flowmeters do not use a constant **B** because chemical instabilities on the electrode surfaces (electrochemical effects) could pollute the signal. The voltages between electrodes due to electrochemical effects are of the order of millivolts. In order to reduce this unwanted effect it is necessary to choose appropriated electrode materials and silver, silver chloride or platinum plated electrodes are commonly used because of their low electrochemical noise.

If a constant **B** were used, high currents would be necessary in order that electrodes might achieve signal voltages above electrochemical noise. This would lead to coils with impractical sizes and excessive power requirements, a big problem for autonomous instrumentation. Fortunately the electrochemical effects change very slowly with time, and this issue can be addressed by modulating **B**. Thus, because noise is of low frequency and **B** of high frequency, signal can be recovered from noise by frequency filtering. For this purpose manufacturers adopt different waveforms for the current in the coils (Fig. 5.7).

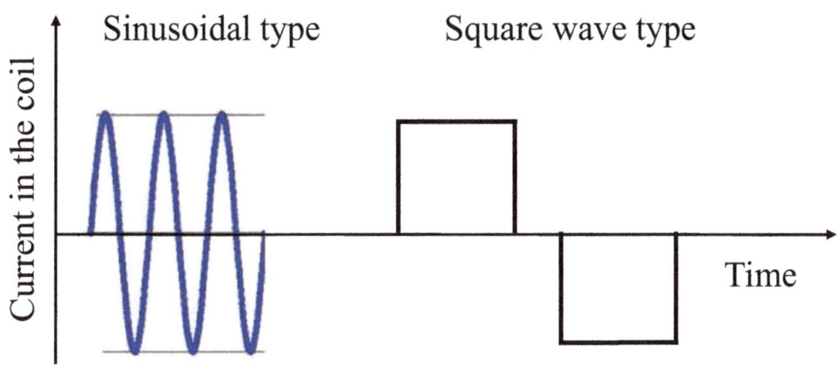

Fig. 5.7: Typical current waveforms in the coils to reduce electrochemical noise effects.

The generated emf (output signal) is a function of the fluid speed, but is independent of temperature, pressure, density, viscosity, turbulence and, to some extent, of the fluid conductivity. Among the limitations of these flowmeters are that the liquid to be measured must be non-magnetic and should have a minimum conductivity level specified by the manufacturer. The minimum conductivity allowed is generally over 1 μS/cm (drinking tap water conductivity is about 50 μS/cm).

5.4 Acoustic Flowmeters

These flowmeters generate a pressure wave of ultrasonic frequency in the flow being measured. They use piezoelectric narrow beam acoustic transducers (Section (4.13.3)) to send and receive the ultrasound wave. The frequencies employed in commercial meters ranges from 0.5 to 2 MHz. Figure 5.8 schematically represents one of the nowadays technologies in which transducers are placed inside the pipe where the flow is measured. The path of the pressure wave forms an angle θ with the axis of the pipe.

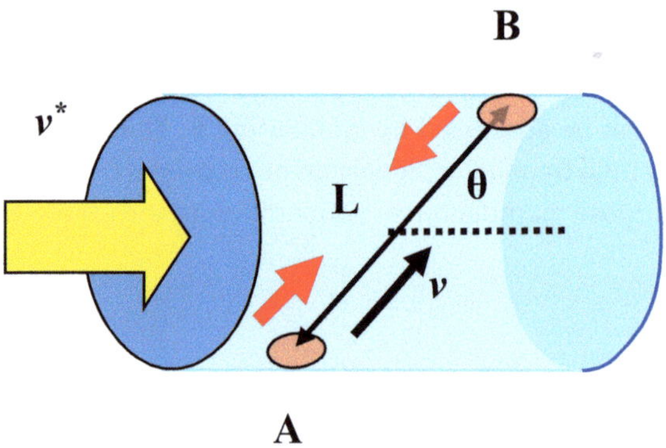

Fig. 5.8: Schematic to introduce the basic concepts on acoustic flowmeters. It represents a pipe full of fluid. A and B are acoustic transducers that send and receive ultra sound pulses. The flow **v*** has a component *v* in the direction of the wave propagation; θ is the angle between both directions.

The flow through the pipe has a velocity **v*** with a component in the direction of the pressure wave path $v = v^* \cos\theta$. If this component has the same direction of the acoustic wave propagation, the speed of the wave will increase. On the other hand, if the acoustic wave has the opposite direction, the speed of the wave will decrease. The fluid velocity can be calculated from the speed of the acoustic wave. For a device such as that depicted in Figure 5.8 there are different ways to measure the wave speed.

5.4.1 Transit Time or Time of Flight Method

5.4.1.1 Direct Measure of Time

This method has some practical limitations but it is useful for introducing the basic concepts on this topic. By means of the transducers an acoustic pulse is sent from A to B, and immediately after, another from B to A. The transit time t_{AB} and t_{BA} are measured and the difference between the times is proportional to the velocity component of the flow in the direction of the wave propagation (v). It also depends on the speed of sound (c) and the distance between sensors (L).

$$t_{AB} = \frac{L}{c+v}; t_{BA} = \frac{L}{c-v}; \quad \Delta t = t_{BA} - t_{AB} = L\left[\frac{1}{c-v} - \frac{1}{c+v}\right] = \frac{2Lv}{c^2 - v^2} \qquad (5.3)$$

In practice, for most fluids $c^2 >> v^2$ and $c^2 - v^2 \approx c^2$ (for example for water $c = 1500$ m/s and $v = 10$ m/s). so that Eq. (5.3) becomes

$$\Delta t = \frac{2Lv}{c^2} = \frac{2\,L\,v^* \cos\theta}{c^2}; \quad v^* = \frac{\Delta t\, c^2}{2\,L \cos\theta} \qquad (5.4)$$

Thus, once the time difference Δt is computed, the flow speed (v^*) can be calculated. Because changes in fluid temperature modify the speed of sound in the fluid due to changes in density, this change should be considered when computing v^* with Eq. (5.4). The volumetric flow rate could be estimated from v^* and the cross-sectional area of the pipe

The time difference (Δt) that an instrument should measure for detecting a change in fluid velocity inside a pipe is small. Suppose that the fluid is water ($c = 1500$ m/s), $L = 0.5$ m and $\theta = 45°$ ($\cos\theta \approx 0.7$). If it is desired to detect a speed change of $\Delta v = 0.01$ m/s, the time difference to be measured will be:

$$\Delta t = \frac{2 \times 0.5 \times 0.01 \times 0.7}{1500^2}\,\mathrm{s} \approx 3 \times 10^{-9}\,\mathrm{s}$$

The simplest way to measure a transit time is to start a digital clock synchronized with the positive slope of the transmitted pulse and to stop it with the positive slope of the arriving pulse. As the acoustic signal arriving to the sensor has noise, the pulse shape is distorted and its slope difficult to detect (Fig. 5.9).

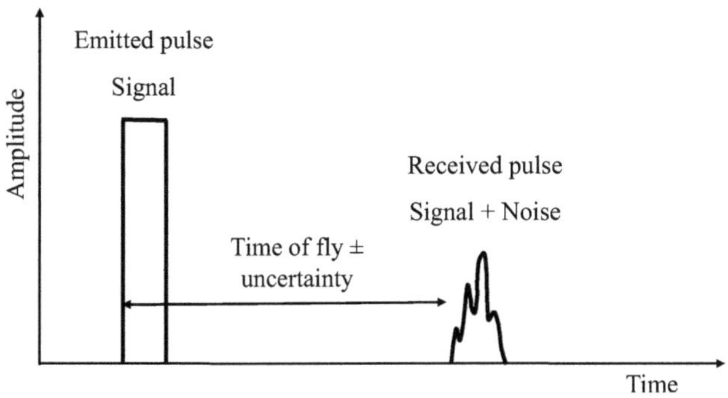

Fig. 5.9: Emitted and received pulses. The received pulse is smaller and noisier; its slope is not well defined, which makes it difficult for the time of flight to be measured.

5.4.1.2 Frequency Difference Measure

This method uses the same physical scheme of Figure 5.8, but the electronics and the processing algorithm are different. In this case, when a pulse emitted from A arrives to B, another pulse is emitted again from A to B. This operation is repeated n times. Then the direction of the transmission is reversed, and other n pulses are sent from B to A. Because v makes the wave velocity increase when traveling from A to B, the first n pulses will be issued in less time than those from B to A. The pulse frequency in the first case is greater than in the second. The difference in frequency between both propagation directions (Δf) is shown in Eq. (5.5), where f_1 is the frequency in the flow direction and f_2 in the opposite direction,

$$f_1 = \frac{1}{t_{AB}};\ f_2 = \frac{1}{t_{BA}};\ \Delta f = f_1 - f_2 = \frac{1}{t_{AB}} - \frac{1}{t_{BA}} = \frac{c+v}{L} - \frac{c-v}{L} = \frac{2v^* \cos\theta}{L} = \frac{2v^* D}{L^2} \rightarrow v^* = \frac{L^2}{2D}\Delta f \quad (5.5)$$

The frequency difference is thus proportional to the fluid velocity. Now the speed of sound does not appear in the equation. This is an interesting property of this method since it becomes independent of the temperature and salinity of the fluid and the velocity of the fluid may be obtained from the distance between sensors (L), the pipe diameter (D) and difference in frequency.

The use of a number n of pulses in each direction of transmission reduces the random noise because sometimes it will be added to the signal and sometimes subtracted from it. Thus, on the average, noise will tend to cancel. It should be noted that all deductions were made assuming that $\mathbf{v}^*$ remains the same while both transducers alternate as senders and receivers, i.e. fluid velocity does not change while 2 n pulses are sent and received.

5.4.1.3 Other Acoustic Flowmeters Characteristics

Acoustic sensors can be mounted so that the acoustic wave follows a direct path through the fluid as in Figure 5.8, or is reflected on the wall of the tube, reaching the receiver upon rebound as in Figure 5.10. In the second case, the accuracy of the method is increased because the acoustic wave travels a longer distance and "feels more flow". Figure 5.10 shows the constituent parts of a simplified industrial flowmeter.

A preamplifier placed near the transducers allows the junction box, from which the instrument is powered, to be separated from the pipe. The communication port can be used to transfer the flowmeter data to recorders, data acquisition systems, transmission units, etc.

If the mass flow is needed, it is necessary to know the pressure (P) and temperature (T) of the fluid to correct for density changes. For this purpose both sensors are placed downstream to avoid disrupting the fluid regime (Fig. 5.10).

Accuracies of ±1% are achieved with this type of flowmeters. They are limited by the ability of the signal processing electronics to determine the transit time. Transducers can be placed inside the tube during the manufacturing process or be fixed externally to the walls of the pipes. The last are known as "clamp-on" transducers. Internal transducers in contact with the fluid (wet transducers) are usually more accurate, but the externals are easier to install

These flowmeters are able to measure bidirectional flow and the cost is almost independent of the pipe size. Because they are not intrusive, pressure drop (head loss) is practically zero, and because they do not have moving parts they require low maintenance. As in most flowmeters, mounting the transducers in an area where the pipe may become partially filled should be avoided, because partially filled pipes will cause erroneous measurements.

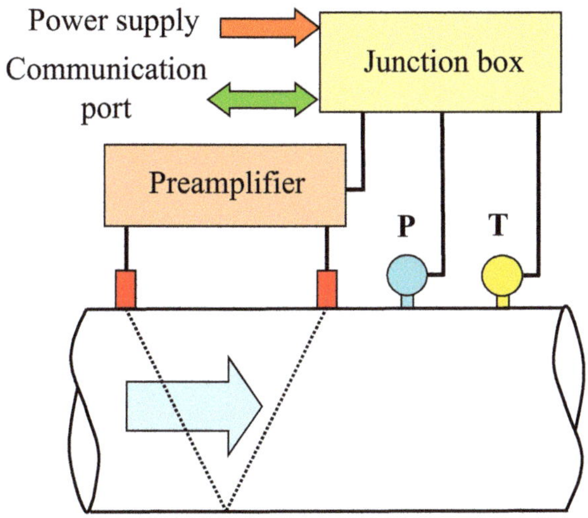

Fig. 5.10: A more complete schematic of a flowmeter shows the pressure (*P*) and temperature (*T*) sensors together with the electrical parts. In this flowmeter the acoustic wave rebounds on the opposite wall of the pipe traveling a longer distance.

Instruments with external transducers must be programmed for each pipeline material, diameter, and wall thickness. The pipe must be constructed of an acoustic conducting material. The largest benefit of the external transducers is obviously the ease-of-installation. They are especially useful for short time measurements or when shutting the system down to install the flowmeter is not possible.

The principle of operation of these particular acoustic flowmeters requires that the fluid be clean because if it is not, the sound wave could be reflected/scattered by the fluid particles and would not arrive at the receiver. Also fluid containing air bubbles or eddies, reflects/scatters the sound making the measurements difficult. It is worth noting that the use of terms 'reflected' or 'scattered' depends upon the size of the 'reflecting particle' with respect to incident acoustic wavelength.

If the size of the reflecting particle is larger than the acoustic wavelength (Fig. 5.11b) it is possible to talk about reflection, but if the size of the 'reflecting particle' is much smaller than a wavelength the law of reflection can no longer be applied, and one must talk about scattering rather than reflection (Fig. 5.11a). In the first case the reflected waves are basically plane waves, whereas in the second case the scattered waves are essentially spherical waves. Although reflection and scattering are present in most cases, the term scattering will be used from now on for the sake of clarity.

For those meters with "clamp-on" transducers, acoustic discontinuities, as those found in concrete-lined or fiberglass-reinforced pipes, may scatter and attenuate acoustic signals, increasing the flowmeter error to ±20%.

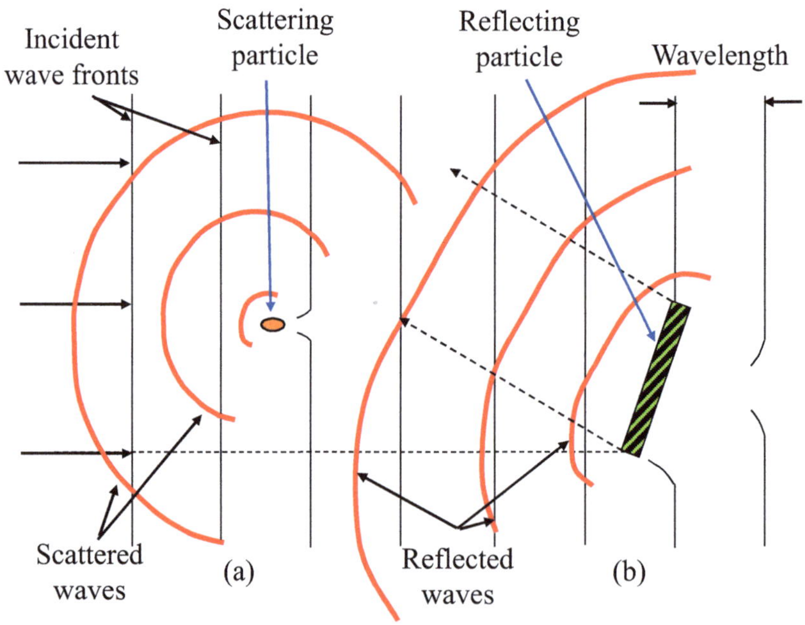

Fig. 5.11: (a) Scattering of sound waves by a particle of size much smaller than a wavelength. (b) Reflection of sound waves by a particle of size comparable to a wavelength.

5.4.2 Doppler Flowmeters

Doppler flowmeters (DFM) use piezoelectric transducers to project an ultrasound acoustic beam in the fluid whose speed is to be measured. The beam is emitted at an angle θ with respect to the direction of the flow and suspended particles or bubbles in the fluid scatter part of the energy received from the acoustic beam (Fig. 5.12). The DFM theory is based on the assumption that particles move at the same speed as the fluid. For this to be true the fluid speed must keep the solids in suspension.

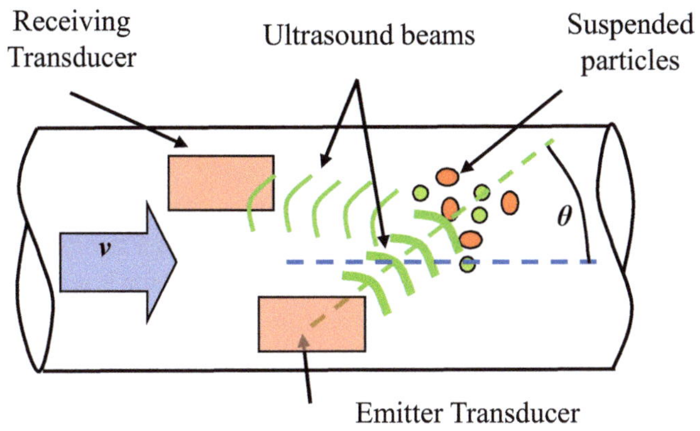

Fig. 5.12: Flow in a pipe. The emitter and receiver transducers are installed such that their beams form an angle θ with the pipe direction. Emitted pulses are scattered by suspended particles that move at velocity **v**. The scatters are received by the opposite transducer.

Suspended particles move with respect to the fixed transducers (emitter and receiver) and, according to the Doppler effect (Section (3.3.1)), the frequency at the receiver is given by

$$\Delta f \approx \frac{2\, v\, \cos\theta\, f_0}{c} \tag{5.6}$$

As before c is the speed of sound, f_0 the emitted frequency, v the flow speed, and θ the angle between v and the beam direction. The electronic circuitry measures the frequency difference Δf and calculates v as

$$v = \frac{c\, \Delta f}{2 \cos\theta\, f_0} \tag{5.7}$$

Reliable measurements require knowledge of c. Because the equation relating temperature to c for different fluids is usually known, measuring the temperature allows correcting c for temperature variations.

5.4.2.1 Installation and Characteristics

Doppler flowmeters should not be mounted on pipes that vibrate excessively. To obtain reliable measurements the installation should be done in sections of the pipe which are always full of fluid, as in vertical upwards flow sections. Sensors should not be mounted at the top and bottom in horizontal sections of the pipe, because fluid's foam (at the top) and deposited sediment (at the bottom) can attenuate the signal excessively.

Weak signals not suitable to obtain reliable measurements may be due to a very clean fluid, low flow speed, pipes with internal coating or large diameter pipes. It is worth noting that Doppler acoustic meters require the presence of particles in the fluid, as opposed to the acoustic current meters described above. Then when signals are not strong enough, the addition of bubbles or solids in suspension upstream, whenever possible, would improve the signal to noise ratio.

DFM have low cost of installation since in many cases are simply mounted on the outside of the pipes. There are portable models which are very handy for quick checks of flow in industrial plants. The portability is generally achieved at the expense of accuracy.

Some manufacturers indicate that for adequate signal reflections a minimum of approximately 100 mg per liter of solids with a particle size of about 75 μm is required. If solid concentrations are in excess of 45 % by weight the output of the reflected signal could be too low to be correctly processed due to the acoustic wave attenuation.

DFM with external transducers cannot be used in pipes of concrete or plastic reinforced with glass fibers, since the waves could be mostly reflected in the external layers. Pipes with internal protective coatings should be avoided because air could be trapped between the external tube and the internal coating, thus preventing the passage of the signal. DFM work well with PVC tubes, instead.

Although DFM can be used with sewage sludge, highly concentrated sludge precludes wave penetration into the center of the flow. Waves are reflected by the slower fluid near the walls where speeds are lower, which results in measures misleadingly low.

Some manufacturers claim that the lower limit of detectable speed for this technology is between 0.1 and 0.6 m/s, the accuracy of the measurements being about ± 2% of full scale. An important advantage of DFM is that they produce no pressure drops because they do not interfere with the flow. The most appropriate piping materials where these flowmeters can be used are iron, steel, PVC, plastic and aluminum.

In order to compute the volumetric flow it is necessary to know the internal diameter of the pipe. In old pipes, however, depositions or inlays make it difficult to know the internal diameter. Additionally, inlays can prevent the passage of the acoustic signal precluding measurements, as is the case of calcium in water pipes.

5.4.3 New Flowmeters (Both Transit Time and Doppler)

As it has been stressed, transit time flowmeters work in clean water but not in water with bubbles or sediment in suspension. On the contrary, Doppler flowmeters require some material in suspension to reflect the sound wave. Manufacturers realized that the methods have similar sensors and hardware but different data processing. Then they designed an instrument capable of measuring in both cases, fluids containing air bubbles and sediments, and clean fluids. They added some intelligence on board the instruments to detect the degree of cleanness of the fluid. According to the fluid condition the instrument switches automatically to measure using one or the other functioning principle.

5.4.4 Acoustic Doppler Current Profiler (ADCP)

The word “profiler” in the name of this kind of acoustic current meter means that it has the ability to measure flow velocity at different distances from the transducer simultaneously. The distances measured by these instruments may range from some decimeters to several meters.

There are diverse shapes of acoustic current profilers with different amounts of acoustic transducers, but all of them work in a similar way. With the aim of simplifying the functioning description of an ADCP, a profiler which measures two-dimensional (2D) velocities along a channel will be illustrated first. The concepts developed in its description will be used to extend the explanation to three-dimensional (3D) profilers.

5.4.4.1 Two-Dimensional Profiler

This profiler was developed to continuously monitor flow in channels, rivers, pipes, culverts, etc. It has three acoustic beams and is mounted at the bottom of the water course. One of these beams points straight up, and the other two point upstream and downstream at an angle of 45° (Fig. 5.13). The first beam measures the height of the surface while the inclined beams measure the water velocity using the Doppler shift as explained in Section (3.3.1) and shown in Eqs. (3.8) or (5.7).

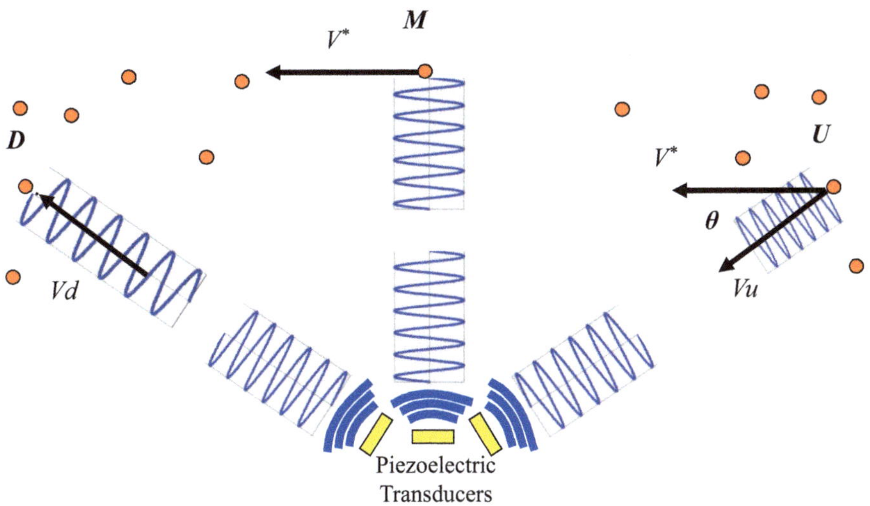

Fig. 5.13: Three piezoelectric transducers emit acoustic waves of the same frequency. Water carries particles in suspension or bubbles at a velocity **V***. Particles U (upstream) and D (downstream) produce different frequency shifts; M (in the vertical) scatters the same sent frequency without shift.

The acoustic transducers used in these meters are piezoelectric ceramics. They are used both for transmission and reception of the acoustic signals, and because of this characteristic the ADCP is called monostatic sonar. Piezoelectric ceramics for these applications produce a narrow beam of sound, so most of the energy is concentrated in a few degrees beamwidth. Sound frequency (f_0) is usually between 1 to 2 MHz.

In the first step of the measuring process, the electronics sends an electric pulse to the transducer which generates a pulse of ultrasound. This pulse propagates through the fluid and is scattered in all directions by acoustic heterogeneities present in the water (suspended material, sediment, bubbles, organic matter, etc). The energy scattered by the heterogeneities that the ultrasound wave finds in the direction of the transducer beam is received back by the transducer and converted into an electrical signal. That is to say, the transducer works, firstly converting electrical pulse energy into vibration in the fluid, and secondly receiving vibrations coming back from suspended particles and converting them into an electric signal. The frequency of the received back signal (f) is compared to that emitted (f_0) and the Doppler shift is calculated $\Delta f = f - f_0$ by the on board electronics, thus the velocity of the fluid is calculated from Eq. (5.7) if c is known.

The three piezoelectric transducers are schematically represented in Figure 5.13. They emit acoustic waves of the same frequency. The dots represent particles or bubbles in suspension. Water is supposed to move at a velocity **V***. Particle U (upstream) has a velocity component *Vu* along the direction of the right transducer beamwidth. Particle D (downstream) has a velocity component *Vd* along the direction of the left transducer beamwidth. Both *Vu* and *Vd* produce frequency shifts. Particle U scatters sound at a higher frequency meanwhile particle D scatters it at a lower one. Particle M (in the vertical) has no velocity component in the direction of the vertical beamwidth, so it scatters the same sent frequency without shift. The maximum reflection received back at the vertical transducer is produced by the water surface.

For known sound speed propagation, the elapsed time since the pulse was emitted and received back allows the distance between the piezoelectric transducer and the scattering particles to be calculated. Therefore, by calculating the Doppler shift for each distance, the velocity profile of the particles at different distances from the transducer can be estimated from the beams pointing upstream and downstream. The central beam with no frequency shift is used to measure the distance from the instrument to the water surface.

In order to process and record velocity information as a function of distance, the transducer's beam has to be discretized at regular intervals. The spatial averaged velocity of the particles on these intervals is estimated. Then, the transducer beam is divided into equal length fractions called cells (Fig. 5.14). Generally, users can select the total number and the length of the cells from a set offered by the manufacturer. Because transducers need a certain time interval to switch from emission to reception, no velocity can be measured in the portion of the beam closest to the transducer. This portion is called the blanking distance.

This blanking distance and the distance to the farthest cell measured from the transducer's surface, depend on the transducer's characteristics and the pulse length. For a real instrument they can be on the order of 0.1 m and 5 m, respectively. The beams should be free from any obstruction, at least for the total measuring distance.

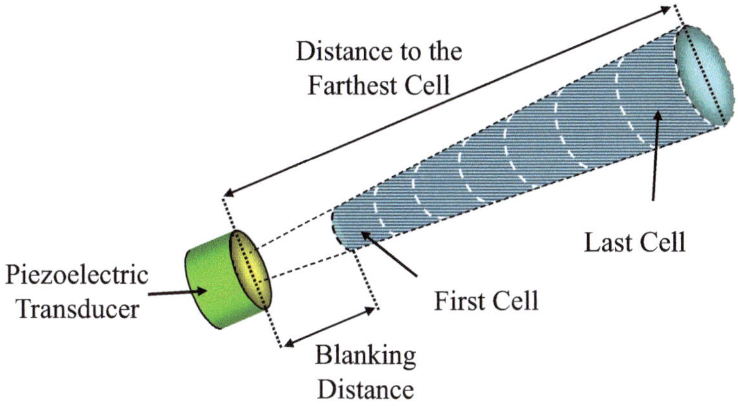

Fig. 5.14: The transducer beam of the piezoelectric transducer is divided into equal length fractions called cells. Average velocities in each cell are measured. No velocity can be measured in the portion of the beam called the blanking distance.

This kind of ADCP must be mounted such that the plane containing the transducer's beams is aligned with the stream. A schematic of an ADCP installation in a natural channel is illustrated in Figure 5.15.

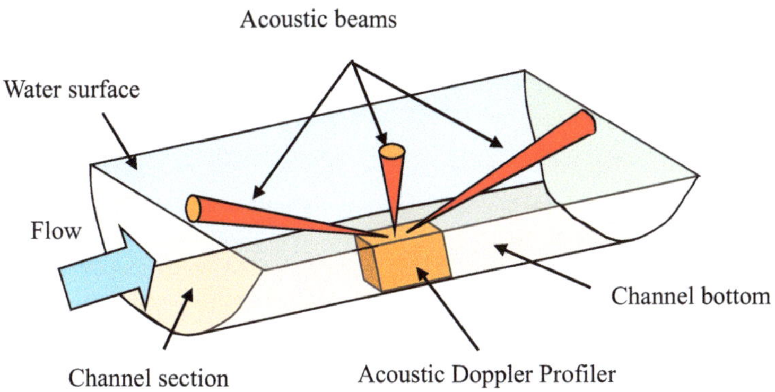

Fig. 5.15: The ADCP is installed at the bottom of a natural channel. The plane containing the three transducer's beams is aligned with the stream.

From the Doppler shifts both beam velocities *Vu* and *Vd* are calculated and converted into the along-channel velocity and vertical velocity through the beam geometry. Under channel or river conditions, the along-channel velocity is the main velocity, the vertical velocity being quite small.

Figure 5.16 shows the installation of an ADCP in a channel where it is expected to find sediment deposits. The stream velocity direction is indicated by the arrow and the three discs (yellow) are the acoustic transducers. Sediment on transducers attenuates the sound wave and hinders measurements to be performed correctly. To avoid sediment deposition on the transducers the profiler was separated from the channel floor. This separation from the bottom has to be considered in flow calculations.

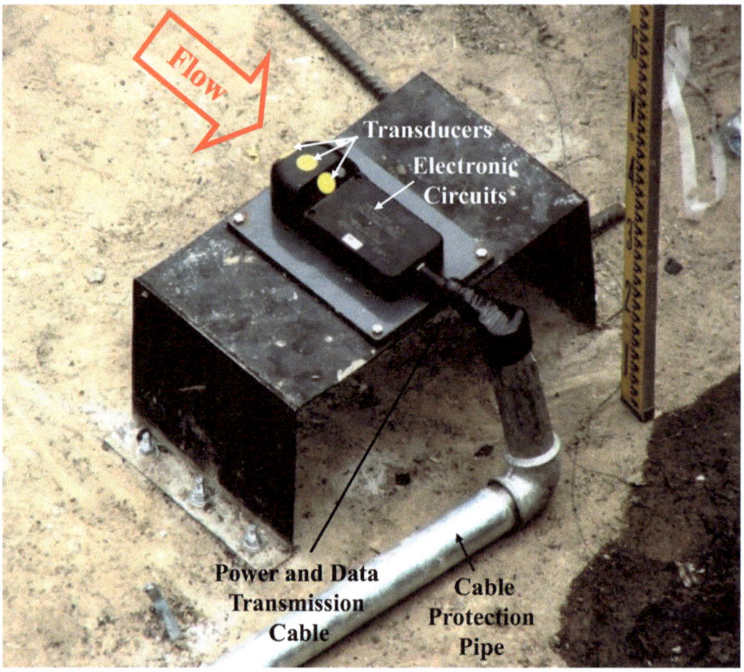

Fig. 5.16: Photography of a two-dimensional profiler installed at the bottom of a concrete channel. The discs are acoustic transducers mounted such the stream velocity is aligned with the plane containing the transducers. Protection to the cable is provided by a pipe. A metal U-shaped mounting device, bolt to the floor, is used to avoid sediment deposition on the transducers.

In this kind of ADCP that measures the water height, the cross-sectional area can be calculated by the instrument if the geometry of the channel is known. Then using the velocity profile data, the volumetric flow rate can be estimated. For this ADCP the manufacturers claim an accuracy of ±1 % of the velocity measuring range (5 m/s).

When the flow to be measured carries solids that could damage the ADCP (such as stones in mountain river or solid waste in a discharge channel of a city), the installation at the bottom has to be avoided. Some manufacturers offer instruments to be mounted in a vertical wall of a channel or on a pile in a river. These ADCP measure the velocity profile in a horizontal plane with two beams and the water height with the third beam (Fig. 5.17).

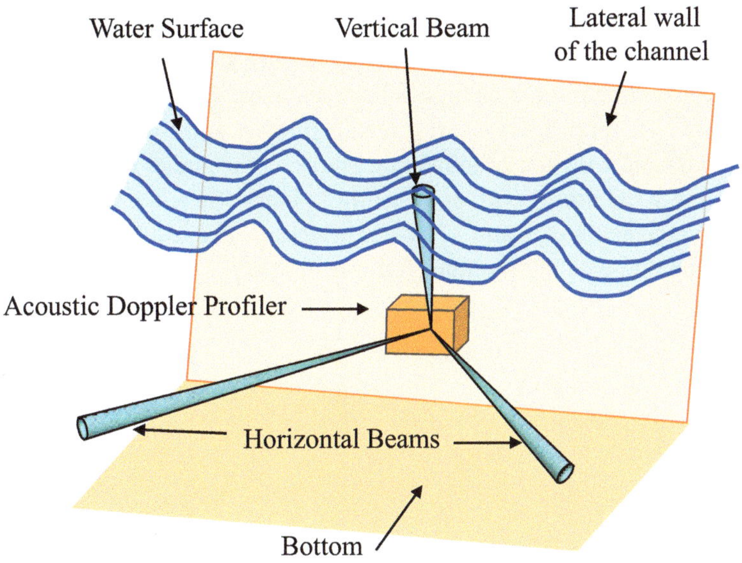

Fig. 5.17: ADCP mounted on the lateral wall of a channel, it measures the horizontal profile with two beams. The third beam points upward to measure the water height. This arrangement could be used when the flow to be measured carries solids that could damage the instrument if mounted on the bottom.

5.4.4.2 Three-Dimensional Profiler

Although, as noted above, there are ADCP with different geometries, all of them can be understood as an extension of the previous 2D profiler because they work on the same functioning principle.

In order to capture 3D information, for instance in open seas, the first idea that comes to mind is to install two 2D ADCP with their vertical measuring planes perpendicular to each other; for example, installing one with its measuring plane in the north–south direction and the other in the east–west direction. Then, combining the information gathered by the two ADCP a 3D current map can be drawn.

In principle, two 2D ADCP as that shown in Figure 5.16 could be combined in one instrument with five transducers, one pointing straight up, and the other four on two orthogonal planes and pointing up with their beams at a certain angle from the vertical (Fig. 5.18).

The first beam measures the height of the surface while the inclined beams measure the water velocity profiles on two perpendicular planes. As previously described, the energy scattered by the suspended particles along the direction of each beam is converted back to an electrical signal, and the Doppler shift calculated in the same way as before.

Profilers similar to this description are used to measure in open waters and they are mounted at the bottom or on a moored subsurface buoy. They may be deployed in 40 m – 180 m of water and can measure directional waves and currents. They use a magnetic compass and a tilt sensor to refer measures on the planes containing the transducers' beams to the north-south and east-west planes. There are some models of ADCP with arrangements of 3 and 4 sensors which do not require the vertical transducer to obtain the same information (Terray et al., 1999).

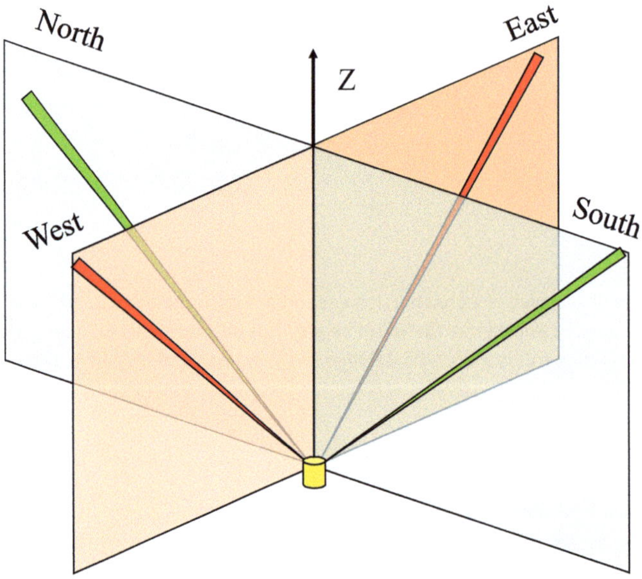

Fig. 5.18: Three-dimensional ADCP, four transducers are place on two orthogonal planes pointing up with their beams at a certain angle from the vertical. They measure the water velocity profiles on two perpendicular planes.

Similar arrangements to those previously described can be mounted looking down on a boat or a buoy for surveying flow in rivers or lakes.

5.4.5 ADCP General Characteristics

With the purpose of having reliable results, some measuring conditions that could limit the utilization of ADCP will be mentioned below and potential users of this technology should verify that their applications are not under the influence of these circumstances.

For a low number of acoustic heterogeneities (low number of particles) in water the energy backscattered to the transducers will be low and the electrical signal weak. Also, for a given number of particles the magnitude of the received backscattered signal decreases with their distances to the transducers. The farthest particles will produce the weakest signals. If the signal strength approaches the noise level, the measured velocity could be unreliable. Usually, low signal to noise ratio indicates less reliable data.

Some manufacturers record the signal to noise ratio so that users may judge the quality of the data. With this information, users could decide if the recorded velocity should be kept or discarded. For each application at hand, users should define the minimum amount of signal to noise ratio that they will accept to consider the data as valid. In order to define this figure, users should perform some previous tests in a tank where measuring conditions are controlled and known velocities could be generated.

In order to correctly measure distances from the transducers to the heterogeneities, the instrument needs to know the actual sound propagation speed in the fluid, which depends on temperature and salinity. ADCP should measure both parameters for sound speed corrections but, in general, they measure only temperature, which has the greatest influence on the speed of sound. Usually users are asked by the instruments to introduce the expected water conductivity at the beginning of the record. This value is used in the internal speed of sound calculations so that to have reliable distances users should have a good estimation of water conductivity. The conductivity should remain constant during the record, or its value changed when appropriate.

Because this method measures the velocity of particles and indirectly estimates that of the water, the assumption that the particles match water velocity has to be accomplished to measure water velocity correctly.

5.4.6 Acoustic Doppler Velocimeters

Acoustic Doppler Velocimeters (ADV) also use Doppler frequency shift produced by particles moving in the water to measure flow velocity, but instead of measuring a profile they give a single measure as representative of a given volume. Generally, these volumes are small, which permits a high spatial resolution to be offered.

Figure 5.19 presents an ADV that measures 3D water velocity in a small volume of fluid. Because of this it is useful to measure velocity in laboratory models. The measuring principle and the technology employed by this instrument have some points in common with Doppler Flowmeters and Acoustic Doppler Current Profilers, but some important differences will be noted.

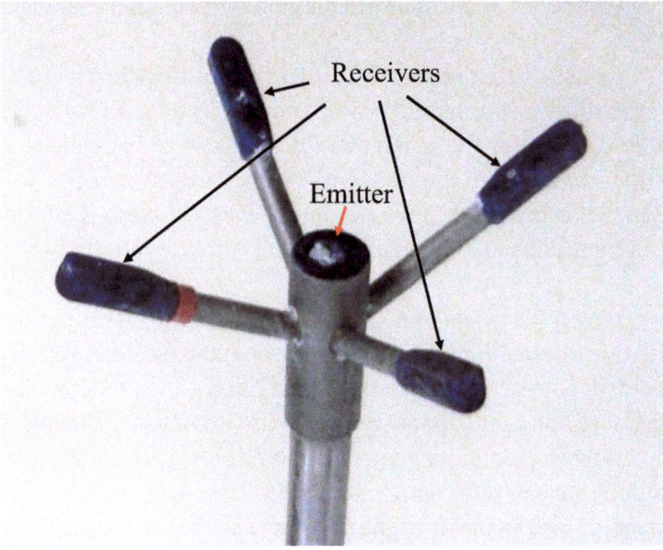

Fig. 5.19: Photography of an Acoustic Doppler Velocimeter. It shows the central emitter and the four receivers forming a cross around the emitter.

To calculate fluid velocity an ADV takes the information from a sample volume of the fluid quite close to the transducers. The distance of the sample to the transducers may range from 50 to 200 mm, depending on the manufacturer. Then an ADV does not measure a profile, but measures in a specific spot of fluid.

ADCP were called monostatic sonars because their transducers are used for both sending and receiving sound waves. Instead, an ADV uses one transducer to send acoustic pulses and three (or four, depending on the model) to receive the scattered acoustic signals from suspended particles passing by the specific spot (Fig. 5.20). Because of this particular characteristic of the transducers ADV are called bistatic sonars. This arrangement of the transducers is sensitive to the velocity component along the direction of the angular bisector between the transmitter and receiver beams. Particles moving at the velocity **V** have components *V*u and *V*d along the angular bisector directions. For this reason, ADV are more sensitive to the velocity component parallel to the transmitter beam, yielding lower measurement uncertainties along this direction.

For the sake of clarity, only the central emitter and two receivers are shown in Figure 5.20. The emitter has a narrow beamwidth (shown vertical in Figure 5.20) that is intersected by the receivers' beamwidth and this spatial intersection defines a "sampling volume". In other words, each receiver's beamwidth is pointed to the same particular volume, and then it senses the acoustic signal scattered by the particles in a specific region of the emitting beamwidth.

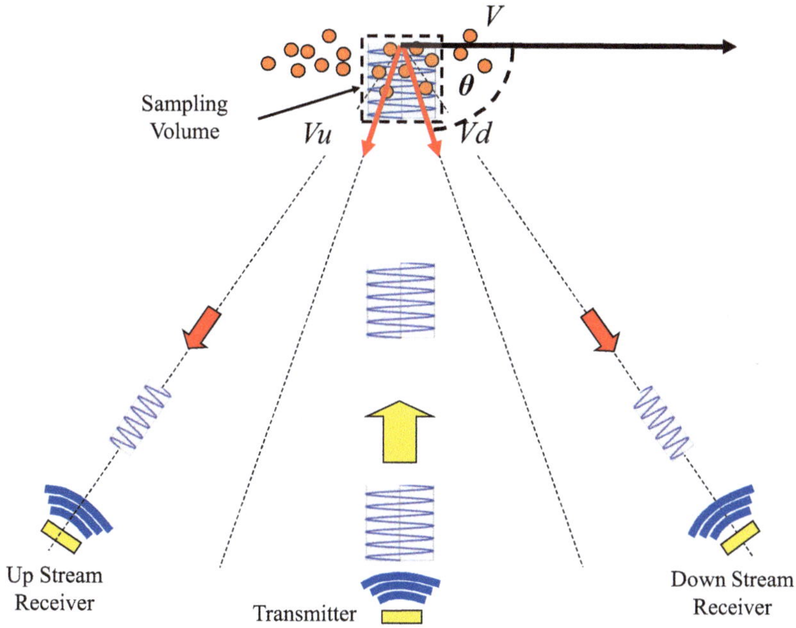

Fig. 5.20: The central emitter and only two receivers of the ADV are schematized. The emitter has a vertical beam that is intersected by the receivers' beams; this spatial intersection defines a "sampling volume". Acoustic signal scattered by the particles in that specific volume arrive at the receivers. The emitter sends short pulses, and the receivers open a sampling time windows to catch the emitted pulse scattered in that volume.

The emitting transducer sends short acoustic pulses, and based on the knowledge of the speed of sound in the fluid, the electronic circuits calculate the time required by the pulse to travel from the emitter to the sampling volume and back to the receivers. A synchronized time window is opened by the electronics to listen and capture only the echo caused by the heterogeneities found in the sampling volume. This technique improves the signal to noise relation.

Water salinity and temperature values are required to permit corrections of sound velocity propagation in water. Usually the first is manually introduced by users and remains fixed while temperature is continuously measured by an on board sensor and used for automatic correction of sound speed.

The spatial resolution of the measurements depends on the beamwidth and frequency of the transducers. With the aim of achieving the above-mentioned high spatial resolution, the transducer's frequency is about 10 -16 MHz, approximately one decade higher than the frequencies used by the ADCP.

Each emitted pulse covers a length of a few millimeters. For example, for one commercial ADV each pulse covers between 3 and 15 mm (users adjustable), which for a 10 MHz frequency in water (c = 1500 m/s) it is about from 20 to 100 periods of the emitted wave. The higher the amount of periods sent (and received) the lower the uncertainties in the Doppler shift calculation because more signal is processed, which improves the signal to noise ratio.

For a given concentration of moving particles, if it were desired to increase the spatial resolution the emitted pulse should be shortened, and so the amount of emitted wave periods decreases. In so doing, the amount of signal received back to be processed decreases, and since the noise is constant, the signal to noise ratio decreases. Eventually the reliability of the calculated velocity decreases.

When a high spatial resolution is required and the signal to noise ratio is so low that the reliability of the calculated velocity is not acceptable a possible solution would be to artificially increase the particle concentration by seeding some particles (for example, adding some insoluble powder to the water such as talc). Because the energy that arrives to the receivers depends on the concentration of scattering particles, sometimes seeding the water with scattering material provide adequate return signal strength. This improves the signal to noise ratio (SNR) resulting in more accurate velocity measurements. Unfortunately, seeding the flow is not always desirable or practical. SNR is a mean of evaluating the quality of the velocimeters measurements and, in general, manufacturers provide this figure, which advice the user about the quality of the data being acquired.

Figure 5.21a shows the velocity of a turbulent flow measured in a laboratory channel downstream of a hydraulic jump where the particle concentration has been increased. It can be observed that the velocity varies around a mean value of 0.3 m/s.

Figure 5.21b shows a noisy record from a similar experimental set up. Mean velocity is around 2 m/s but spikes of about ± 2 m/s due to low particle concentration are observed. If only the average velocity is needed spikes could be filtered, but when high frequency hydraulic phenomena are analyzed spikes have to be avoided.

The emitted pulse which contains several periods of the emitted wave is usually referred to as a ping; each ping permits the water velocity in the sampling volume to be estimated. Many pings are averaged together to reduce measure uncertainty giving as a result just one data. The standard data collection rate is about 25 data per second, but some ADV models collect 50 and 200 data per second.

Because the instrument is more sensitive to the velocity component parallel to the transmitter beam, it should be ideally mounted with the transmitter beam parallel to the expected mean velocity of the fluid; therefore, it would yield lower measurement uncertainties. But, if installed in this way, the instrument itself would disturb the flow being measured because transducers would stand in the flow way.

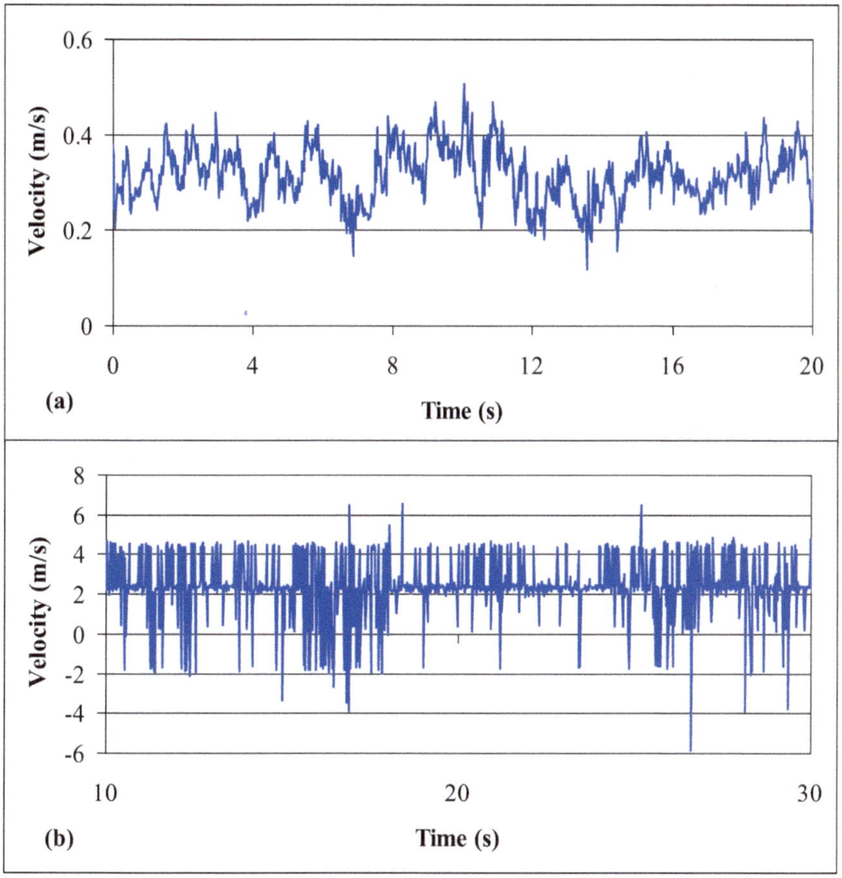

Fig. 5.21: Records of velocity measured with an ADV. (a) Increasing particle concentration improves data quality. (b) Low particle concentration produces noisy records. (Courtesy of Juan Parravicini and Mariano De Dios, Laboratory of Hydromechanics, Hydraulics Department, Faculty of Engineering, National University of La Plata, Argentina).

From the hydraulic point of view, the minimal flow disturb (best data quality) is achieved when flow direction is perpendicular to the transmitter axis because flow perturbation due to the instrument itself is minimized.

By virtue of the foregoing, there is a compromise in the way the instrument should be mounted with respect to the main flow direction because the direction of maximum sensitivity is that of the maximum hydraulic perturbation. The manufacturer advises avoiding flow measurements into the transmitter axis to prevent flow disturbance, but when the concentration of particles in water is low, flow perpendicular to the transmitter axis could result in low sensitivity and high noise.

5.5 Flow Measurements Based on Heat Transference

Two kinds of thermal mass flowmeters will be described below. One of them supplies a constant heat to the moving fluid and measures a **differential temperature** which, under some conditions, is proportional to the **mass flow**. The other computes the amount of electrical power required to maintain the temperature of a heater exposed to the flow stream. Here, the **injected power** is related to the **mass flow**.

There is another instrument based on heat transference, called hot wire anemometer, which is used for high flow speed measurements in which the **electrical power** to keep sensor's temperature constant is representative of **fluid velocity**.

Although the theoretical models expressing the thermodynamics operating principle for each type of instrument have some equations in common, these equations are, however, somewhat different and they will be developed only in a simplified and conceptual way.

5.5.1 Introduction to Heat Transfer

A general explanation of some of the heat transference processes used in these instruments is presented below with the aim to help the reader understand how these meters work.

When the temperature of a body changes it is said that the body has received or given a certain amount of heat. Heat is energy, and although sometimes expressed in calories (cal), its SI unit is the Joule (J) (1 cal = 4.186 J). The amount of heat required to heat a body is proportional to its mass.

The amount of heat Q (J) necessary to increase the temperature of a mass m (kg) from T_1 to T_2 (°C) is given by

$$Q = mc\,(T_2 - T_1) \tag{5.8}$$

where c the specific heat of the body (J kg^{-1} °C^{-1}). Equation (5.8) is valid for a certain range of temperature, since for large temperature variations the specific heat capacity is not constant.

Equation (5.9) is the equation for convection of heat (natural and forced) through a surface of area A.

$$q = h_c A\,(T_2 - T_1) \tag{5.9}$$

where q is heat transferred per unit time, h_c is the convective heat transfer coefficient, T_2 is the temperature of the surface and T_1 the temperature of the fluid. The convective heat transfer coefficient depends on the type of media (gas or liquid) and the flow properties such as velocity and viscosity, among others.

The amount of heat dissipated during a time interval Dt in an electrical resistance R connected to a voltage source E and through which an electric current I flows is given by

$$Q = R\,I^2\Delta t = \frac{E^2}{R}\Delta t \tag{5.10}$$

If the resistance R is placed into a fluid, the energy dissipated in the resistance may be used to heat the fluid, and the fluid temperature rises from T_1 to T_2 according to Eq. (5.8). Then, if the fluid is moved, it will transport heat from the resistance as expressed by Eq. (5.9). These three simple equations are combined to explain the working principle of the above mentioned instruments.

Units used in the above equations are presented in Table 5.2.

Table 5.2: Units of heat transfer equations

m	I	E	T	R	Q	A	c	h_c
kg	A	V	°C	Ω	J	m²	J kg⁻¹ °C⁻¹	W m⁻² °C⁻¹

5.5.2 Mass flowmeters

There are two types of thermal technologies that measure fluid mass flow directly. They are based on the principles of heat transfer. Both have a heated surface that transfers heat to the molecules of the flowing fluid (McMahon & Rouse, 2008). Because the types of thermal mass flowmeters have different physical implementations, their respective theoretical thermodynamic models are not the same. In one of the methods the thermal energy is used to heat the fluid, whereas in the other the energy heats a cylinder immersed in the fluid. In order to present the functioning principles simplified models will be adopted.

5.5.2.1 Measuring Temperature Difference

These flowmeters are based on the heat transfer between a hot body and the fluid that passes by and cools it. An outline of the principle is seen in Figure 5.22, where flow in a pipe is being measured. The fluid is heated at the midpoint between two temperature sensors. When the fluid is at rest heat spreads symmetrically and $T_1 = T_2$.

When a fluid begins to move, as indicated in the figure, a certain amount of heat is transported from point 1 to point 2, so that T_1 decreases and T_2 increases with respect to the previous situation. Within a particular range of flow speeds, and for certain fluids, the temperature difference is proportional to the mass per unit time flowing through the pipe,

$$\frac{\Delta m}{\Delta t} = K\left(T_2 - T_1\right) \tag{5.11}$$

If the speed of the fluid increases beyond a certain point the relation between mass flow and temperature difference is no longer linear.

Equation (5.11) shows that injecting heat by means of a heater and measuring temperature upstream and downstream the injection point, the mass flow may be estimated. The constant K is the slope of the flowmeter transfer and can be obtained through a calibration process.

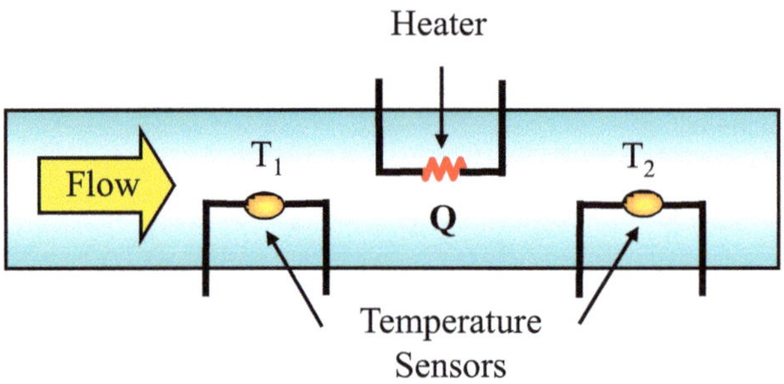

Fig. 5.22: Flow through a tube is heated at the midpoint between two temperature sensors (T_1 and T_2). Fluid transports heat from point 1 to point 2.

There are other designs based on the same principle (Fig. 5.23) where two heated temperature sensors are used. Sometimes they simply consist of two platinum resistance temperature detectors (RTD) (Section (4.7.2)). They are identical and equally heated by means of an electronic circuit so that with null flow both sensors will be at the same temperature. When fluid moves as in the figure heat is transported by the fluid mass, resulting $T_2 > T_1$ and Eq. (5.11) is met as in the previous design.

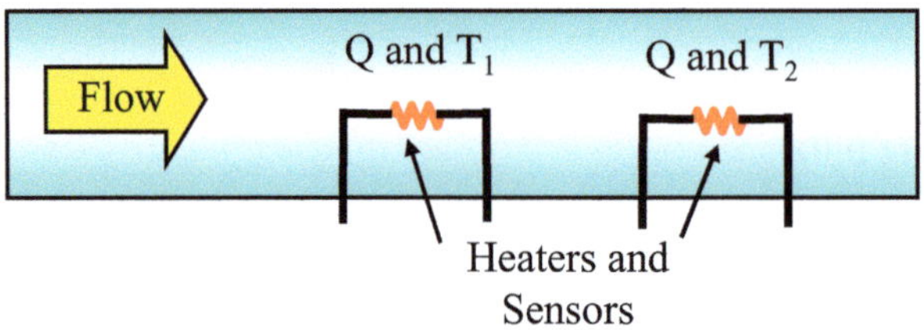

Fig. 5.23: Two heated temperature sensors are used to heat the fluid and to measure its temperature at two points. Sensors may be resistance temperature detectors (RTD).

The measurement schemes presented in Figures 5.22 and 5.23 are used for small pipe diameters and small flow rates. Otherwise, the amount of heat that must be provided to generate measurable differences in temperature would be very great. With the purpose of using this technique in larger pipe diameters its practical implementation has to be modified as in Figure 5.24.

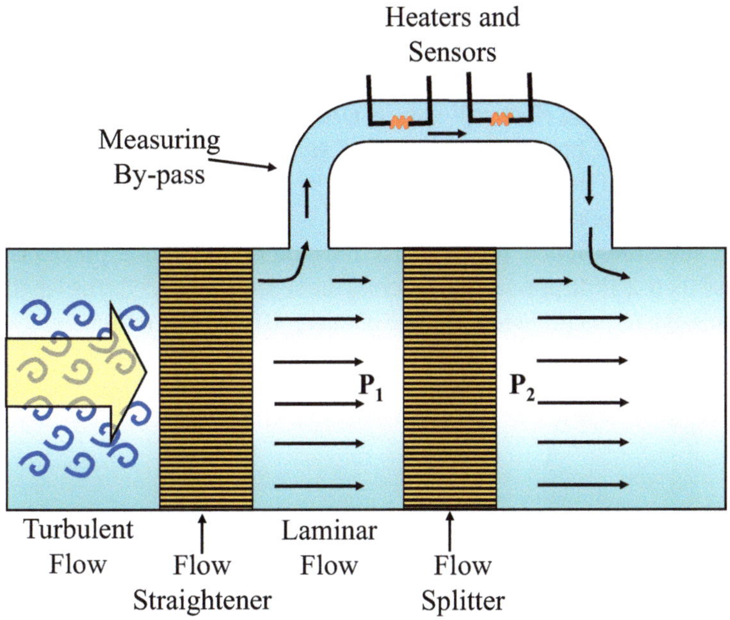

Fig. 5.24: The technique is modified to be used in larger pipe diameters. The flow straightener converts turbulent flow into a laminar one. A flow splitter produces a differential pressure ($\Delta P = P_1 - P_2$) which derives the fluid into the by-pass where the measuring system is installed. Flow through the by-pass is proportional to that passing through the main tube.

Figure 5.24 shows that by means of a flow straightener the turbulent flow at the input is converted into a laminar one. Downstream a measuring by-pass, a hydrodynamic resistance (flow splitter) is introduced to produce a differential pressure ($\Delta P = P_1 - P_2$) which derives the fluid into the by-pass (http://sierratechsupport.com). The amount of fluid that goes by the measuring system (heaters and sensors) is proportional to that passing through the main tube. Therefore, measuring the differential temperature in the smaller tube allows the total mass flow through the pipe to be calculated.

This technology requires simple electronic circuits, since they basically need measuring temperature differences. This type of flowmeter is suitable for measuring in air, nitrogen, hydrogen, oxygen, helium, ammonia, argon, carbon monoxide, carbon dioxide, hydrochloric acid, ethane, ethylene, methane, etc. These flowmeters are not useful to measure in fluids whose *c* (specific heat) varies with temperature or time (such as could occur in mixtures). It is a very useful technique for detecting slow fluid movements such as leaks.

5.5.2.2 Measuring Electrical Power

Another way of measuring mass flow is by means of two precision platinum-resistance temperature sensors completely immersed into the flow stream. One sensor measures the temperature of the fluid, while the other is heated at a higher temperature than that of the fluid by means of an electronic control circuit (about 40 °C hotter). When the fluid begins to flow, the hotter sensor is cooled as the fluid passes taking heat away from it. The amount of heat removed is replaced by the electronic circuit so that the temperature of the hotter sensor is kept constant.

This method measures much higher mass flow rates that those previously explained because there is little pressure drop across the sensors (Fig. 5.25). The length (L) of the sensors is of a few centimeters and the diameter (D) of a few millimeters.

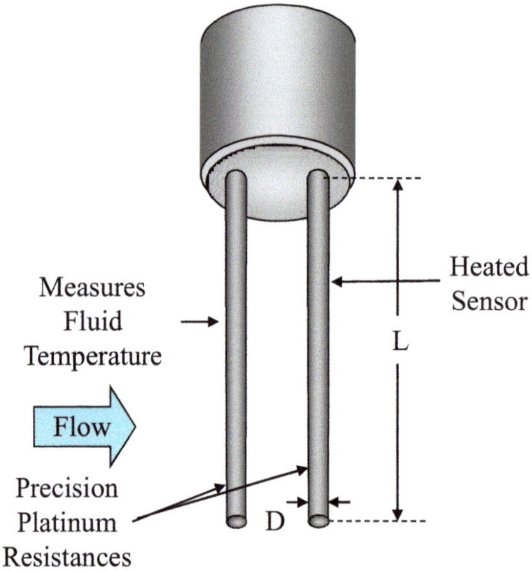

Fig. 5.25: Two precision temperature sensors are immersed into the flow stream. One sensor measures the temperature of the fluid and the other is heated. D and L are the sensor diameter and length respectively.

Assuming steady-state operation conditions, and disregarding heat losses by conduction and radiation, the power supplied by the heated sensor (Eq. (5.10)) equals the heat taken away by the fluid flow due to convective heat transfer (Eq. (5.9)), then

$$RI^2 = h_c A\,(T_2 - T_1) \tag{5.12}$$

The left member in Eq. (5.12) is the electrical power supplied to the resistive sensor, which can be accurately measured, and the right member is the heat transferred from the heated sensor to the fluid due to natural and forced convection. In this case A is constant ($A = \pi\, DL$), D and L being the sensor diameter and length respectively. The temperature difference is kept constant by the electronic circuit, but h_c is variable and depends on the mass flow.

The mass flow of the fluid cannot be found analytically, but it can be known empirically by a calibration curve relating the power input to the mass flow passing by the sensor. Calibration has to be done with the same fluid in which the flowmeter will be used, and it should be performed in a pipe of similar characteristics. The calibration is stable as long as the coefficient h_c does not change with time.

5.5.3 Hot Wire Anemometers

Hot wire flowmeters (also called hot wire anemometers) are based on the measurement of the heat transferred from a heater to the environment that surrounds it. The sensor must be heated to a certain temperature greater than that of the environment. When the sensor is exposed to a fluid stream it loses heat by forced convection. This transfer increases with fluid velocity.

These flowmeters measure flow using the relation that exists between the electric power supplied to the heater and the velocity of the fluid in which it is immersed. Heat transfer from a resistive wire (probe) to the fluid in motion is given by

$$A + BV^n = I_s^2 R_s \tag{5.13}$$

where V is fluid speed, A, B and n are calibration constants, I_s is the electric current through the sensor and R_s is the sensor electric resistance. In some cases it may be assumed $n \approx 0.45$, but n should be determined along with A and B by calibration. Assuming R_s invariable, in order to get these calibration constants, I_s must be measured for a number of known flow velocities and a least squares fit has to be performed for the values of the constants which produce the best fit to the data.

Therefore, once the relation between V and I_s has been established the flow speed can be estimated by measuring the electrical current I_s. This kind of anemometer is frequently used in wind tunnels and fluid mechanics laboratories because they are able to measure very fast fluctuations of fluid velocity.

It should be underlined that unlike thermal mass flowmeters, I_s is related to fluid velocity. If the velocity were measured inside a pipe, the pipe area (A) should be known in order to compute the **volumetric flow rate**. If the fluid density (ρ) is known, the mass flow can be computed from Eq. (5.14); but changes of density with temperature have to be taken into account to compute the **mass flow rate** correctly.

$$\frac{\Delta m}{\Delta t} = \rho V A \tag{5.14}$$

Probes using this method are basically of two types: hot films and hot wires. The first are employed to measure in water and gases, and the last to measure in air.

5.5.3.1 Hot Film

Platinum films stuck on a quartz substrate are sometimes used as sensors (Fig. 5.26b) and are known as **hot films**. They are protected with alumina for use in gases and with quartz for use in liquids. They present an electrical resistance which is used to produce heat. The hot film has electrical contacts at the point where the wires are soldered to connect them to the electronic circuit. The size of these probes is small, for the example shown in the figure the diameter D may be on the order of 0.06 to 0.15 mm.

The film resistance (R_s) varies with temperature as

$$R_s = R_r\,[1+\alpha\,(T_s - T_r)] \tag{5.15}$$

where R_r is the resistance of the sensor at a reference temperature T_r, and α is the temperature coefficient of the metal film resistance.

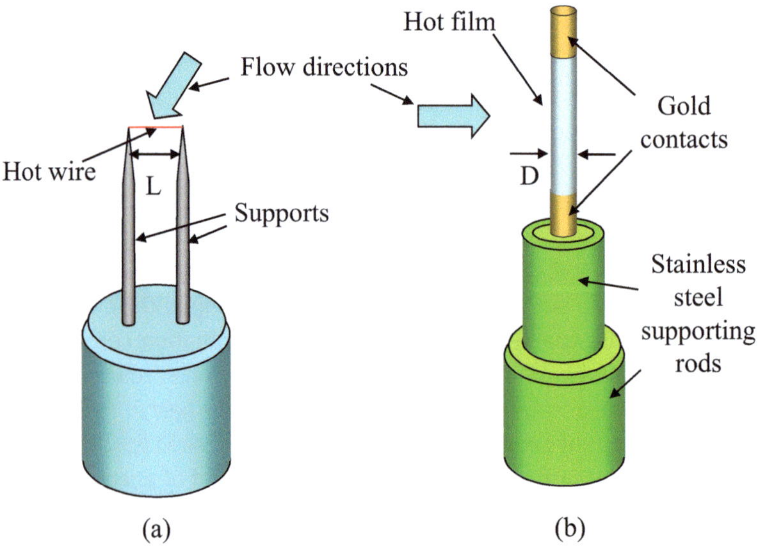

Fig. 5.26: (a) A hot wire probe. (b) A hot film probe.

Usually, the sensor is part of a feed-backed Wheatstone bridge which keeps T_s constant, independently of the flow cooling the sensor. This is achieved by controlling the current through R_s such that R_s = constant. Then, measuring I_s, the fluid velocity V can be calculated from Eq. (5.13).

5.5.3.2 Hot Wire

It is another type of sensor used with the same operating principle but which has a different field of applications (Fig. 5.26a). Hot wire anemometry is an excellent tool for studying turbulent flow in air. It is useful to acquire velocity time series with high time and spatial resolutions. For these applications sensors are made of very small diameter wires. Probes are usually made of tungsten wires of a few millimeters in length and a few micrometers in diameter. Probes with $L = 1$ mm and 5 μm in diameter, mounted on two needle-shaped prongs are reported (http://www.dantecdynamics.com). Because these devices and the associated electronic circuits respond very fast, the upper frequency of the transference bandwidth reaches 100 kHz.

These probes are sensitive to directional flow so that when two or three probes are orthogonally arranged they give information on the flow velocity components, but a directional calibration is required to determine the complete transfer of the instrument.

5.6 Coriolis Mass Flowmeters

5.6.1 Introduction

The Coriolis mass flow meter is perhaps one of the most anti intuitive measuring devices, and at the same time it is one of the most useful flowmeters. The fact that they measure mass flow directly is a characteristic offered by few flowmeters, and there are some applications where this kind of flowmeter is the most adequate option. If it were said that by measuring time it would be possible to estimate mass flow through a pipe, it would sound like a "magic trick". To some extent this instrument is the result of some wizard engineers which combines mechanical and electronic knowledge to produce a work of art. Therefore, to explain this ingenious device requires the use of several pages and a number of approximations. This explanation will follow the ideas found in the technical literature (Micro Motion Inc., 1996). The explanation requires beginning with the Coriolis force concept and how it appears. This concept is extensively developed in Section (3.10).

5.6.2 The Coriolis Flowmeter

A flowmeter that uses the Coriolis force as its operating principle is shown in Figure 5.27a. The development of a simplified explanation will be based on the geometry shown in Figure 5.27b.

Figure 5.28 shows a simplification of the U-shaped tube of the flowmeter that is located on a plane which, for the sake of simplicity, is assumed to be horizontal. The plane is made to oscillate vertically by means of electromagnets. The physical principle by which the vibration is produced is similar to that shown for a tuning fork oscillator (Section (4.5.1)). In the flowmeter case the fluid-filled metal pipe represents the mass, and the elasticity of the metal, the spring. In practice the total vibration amplitude of the tube is small and the frequency low (e.g. 3 mm and 80 Hz).

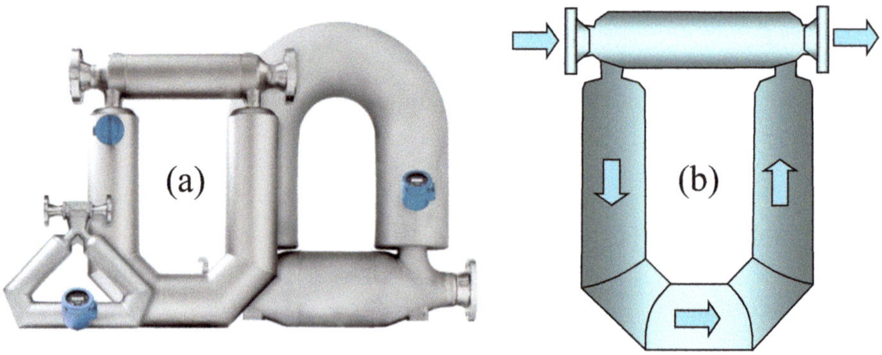

Fig. 5.27: (a) Coriolis Flowmeters. (Courtesy of Emerson Process Management - Micro Motion, Inc.). (b) Schematic of the flowmeter; arrows indicate flow path.

We will follow an intuitive introduction to the phenomenon to develop the working principle. At the beginning no flow circulates and the tube vibrates as in Figure 5.28a. It must be recalled that the flow passing through the fixed portion of the tube (Rotation Axis) has a certain momentum. Suppose that in passing the curve, when the fluid is in the first arm of the "U" tube, the "U" tube is moving upwards, so that the fluid is forced to acquire the vertical movement of the vibrating tube. The fluid is thus compelled to change its momentum, and because it refuses to be moved upwards, it produces a thrust down onto the wall of the tube Figure 5.28b. When the fluid is in the opposite arm of the "U" tube it is forced to decrease its vertical momentum and it resists pushing the tube up. These two forces in opposite directions produce a torque on the tube.

It is interesting to note that in the part of the tube that is parallel to the axis of rotation, $\mathbf{V}$ is collinear with $\boldsymbol{\omega}$ ($\sin\theta = 0$), so on that portion of the tube there is no

Coriolis force when the "U" is horizontal. At the central point there is a node of the twisting motion.

The amount that the tube is twisted depends on the mass flow. Applying Eq. (3.31) and the rule of the corkscrew for the vector product, the directions of the Coriolis forces are obtained and displayed in Figure 5.28b along with the torque. Obviously, when the tube stops moving upwards and begins to move down, the direction of the torque is reversed.

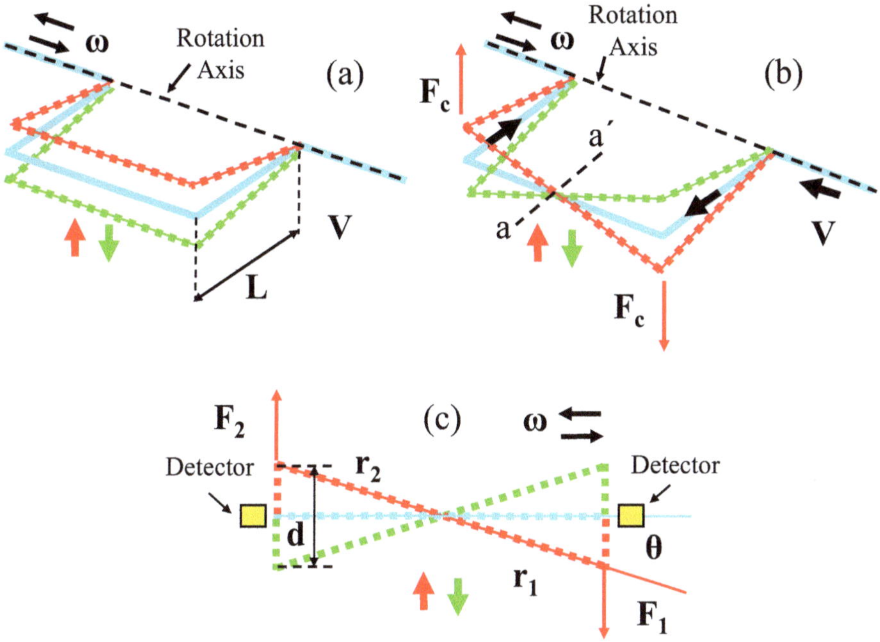

Fig. 5.28: (a) The U tube is horizontally placed and vibrated up and down about the rotation axis. Flow is null. (b) Flow circulates as indicated by the arrows; forces F_c appear. (c) Forces F_c create an oscillating momentum. The detectors permit the difference in time between the tubes to be known.

Some relationships based on the Coriolis force (Eq. (3.31)) will be developed to show that the mass flow can be measured with this device. Several simplifications are made for this purpose. The oscillating movement up and down, which is an accelerated movement, has to be assumed as a rotating motion of constant angular velocity value (ω) whose axis of rotation is the fixed part of the pipe (Figs. 5.28a and 5.28b). Therefore, the direction of the vector $\boldsymbol{\omega}$ agrees with the rotation axis. Also, it has to be assumed that the velocity **V** of the fluid has the direction of the pipe in each section of the "U" tube; and its direction is as depicted in Figure 5.28b.

Because of the symmetry of the tube, (Figure 5.28c), the forces $\mathbf{F}_1$ and $\mathbf{F}_2$ exerted by the fluid in each of the branches are equal in magnitude but opposite in direction. When the tube vibrates, the forces create an oscillating momentum **M** around the axis a-a´, with radii $\mathbf{r}_1 = \mathbf{r}_2 = \mathbf{r}$ approximately given by Eq. (5.16), because the forces and the radii are nearly perpendicular.

$$M \approx r_1 F_1 + r_2 F_2 \tag{5.16}$$

From Eq. (3.31), and taking into account that the Coriolis forces F_1 and F_2, and the distances r_1 and r_2 are equal, we get

$$M = 2\,r\,F = 4\,m\,V\,\omega\,r \tag{5.17}$$

As we are interested in the magnitude of the Coriolis force we will disregard the negative sign in Eq. (3.31). Then, due to the mass flow rate passing through the tube and the rotation imposed by the vertical vibration, there appears an oscillating momentum applied to the tube. It should be noted that the momentum is proportional to the mass ($m = \rho\ Vol$, ρ being the fluid density and *Vol* the fluid volume).

It is interesting to note from Eq. (5.17) that once the instrument is built (i.e. $\boldsymbol{\omega}$ and **r** being fixed), and for a given fluid velocity *V*, the measured momentum depends only on the mass of the fluid. This means that the instrument will produce lower momentum with low density fluids, resulting less sensitive for this kind of fluid.

The speed **V** is replaced by the distance *L* traveled by the fluid in time *t*. The mass flow rate Q_m is defined as the mass *m* that passes through a given section of the pipe in time *t*, thus Eq. (5.17) is rewritten as

$$M = 4\,m\,\omega\,r\frac{L}{t} = 4\,L\,\omega\,r\frac{m}{t} = 4\,L\,\omega\,r\,Q_m \tag{5.18}$$

Figure 5.28c shows that the momentum **M**, which is proportional to mass flow rate, causes an angular deflection θ about the axis a - a´. A resistant torque **T** due to the elasticity of the tube (k_s) opposes to **M**. Equating both torques and solving for the mass flow rate, we get

$$T = k_s\,\theta = M = 4\,L\,\omega\,r\,Q_m; \qquad Q_m = \frac{k_s\,\theta}{4\,L\,\omega\,r} \tag{5.19}$$

This equation states that the mass flow can be known if the angular deviation θ is known. For this purpose two detectors are placed as in Figure 5.28c. Each detector measures the instant at which each tube arm passes by the midpoint of the swing.

When the liquid is at rest, there is no bending of the tube and both arms pass by the midpoint at the same instant (Fig. 5.28a). As the fluid speed increases, the angle θ begins to increase, which increases the time difference (Δt) between the crossings of both arms of the tube by the detector's positions. It is possible to relate the distance traveled by the extreme of the tube (*d*) with the angle θ. This distance can be estimated from the average tangential velocity of the extreme of the tube (v_t) and Δt. For small θ, $\sin\theta \approx \theta$ and $v_t = \omega\,L$,

$$\sin\theta \approx \theta = \frac{d}{2r} = \frac{v_t\,\Delta t}{2r} = \frac{\omega L\,\Delta t}{2\,r} \tag{5.20}$$

From Eqs. (5.19) and (5.20),

$$Q_m = \frac{k_s\,\Delta t}{8\,r^2} \tag{5.21}$$

Expressed in words, Eq. (5.21) states that **the mass flow is proportional to the constructive constants and to the time interval Δt, and is independent of the vibration frequency.**

All this simplified mathematics was done to show that in order to measure the mass flow it is only needed to know the time difference Δt. In practice this time difference is measured by means of magnetic detectors placed as in Figures 5.28c and 5.29.

Figure 5.29 shows a schematic diagram of the instrument; it depicts a real flowmeter in which the flow is distributed into two vibrating tubes. The theory remains the same for each tube, but flow splitting has some practical advantages. One advantage is that the external vibration (due, for example, to the pumps that impel the fluid in the main pipe) which moves both tubes in a similar way is a common mode noise, affecting both measures in the same way. Thus, measuring the relative motion between both "U" tubes, known as a differential signal, greatly attenuates the common noise introduced by the external vibrations. The figure also shows the location of the driver, used to produce the vibration of both tubes, and the detectors, used to measure the time difference Δt.

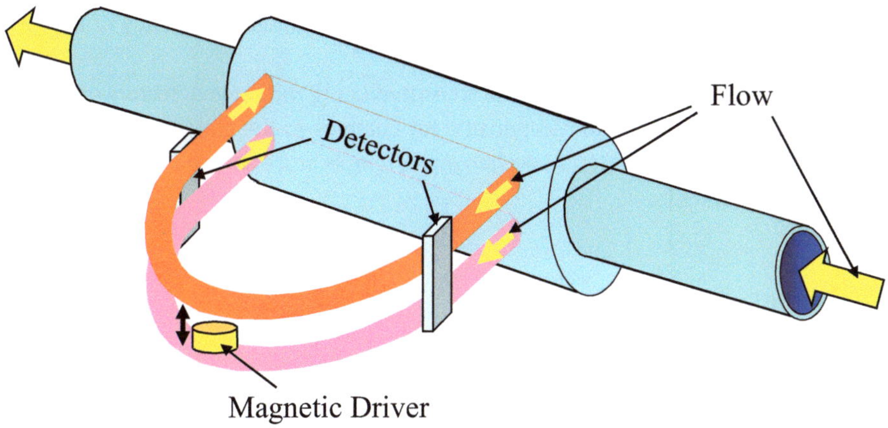

Fig. 5.29: The flow to be measured is driven into two vibrating tubes. The magnetic driver makes the tubes vibrate 180° out of phase. Detectors measure the differential time between crossings of both arms of each tube.

5.6.2.1 Some Characteristics of Coriolis Flowmeters

This type of flowmeter is not affected by changes in the fluid temperature, pressure, viscosity, density or velocity profile. It is also insensitive to inlays that change the internal section of the tube. They can be used in almost any type of fluid: mud, multiphase fluids, dry gases, steam, pulsating flow, etc. In the case of gases, their densities must be sufficiently high so that the produced torque is detectable. They are manufactured for several pipe diameters. The errors are less than ± 0.2% for dynamic ranges of 1: 25 or greater.

As disadvantages it should be mentioned that they produce moderate load losses and are large in size compared to other technologies (e.g. electromagnetic). Since the flow is separated into two tubes of smaller sections than the inlet pipe, cavitation or secondary phases could be generated which can clog the tubes, interrupting the measurements.

5.6.2.2 Measuring Density

Each tube of the flowmeter may be considered as a mass and spring set which can vibrates at a certain resonant frequency (Section (4.5.1)). This frequency depends on the mass of the tube plus the mass of the fluid, as well as on the tube elasticity. In order to bring the tube to its resonant frequency, a driving coil and a feedback circuit is used and then the tube always oscillates at resonance.

The mass of fluid filling the pipe depends on the fluid density and the internal volume of the pipe ($m = \rho\ Vol$). Let us assume the internal volume of the tube, the tube's elasticity and the tube mass to be constant. Then the resonant frequency will depend only on the fluid density. Thus knowing the resonant frequency of the tube the density of the fluid may be inferred.

The frequency at resonance can be measured using the same detectors employed to measure the mass flow. To compensate for changes in the elastic modulus of the pipe's material due to temperature changes, a temperature sensor is used to correct density measurements.

References

McMahon, M. & Rouse, S. (2008). A Tale of Two Thermals: Capillary & Immersible. *Core Technology White Paper Series, Part 1: Thermal.* Sierra Instruments.

Micro Motion Inc. (1996). *Micro Motion Mass Flow Meter Manual*, Boulder, CO, USA.

Miller R.W. (1996). *Flow Measurement Engineering Handbook.* New York: McGraw-Hill.

Terray, E. A., Brumley, B. H., & Strong, B. (1999). Measuring waves and currents with an upward-looking ADCP. *Proc. Sixth Working Conf. on Current Measurements*, San Diego, CA, IEEE (pp. 66–71).

A Tale of Two Thermals: Capillary & Immersible. *Core Technology White Paper Series.* Sierra Instruments. Available at: http://sierratechsupport.com/news_letter/newstrak/a_tale_of_two_thermals.pdf

http://www.dantecdynamics.com/Default.aspx?ID=1057

6 Water Level and Groundwater Flow Measurements

In the first part of this chapter we discuss instruments for measuring water level in dynamic and quasi-static conditions; in the second part, modern methods for measuring ground water flow are presented.

6.1 Water Level Measurements

As stated in Chapter 2, requirements for water level measurements in environmental problems may be classified into two broad categories, one associated with static or average levels which requires low spatial and temporal resolution, and another related to fast changing levels which needs measuring in small areas and short times. Measurements of water table in aquifers, water level in rivers and tides correspond to the first category. In these cases, spatial integration of several meters and sample periods of a few minutes result adequate for most applications. The second category is represented by wave measurements in lakes, rivers, oceans, wave tanks and hydraulic physical models. In these cases the time scale imposes that several samples per second be taken, and the spatial scale is in the range of centimeters for field measurements and in the order of millimeters for models.

Because in water level measurements there exist different applications with dissimilar requirements there are several instruments and distinct operating principles. Some of the most familiar instruments used for level measurements will be presented below. The basic ideas supporting a number of operating principles used by instruments described in this chapter have already been explained in previous chapters and therefore, the reader will be referred to those sections for the sake of synthesis.

6.2 Wave Measurements

Wave measurements in the sea and in hydraulic models have many characteristics in common, although different spatial and temporal scales. Because of the size of the waves in models they need better resolution in both scales. Some operating principles used in wave meters are similar for both applications, but others only measure properly in one of the scales.

Ocean waves are a critical variable for many marine problems that can be found in basically two quite different scenarios: near-shore and open oceans. Users of wave meters in each one of these cases have different measuring needs. In general, temporal and spatial resolution requirements increase from the open ocean to the coast (Alliance for coastal technologies, 2007).

Tests in hydraulic models and wave tanks require measuring waves with smaller amplitudes and sampled at higher frequencies than sea waves, calling for instruments with higher spatial and temporal resolution. Increasing temporal resolution is not a problem because it is simple to get more samples per second, but to increase spatial resolution requires that some of the methods used for sea waves be discarded and to search for new ones. Basically it is necessary to reduce the size of the sensors to minimize the spatial sample over which the instrument takes the wave information. In other words it is desired to reduce the spatial integration introduced by the sensor itself.

Some of the modern technologies identified for wave measuring (Alliance for coastal technologies, 2007) will be described below, indicating which are the most suited for each scenario.

6.2.1 Submerged Pressure Sensors

Pressure sensors are mainly used in field scale (seas, rivers, lakes, etc.). They are of little use for dynamic measurements in wave models, although they may be used for average level estimations. These sensors were described in Section (4.8.2) and their application to water level measurements were discussed as examples of practical uses of sensors in Section (4.8.4). When pressure sensors are used to evaluate field waves they sense the dynamic pressure generated by the change in water height produced by the passing wave plus the average hydrostatic pressure due to the water column below the still water level, as in any static fluid. Therefore, they have to be mounted at a constant depth and the sensor mounting plays an important role in the quality of the measured data. Sensors may be installed on platforms fixed to the sea floor or attached to subsurface mooring. The first are employed for long term wave data collection in near-shore places; the second are needed in deep waters or when the cost of fixed platforms is off-budget.

The subsurface mooring is frequently composed of an anchor, a subsurface float and a line which links the anchor and the float. In general, pressure sensors are attached to the line. In order to pull the line and minimize movements due to waves, the float has to be always some meters underwater, even at low water.

Because sensors are at a certain depth, fast changes in sea surface height are low-pass filtered by the water column. Thus, sensors record a version of the actual signal whose higher frequencies are attenuated. The characteristics of the low-pass filter depend on the sensor depth and can be predicted approximately.

Using linear wave theory and knowing the depth of the sensor, the time series of pressure recorded at some depth can be converted into sea surface elevation (Van Rijn et al., 2000). This conversion has some limitations because the problem is nonlinear, errors in the depth produce errors in the calculated water surface elevation, and there is also a maximum frequency for which the conversion equations may be applied.

In the case of platforms fixed to the bottom in places where the average sea surface varies by several meters a day due to the tide, the characteristics of the filter varies. If the mooring depth used in the reconstruction equations is considered constant some errors will be introduced in calculating the wave height. Underwater pressure sensors also measure variations due to changes in the atmospheric pressure, which should be taken into account. In subsurface mooring, the movement of the line due to wave forces on the float will introduce lateral and vertical motions of the sensor, producing additional errors.

In general, submerged pressure sensors process and store wave information on board for a later retrieval, but some instruments placed close to shore may have a cable for power supplying and data pickup. The onboard processing generally consists of the wave spectral analysis and the computation of some parameters representative of the wave characteristics in the time domain.

The installation of three or more submerged pressure sensor adequately located may be used to obtain the wave directional spectrum.

A more frequently way used to attain wave direction is to combine one pressure sensor with a current meter similar to those described in Sections (5.3.2) and (5.4.6). The current meter measures two orthogonal components of the horizontal water particle orbital velocity, and by means of a magnetic compass or a magnetometer these components are converted into the well-known u and v components. With this information the directional spectrum referred to the geographical coordinates is obtained.

6.2.2 Buoys

Buoys are often used for field studies and they are well suited to measure in deep waters because they do not need a platform or a pier fixed to the bottom as was the case of pressure sensors. Floating buoys may follow, up to some extent, sea surface movements. Measuring the buoy displacements allow wave characteristics to be known. Usually, two kinds of buoys are referred to according to their operating principle and size (Barstow et al., 2003). Some of the buoy types are known as pitch, roll and heave buoys or wave slope buoys. They are medium size buoys used for the installation of oceanographic and meteorological instruments in the sea. They usually have a disc shape some meters in diameter and, due to their hydrodynamic characteristics, they are slope followers. The other types are called displacement buoys or orbital following buoys. They are spherical buoys with a diameter smaller than 1 m and because their shape, size, weight and mooring, they follow the sea surface water particle orbit (Lawrence, 2012). Sensors frequently used to measure buoy movements include accelerometers as those described in Section (4.11).

6.2.3 Displacement Buoys (Orbital Following Buoys)

In order for a displacement buoy to follow the waves the mooring must allow the buoy to move freely. Therefore, the installation requires a slack mooring line attaching the buoy to the anchor; usually an elastic rubber cord is used as parte of the mooring line.

The early displacement buoys measured wave parameters by means of only one accelerometer which measured the vertical displacement of the sea surface. Buoys transmitted wave information to a shore station as an analog signal modulating a radio frequency carrier. The wave analog information was recorded at the receiver unit at regular intervals over certain periods for later analysis. With this information non directional wave spectra were obtained. This way of data transmission limits the maximum distance from the buoy to the coast to a few tens of kilometers.

Buoys that follow the wave movement and use only one accelerometer calculate wave height by integrating the vertical acceleration twice (Fig. 6.1). To avoid unwanted acceleration due to roll and pitch of the buoy, the accelerometer has a special gimbals and pendulous mechanism, also known as a gravity stabilized platform. The platform is formed by a suspended disk in a fluid of equal density and the disk is made sensitive to gravity force adding a small weight. The set forms a pendulum with low natural frequency of oscillation and the pendulum is mounted in gyroscopic gimbals. This platform remains almost horizontal because it works as a mechanical low-pass filter for the higher frequencies existing at sea. The accelerometer axis is mounted perpendicularly to the horizontal platform to measure only the vertical acceleration (Datawell, 2009).

Fig. 6.1: A buoy that follows wave movement is being deployed in the sea. The top white pole is the transmitting antenna.

A further extension of the previous idea permits wave direction to be recognized, thus giving directional wave spectra. The information from the vertical sensor mounted on the stabilized platform is combined with data from two orthogonal horizontal accelerometers, pitch and roll sensors, and a three axis fluxgate compass. The latter is used to refer the spectra to the geographic coordinates (Barstow et al., 2003).

Buoys following the wave movement can also use accelerometers installed on three orthogonal axes on board the buoy to sense buoy displacements. By integrating the output of these accelerometers (Eq. (4.16)) it is possible to describe the motion of the buoy in space without a reference platform. If the assumption that the buoy follows the wave is correct, the movement of a water particle as a function of time can be calculated from the buoy displacement. The vertical displacement of the buoy permits wave height and period to be estimated, whereas the orbital motion of the buoy allows knowing wave direction. By means of a three axis fluxgate compass the directional spectra are referred to the geographic coordinates.

6.2.4 Pitch, Roll and Heave Buoys (PHW) (Wave Slope Buoys)

Other types of buoys, generally bigger buoys designed for long time unattended operation in the open sea, use a single-axis accelerometer with its measurement axis perpendicular to the deck of the buoy (NDBC, 1996). For perfect slope following buoys, the accelerometer axis is perpendicular to the wave surface. Spectral analyses are derived from the acceleration time series. Measured accelerations are corrected for frequency-dependent responses of the buoy hull and its mooring. In order to estimate directional wave spectra, buoy pitch and roll time series are measured and the buoy slope time series is referenced to east-west and north-south direction by means of a magnetometer which obtains the azimuth of the buoy from measurements of the Earth's magnetic field. Some PHW buoys also use an accelerometer with a stabilized platform as described for displacement buoys (NDBC, 1996; Lawrence et al., 2012).

Advances in electronics resulted in small integrated sensors that permit lighter, smaller and lower cost sensitive buoys (Teng et al., 2009). The motion detection of these buoys is achieved by sensors from different manufacturers: three single-axis angular rate sensors, two dual axis accelerometers, and a single 3-axis magnetometer. The result is an unprecedented amount of motion information on nine measurement channels: acceleration, angular rate and magnetic flux density along three orthogonal axes at a rate of 35 hertz. By digitally processing the angular rates they are converted into pitch and roll angles, and then the directional wave spectra and other wave parameters are extracted.

Modern buoys have the possibility of analyzing the information on board, so that only the results of the analysis are transmitted. In this way the information can be sent through a satellite link (as will be seen in Chapter 9). In general, the results from the onboard analysis contain information in the time and frequency domains. Among other wave characteristics reported are the period corresponding to the frequency where the non directional spectrum is a maximum, the average and zero-crossing periods, the significant wave height, the full directional wave spectra, the mean wave direction and parameters which describe the directional spreading about the main directions.

6.2.5 GPS Buoys

These are displacement buoys with no sensor on board, but only a GPS receiver and some electronics. GPS stands for Global Positioning System; this system will be described in detail in Chapter 9. It is a system based in 24 satellites for the positioning in real time of military and civilian platforms. It works 24 hours a day and under any climatic conditions. The platform must have a special receiver to take information from several satellites simultaneously in order to calculate its position.

Buoys based on GPS receivers can measure directional and spectral wave data and they can be deployed free-floating or moored. Because the receiver and the associated electronics are of reduced size and consume little power, they can be housed in small buoys with diameters smaller than half a meter. As GPS signals do not break through water, signals may be blocked by high ocean waves and sometimes data may be lost. Some manufacturers claim that resolution direction of their buoys is 1.5°.

Some methods of buoy positioning require a differential GPS strategy which needs an additional GPS reference station on shore. Thus, the use of this technology limits the applications to near-shore buoys. Other buoys use a single GPS receiver; they are based on the determination of the moving speed of the buoy as computed from the Doppler shift of the frequency received at the buoy. The buoy speed is then integrated to estimate its motion.

There are other buoys that also use a GPS receiver to obtain information on the wave movement, but the velocity information output is directly derived from the receiver (Doong et al., 2011). The authors of these studies express that the main concept behind this method is to transfer the velocity spectrum to the vertical displacement spectrum to obtain the water surface elevation. Velocities in three-directions were used to derive the directional wave spectrum. Authors found, after a field test comparing simultaneously buoys equipped with accelerometers and equipped with a GPS receiver, that the one-dimensional spectra and the directional wave spectra are similar for both methods. Wave parameters compared between both types of buoys showed a correlation coefficient higher than 0.95.

6.2.6 Acoustic Level Measurements

Acoustics principle for level measuring is the same either in field (Fig. 6.2) or laboratory (Fig. 6.3) studies. Sometimes the same manufacturer offers two models, each one suited for each application.

Fig. 6.2: Acoustic sensor measuring ocean waves (Courtesy of General Acoustics)

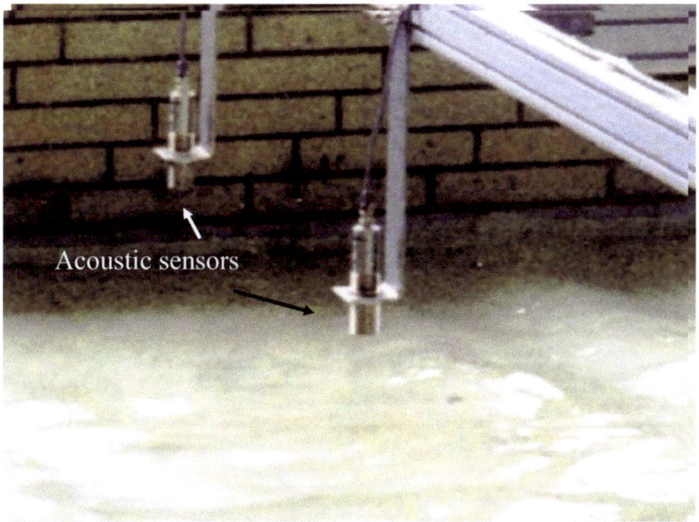

Fig. 6.3: Acoustic sensors measuring small waves (Courtesy of General Acoustics)

In structures close to the shore or in platforms fixed to the seabed it is possible to install an acoustic transducer as shown in Section (2.3.3). The transducer must be installed a few meters higher than the maximum level reached by the waves. The acoustic transducer emits pulses downwards and measures the transit time (or time of flight) (Section (5.4.1)). By knowing the speed of sound in air, the distance to the sea surface is calculated. Taking several samples per second, a time series with the wave information is acquired. In order to have a good spatial resolution (measuring on a small area of the wave), the transducer beamwidth must be narrow (should have a few degrees).

Because the speed of sound in air varies with temperature, a correction to the calculated distance has to be done by measuring air temperature or by a direct measure of the sound speed. The last is done by determining the time of flight of an acoustic pulse to an object placed at a known distance.

Acoustics measurements of sea surface level can also be done offshore by using submersible acoustic transducers pointing upwards and placing them at a fixed distance from the seabed. Now sound travels through the water to the surface and is reflected backwards by the water-air interface. Again, the time of flight is measure and the distance to the surface calculated. The same requirements for a narrow beamwidth have to be met.

Mounting of transducers to seabed fixed platforms is similar to that shown for acoustic profilers (Section (5.4.4)); in this case there is just one beam. In sand moving seabeds, precautions should be taken to avoid sediment deposition on the transducers. Also, in deep waters transducers can be attached to the line of a subsurface mooring as described above (Section (6.2.1)).

There are also miniature acoustic instruments to measure distances in the laboratory. They can measure the surface elevation of the fluid or the moving sand bottom in models. They work by sending an ultrasonic pulse and measuring the travel time (Section (5.4.1)). The acoustic sensor is a piezoelectric transducer (Section (4.13.3)) that works at frequencies about 1 Mhz.

Some commercial equipment has an accuracy of 1% of the measuring range and a maximum resolution of 1 mm. The sampling rate may usually range from 10 to 100 samples per second. Some manufacturers claim that the footprint of the sensor beam on the water surface (spatial resolution) is less than 10 mm for a distance of 1 m. Also, corrections for changes in temperature have to be done, either manually or automatically.

6.2.7 ADCP

Acoustic Doppler Current Profilers, already described in Section (5.4.4), are also useful to measure waves. The method provides measures of both directional and non-directional waves. The principle behind this application is that because ADCP can measure the velocity profile, they are also able to estimate the wave orbital velocities below the surface (Lawrence et al., 2012). Remember that the velocity is measured for several cells at different depths, then processing the data of the cell array, directional information is obtained. Linear wave theory is used to translate the data from velocity spectra to surface displacement.

ADCPs for directional wave measurement purposes use a magnetic compass and a tilt sensor to refer measures velocities to the north-south and east-west planes and include a pressure sensor. The pressure sensor is used to measure the mean water depth, which increases accuracy in applying the linear theory. Also, it provides a second source for non-directional wave statistics

6.2.8 Radar

The use of radar as a remote sensing system in environmental sciences will be addressed later in Chapter 8 with some detail. Here radar application to the particular case of local (at a point) sea level measurements is introduced briefly, the operating principle being similar to that of the acoustic equipments described above.

The installation of a radar level meter, as in the acoustic sensors case, requires also a structure on the shore or a platform fixed to the seabed, but the electromagnetic wave only travels in air and cannot be used from underwater to the surface as in the acoustic case. Radars cannot be used underwater because due to the electrical properties of water the electromagnetic wave results attenuated within short distances.

Radars use as transducer a radiofrequency antenna pointing down to the sea surface. The radar antenna emits pulses downwards to measure the distance from the antenna to the sea surface by means of the time of flight. In this case the radar emits an electromagnetic wave whose traveling speed is much higher than that of the sound in air. Then, because the time of flight is very short, the electronics involved in the process of measuring the water height has to be more sophisticated than in the acoustic case.

6.2.9 Resistive (Conductive)

These sensors are used to measure waves in models and wave tanks. In the literature, these sensors can be found as resistive or conductive gauges. The wave sensor is composed of two vertical parallel conductive rods or wires, generally made of stainless steel (Fig. 6.4). The sensor has a head containing the electronic circuits; the head may include a transmission link to a central station which is responsible for acquiring the signal from several sensors. The spacer, at the lower end, keeps the rods at a constant distance from each other because it is essential that the geometry of the staff remains unchanged.

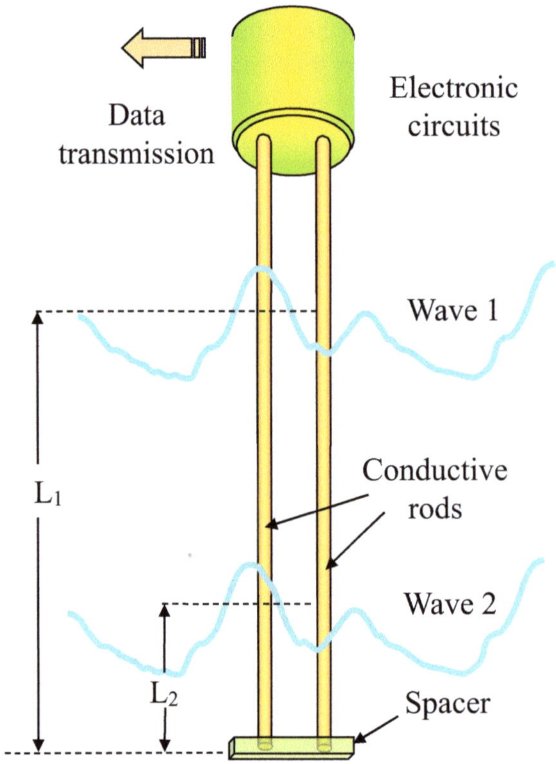

Fig. 6.4: Schematic of a resistive probe

The rods are submerged at a certain depth and the electronic circuit measures the resistance of the rods plus the water. Because the resistance of the rods is low and constant, the change in resistance as "viewed" by the circuit is due to the change in water height. The measured water electrical resistance is given by

$$R = \frac{C}{L\sigma} \tag{6.1}$$

where L is the length that the rods are submerged, σ is the conductivity of the water and C is a constant which depends on rod's diameter and distance between rods. This equation is valid from a certain distance from the lower end of the sensor, for example 5 to 10 % of the rods length. Closer to the lower end the relation between resistance and water height is no longer linear.

The change in resistance measured by the electronic circuit is inversely proportional to the water height. When the sensor measures "Wave 2" (Fig. 6.4), the resistance of the water "seen" by the electronic circuit is R_2; when the water level increases to "Wave 1" the resistance is R_1, and $R_1 < R_2$. The decrease in resistance is because when the rods are more submerged, some resistance is added in parallel to the initial R_2. As explained in Section (3.7.6), the value of the equivalent resistance of two resistors connected in parallel is always less than the smaller of the two resistors; this is why the resistance decreases with the increase in water level.

The measured resistance, and then the voltage output, is an average of the wave height between the rods (Fig. 6.4). Therefore, when the wavelength of the water wave is comparable to the distance between the rods, the sensor works as a low-pass spatial filter for the measured waves.

If sensors were used in a tank where waves travel in just one direction, one way to minimize sensor spatial averaging is to install the plane containing the rods parallel to the wave front. In a bi dimensional hydraulic physical model where waves can travel in any direction, one way of reducing filtering is by decreasing the diameter of electrical conductors and their separation distance. Wires are thus used for replacing rods. Wires require a tension to be exerted between both ends to keep them parallel, for example, fixing the lower end to the bottom of the model.

The dynamic response of the wave staffs can also be limited by the water meniscus on the probe at the water-air interface. Also, when the water falls down around the staff, there is a physical phenomenon developed on the probe surface called *wetting* of the probe which can be more extended in length that the meniscus itself (Clayson, 1989). This physical behavior also produces some kind of low-pass filtering of the wave. In simple terms, wetting is a phenomenon that results from the action of cohesive and adhesive forces acting upon the molecules of a liquid near a solid surface (the walls of the recipient containing the liquid or any structure submerged in it). Cohesive forces are exerted by other molecules within the liquid, whereas adhesive forces are exerted by the molecules of the solid surface. Since liquids cannot maintain any shear stress, the resultant force between cohesive and adhesive forces must be normal to the surface of the liquid at the point of contact with the solid surface. The angle that the tangent line to the liquid surface at the point of contact makes with the solid surface is called the *contact angle*. On the whole, if the adhesive forces are greater than the cohesive ones, the contact angle is small, and it is said that the liquid *wets* the surface. Conversely, if the contact angle is large, the liquid *does not wet* the surface. Water wets clean glass (contact angle near zero), whereas mercury does not (contact angle about 140°). It is worth noting, however, that small quantities of impurities, such as humectants, detergents and waterproofing agents, can produce large variations of the contact angle (Sears & Zemansky, 1964).

Because the resistance is inversely proportional to wave height, in order to have an output voltage directly proportional to the height some simple electronic circuits have to be used. In general alternating signals in the audio range are used in these circuits.

Generally, wave tanks and hydraulic models use fresh water and the conductivity of the water could be considered constant during short time tests. Thus, changes in σ can be controlled by calibrating before and after each measuring cycle. In some cases where σ changes during the test, a compensating circuit that measure the conductivity of the water simultaneously with the wave height is used.

Calibration of the probes is achieved by raising or lowering them a known amount and simultaneously measuring their output voltages. Contaminants in the water, such as oil, may have an important effect on the calibration.

6.2.10 Capacitive

Devices known as capacitive gauges are used in model and field measurements. Field use requires a shore construction, a platform, or a pier fixed to the bottom.

They have a sensor consisting of a metallic rod (or wire) covered by a sheath of dielectric material (electrical insulator) (Fig. 6.5). Usually, the wire is made of copper, and the dielectric of Teflon. In order to have a linear transference, the dielectric thickness must be constant along the wire length. As in the case of the resistive gauge, if a wire is used, some mechanism to tense the wire should be used.

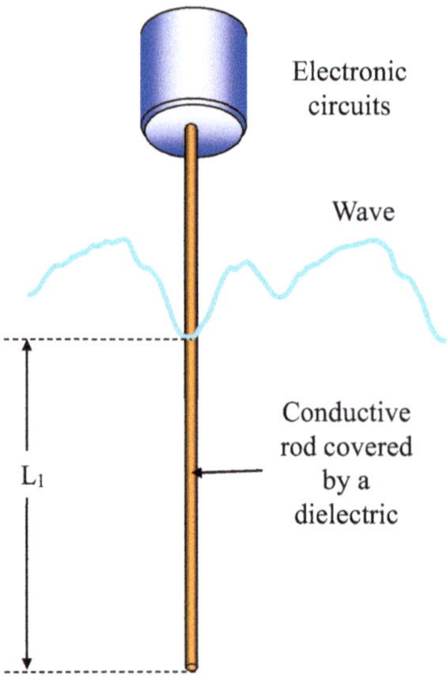

Fig. 6.5: Schematic of a capacitive probe

The internal rod is one of the electrodes of a capacitor. The other electrode may be a second, non-insulated rod, placed close to the first, or may be the water itself, which is at the electrical ground potential of the tank (Payne, 2008). In the second case, the capacitance C between the internal electrode and the water is given by (Clayson, 1989).

$$C = K\,h \tag{6.2}$$

where K is a constant which includes the diameter of the rod and the dielectric sheath and the permittivity of the dielectric, and h is the immersed depth of the sensor. Thus, the capacitance is proportional to the water level. Some electronic circuits, appropriated for measuring capacitive impedance are used to convert changes in the capacitance to voltages. Thus, output voltages represent changes in the water height. In general, signals in the range of radio frequencies are used to excite the sensors.

The capacitive sensor will average the level measured to a certain area around the wire (or rod). The size of this integrating surface and so the spatial averaging was not found in the manufacturers' specifications. Theoretically, it depends on the spatial distribution of the electromagnetic field in the real distributed capacitor, which depends on where the connection of the outer electrode is made, the impedance of the water and the reactance of the capacitor.

In order to achieve high capacitance values, which are less prone to noise, the dielectric material cover is usually made thin and can therefore be easily damaged. Manufacturers of capacitive gauges recommend that the sensing wire does not come into contact with sharp objects, and when not in use, care should be taken to avoid putting any weight on the sensing wire. Also, probes should not be exposed to very hot or cold temperatures (RBR Limited, 2011).

These capacitive staffs also suffer from the problem of meniscus and wetting (already mentioned for resistive sensors) and to have a constant sensitivity the temperature must be held constant (Lawrence et al., 2012). They can be used either in fresh or salt water and the calibration process is similar to that for resistive probes.

6.2.11 Optical

It would be very useful to have an optical method to measure water waves in models without any contact with the water, and with high temporal and spatial resolution (Mulsow et al., 2006), and some efforts have been made in this sense (Mulsow et al., 2006; Payne et al., 2009). The operating principles of the methods are simple, but the recording and processing of the acquired images are complex.

One of the systems consists of a laser source whose beam is vertically installed and pointing downwards at the water surface, and a camera whose line of sight forms a certain angle with the beam (Fig. 6.6). The relative position of the laser and camera is kept constant by mounting them on the same frame. The laser beam produces a spot on the water surface and the camera digitally records images of it as the water moves. Unfortunately, because the beam light is not only reflected by the surface but also scattered by subsurface water, the recorded spot image will not be sharp but diffused. After some statistical image processing, the surface spot is estimated.

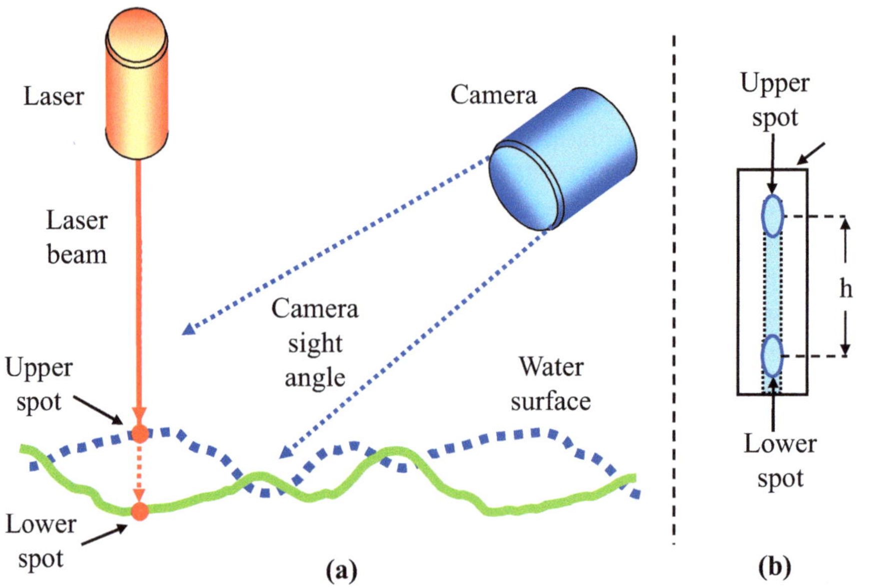

Fig. 6.6: (a) Laser and camera setup. (b) Camera registers. Water surface in dashed and continuous line corresponds to two different instants. The same goes for the upper and lower spot showed on the camera register. When the spot on the surface at both instants is obtained by statistical analysis from the register, the change in water level (*h*) can be estimated.

The calibration of this measuring system is achieved by displacing the laser/camera set vertically on a motionless water surface. The set displacement is measured and compared with the position of the spot on the recorded image and from these values the transference of the system is obtained.

Another optical wave meter uses a slanted laser beam which forms a light sheet that is reflected on the water surface and projected on two parallel vertical surfaces (Mulsow et al., 2006). The images produced by the reflected laser beam on these surfaces are recorded by a camera. By processing the image on the two vertical planes it is possible to separate water-level induced effects from slope-induced effects. The data processing and calibration procedure is more complicated than in the previous measuring system.

Optical systems have the advantage that they do not perturb the water surface as in the case of the conductive or capacitive staffs. But they depend on surface reflection; any floating material or surface turbulence could cause measurement errors. Also, they require considerable data processing.

6.3 Quasi Static Level Measurements

6.3.1 Mechanical Systems for Measuring Level

The old instruments for recording tidal level used a float and counterweight to move a mechanical system, thus converting linear level variations into rotational motion. This rotational motion was associated with a pencil and a recording paper chart. The chart was advanced by means of a mechanical clock. The pencil registered on the paper chart a distance proportional to the water level change.

The same principle is still in use in some instruments to measure slow water level changes such as tide, water table and liquid level in tanks. Nowadays instruments have great modifications of the method through which the information is recorded, in the clock and in the way the mechanical information is transformed into an electrical signal. The rotation produced by the float and counterweight on a pulley is used to move a variable potentiometer or an optical sensor as those described in Section (4.14). Because mechanical systems are easy to understand and their description can be found in many reports (GWPD 14, 2010) they will not be described any further here.

6.3.2 Radar and Acoustic Meters for Low Spatial and Temporal Resolution

Radar and acoustic meters as those explained before for measuring waves are also used to measure slow level changes. Because these technologies were already described they are only mentioned here. These kinds of meters have similar operating principles and installation requirements when they are used either for measuring fast or slow changes in water surface elevation. The main difference is that in low spatial resolution problems, the beam width of the transducers and antennas is not required to be narrow. Also, level time series are low-pass filtered to remove all undesired perturbations.

6.3.3 Pressure

Pressure sensors are well suited for measuring slow changes in the water level and their applications were explained in Section (4.8.4). The more appropriate type of pressure sensor for each case was described in detail, stressing the convenience of using vented gauges in some applications.

6.3.4 Applications of Quasi Static Level Measurements in Hydrology

Quasi static levels in hydrology are extensively measured for multiple purposes. Among them are: tests to evaluate hydraulic conductivity, monitoring of aquifers exploitation and determination of groundwater flow.

The methods used to estimate groundwater flow require monitoring a network of at least three wells and to know the hydraulic conductivity of the soil. With this information water speed is calculated using Darcy´s law and water direction is estimated from the gradients inferred from the piezometric heads.

In general, these methods have low spatial resolution and are prone to errors because the hydraulic conductivity distribution is not well known, as pumping or slug tests do not extend a great distance from the wells. In order to have good data quality it is necessary to increase the number of wells involved in the measurements, which increases costs and efforts employed for each test.

A different concept has been developed in the last two decades to evaluate groundwater flow. Thus, the so-called direct or point-measurement methods that measure the water velocity in just one point of the aquifer, giving local results emerged. At present, there is not a prevalent method that stands out among the few existing; on the contrary, most of them are still under development. These methods will be described in the rest of this chapter. They seem to be methods with a promising future because they require only one borehole and a short time to perform a measurement. Also, as technology evolves they will probably allow collecting data at lower costs than traditional indirect methods.

Perhaps the material here presented on groundwater velocity measurements could be considered too meticulous for an introductory work, but it was done this way for two main reasons: first, this subject was not found addressed in other books, and second, the detailed explanation of these operating principles is necessary to advise the reader about instruments limitations and operational problems that these technologies still entail.

6.4 Measuring Groundwater Velocity

6.4.1 Introduction

Instruments for measuring currents in channels, rivers and seas have reached such a degree of development that almost any research requirement is fulfilled by commercially available equipment (Chapter 5). The situation is quite different with groundwater measurements because flow is very slow and in some circumstances its direction changes in short distances. The selection of instruments for this application should be of special concern because no method has yet been adopted by researchers as standard.

6.4.2 Review of Direct Methods

For the sake of a more structured description, direct methods will be grouped according to their hydrological applications. For this reason, an explanation follows on some quite different potential hydrologic applications where these methods could be employed. The application characteristics taken into account to form these groups are: the range of groundwater speed, the relevance of measuring flow direction, and the environment surrounding the sensor. Obviously, as in most cases, this classification is arbitrary and some measuring methods can be used for more than one application.

6.4.3 Some Quite Different Flow Measurements Needed in Hydrology

6.4.3.1 Vertical Flow

The study of vertical flow in boreholes during pumping tests is a classical application of flowmeters in hydrology. In some applications the instrument for measuring pumping rates has to be lowered at great depths and should withstand large water pressures. The range of speeds for this application is from 0.001 to 0.1 m/s. Flowmeters must work in direct contact with water, but they do not need to measure flow direction (Hess, 1986). From the point of view of instrumentation complexity, this is the simplest case to solve because velocities are still high enough to be measured with simple technological solutions and direction is of no interest.

6.4.3.2 Horizontal Groundwater Flow

The magnitude and direction of horizontal groundwater velocity is a very important hydrological parameter that can be measured in direct contact with the soil or from inside boreholes. In the first case, instruments are permanently buried in "ad-hoc" drilled boreholes and relocation of the instrument is difficult. The second case allows using the same instrument in different wells, or at various depths in the same well. Moreover, already existing boreholes could be used. For this application, instruments can be used at hundreds of meters from the surface. The flow speed ranges approximately from 1×10^{-6} to 8×10^{-4} m/s and flow direction is of great interest. The higher velocities correspond to mountain recharge zones or fractured rock soils, and the lower ones to soils in plain areas.

6.4.3.3 Seepage Flow

Instruments for measuring submarine groundwater discharge are known as seepage meters. They use a funnel inserted into the bottom sediment to collect groundwater. In automatic seepage meters the funnel has a discharge outlet connected to a flowmeter.

The flowmeter sensors are in direct contact with water. The outlet area is much smaller than the funnel mouth area so that flow at the outlet is about 10^3 higher than at the mouth, which simplifies the flowmeter design. A measurement range of seepage velocity from 2×10^{-7} to 5×10^{-6} m/s has been reported (Taniguchi & Fukuo, 1993).

6.4.3.4 Flow in Remediation Works Where Velocities Are Very Low

The monitoring of waste remediation activities and post-closure performance of remediate waste sites requires measuring fluid velocities (in magnitude and direction) as low as 3×10^{-8} m/s. Sometimes the fluid is biphasic (water plus air) and has some degree of corrosive components. Instruments have to be installed in direct contact with the soil and should measure, hopefully, during years without recalibration. In general they are installed close to the surface and should record continuously.

6.5 Direct Methods and Their Hydrological Applications

6.5.1 Vertical Flow in Boreholes

6.5.1.1 Thermal-Pulse Flowmeter

When vertical velocities are small (below 10 mm/s) or a great resolution is required conventional spinner flowmeters cannot be used, and an alternative is to use a method developed by the U.S. Geological Survey (USGS) based on a thermal-pulse flowmeter (Hess, 1986). The device consists of a heater and two temperature sensors separated a distance $d = 40$ mm, arranged as shown in Figure 6.7. It can measure in a flow range from 1 to 100 mm/s. The heater grid is contained in a plane perpendicular to the flow. Sensors are glass bead isolated thermistors, separated 20 mm from the heater.

Flow velocity (V) is measured by recording the time elapsed between the electrical pulse is applied to the heater and the response peak of the sensors (Δt).

The volume flow is obtained from calibration charts developed in the laboratory using pipes of similar diameters to those used in the borehole. The thermal pulse response time may range from 1 to 60 s. Water is heated about 0.1 °C above the ambient temperature. The buoyancy of the heated water causes some asymmetry in the calibration curve for up and down flows.

Because flow is obtained from time measurements, and $V = d/\Delta t$, the transference curve which relates V and Δt is quite non linear (flow is inversely proportional to the elapsed time).

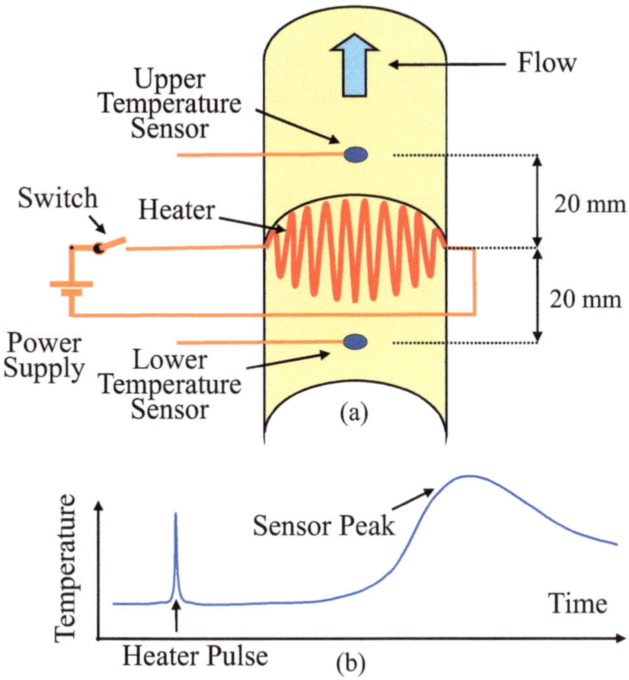

Fig. 6.7: (a) Schematic of heater and sensors. (b) Sensor heat pulse response showing the instant when the electric pulse is applied.

6.5.1.2 Electromagnetic Flowmeter

An electromagnetic flowmeter based on Faraday's law (Section (3.7.12)) was reported for this same application (Moltz & Young, 1993). Authors claim that it is possible to measure velocities as low as about 1 mm /s with errors less than 10%. They use two electrodes and an electromagnet (coil) as the operating principle already mentioned in Section (5.3.3). The main differences with that described before are that the size and power consumption of these flowmeters are smaller, and that they are designed to work at several hundred of meters underwater.

6.5.2 Flowmeters for Horizontal Groundwater

6.5.2.1 Colloidal Borescope (CB)

The instrument, described by Kearl (1997), attempts to measure groundwater velocity by observing the motion of particles inside a well. It consists of two CCD (charged-couple device) cameras, a ball compass, optical magnifying lenses, an illuminating source, and housing (Fig. 6.8). This device is about 0.9 m long and 0.044 m in diameter.

When placed inside a borehole it is capable of transmitting amplified images of the particles to the surface. Due to the insertion of the instrument in the well the flow is disturbed, but after half an hour laminar horizontal flow predominates. Particles illuminated by the back-light source, as in a microscope, are zoomed by one of the cameras while the other focuses the compass to reference the images to the magnetic north. A digitizing system records video frames at selected intervals and a software compares them, gathering information about average particle size, number or particles, speed and direction.

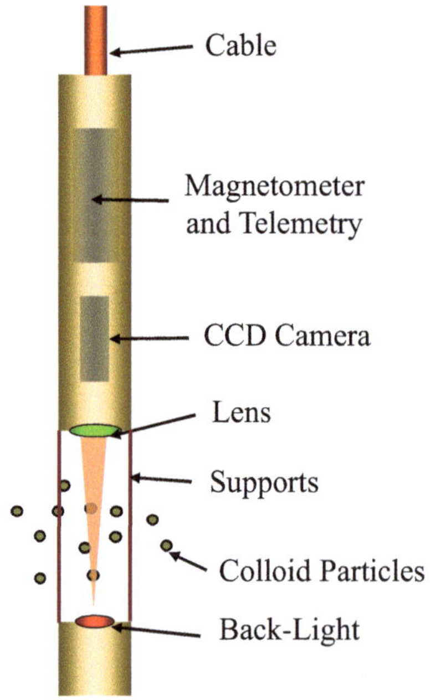

Fig. 6.8: Scanning colloidal borescope flowmeter (SCBFM).

An improved version of this instrument (James et al., 2006), called a scanning colloidal borescope flowmeter (SCBFM), allows investigating three-dimensional flow on an interval of 0.5 m depth inside the borehole without relocating the instrument. The best results are for particles that move at a rate below 12×10^{-3} m/s. A U.S. patent (Foster & Fryde, 1990) presents a similar idea to that described above, but with some technological differences.

6.5.2.2 Groundwater Laser Velocimeter (GLV)

This instrument measures the motion of the particles suspended in the groundwater of a borehole (Momii et al., 1993) as the CB flowmeter, but with a different technology. For this purpose a laser beam is split into two resulting beams that are focused by a lens on a point where interference fringes are formed with a spacing *d*. Figure 6.9 shows the illuminated volume with the interference fringes. The moving particles passing the fringes reflect light from the regions of constructive interference. The illuminated volume is a cylinder of about 100 μm long and 30 μm in diameter; very small particles (a few microns in diameter) are thus detected. The reflected light is received by a photo detector, and by measuring the Doppler frequency shift (*fd*) of the scattered light it is possible to calculate the velocity of the tracer particles, and hence to estimate the water velocity. The velocity of the particles (*V*) is related to the Doppler frequency and the distance between interference fringes (*d*) by Eq. (6.3). No laboratory calibration of the instrument is thus needed.

$$V = d\,fd \tag{6.3}$$

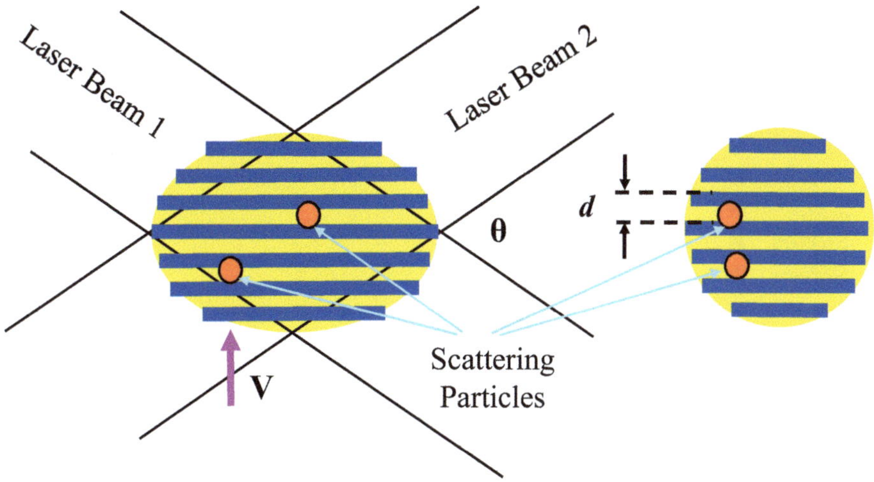

Fig. 6.9: The groundwater laser velocimeter illuminates a volume of water with interference fringes.

The distance *d* is a function of the angle θ between the two laser beams, and the laser wavelength λ. Due to its high spatial resolution, this method needs short observation times, but as in the CB flowmeter, when introduced in the borehole the flow is disturbed and some time is required to restore laminar flow conditions. This instrument cannot be used when the flow is turbulent.

The working speed range of this instrument is from 3×10^{-7} to 1.4×10^{-4} m/s. At the low end of the range, errors are of about 8%, decreasing to 1.4% at the highest velocities.

6.5.2.3 Thermal Flowmeters

6.5.2.3.1 Horizontal Heat Pulse Flowmeter (HHPF)

The most widespread heat pulse flowmeter to measure horizontal flow consists of a probe containing a heater surrounded by a circular array of thermistors. The U.S. patent of Kerfoot & Skinner (1983) describes an array of 8 thermistors; other authors (Melville et al., 1985), an array of 10. The manufacturer recommends placing a nylon mesh end cap at the end of the probe. The external diameter of the nylon mesh end cap matches the internal diameter of the well casing and is filled with glass beads (Fig. 6.10). Thus, when it is lowered into the well and immersed in water, the probe is surrounded by a saturated porous media. When a heat pulse is applied a transient temperature field is generated. While the heat diffuses radially, the water displacement produces some advection of the heated water.

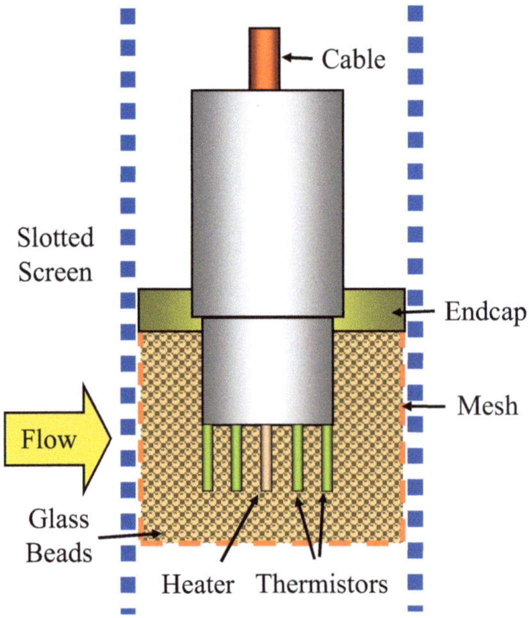

Fig. 6.10: Horizontal heat pulse flowmeter. The instrument is shown inside a slotted screened borehole and the space between the screen and the heater/thermistor array is filled with glass beads which create a saturated porous media around them.

For stagnant water heat diffusion produces a symmetric distribution of heat, so opposite thermistors measure the same temperature. When water moves, an asymmetric distribution of heat causes temperature differences between thermistors. These differences are supposed to be proportional to the component of the water velocity along the direction defined by opposite thermistors. Water velocity is estimated by calibrating the temperature difference vs. water velocity. The probe is hold in the borehole from the surface by means of connecting rods. In order to relate measured velocities to the magnetic north, a particular pair of thermistors is referenced by placing a compass at the top of the connecting rods.

The power pulse applied to the heater is about 15 W, it lasts for approximately 30 s and its effects on the thermistors are measured during 3 minutes. The diameter of the probe is 0.044 m. The velocity range is from 0.35×10^{-7} to 0.35×10^{-5} m/s (Melville et al., 1985). In order to get reliable results the instrument must be calibrated under similar conditions to those of real use.

6.5.2.3.2 Rotary Device Probe (RDP)

It has been developed for measuring groundwater in recharge zones (Guaraglia et al., 2009). The thermal rotary device probe consists of a central heater and four thermistors symmetrically placed around the heater, forming two orthogonal axes (Fig. 6.11a). When the probe is lowered in a monitoring well to the desired depth, a temperature step is applied to the heater and its temperature kept constant. Then the probe is slowly rotated 360° clockwise and counterclockwise. The average power supplied to the heater and the thermistor temperatures are recorded.

If flow is not null the thermistor temperatures increase downstream and decrease upstream. As the thermistors rotate, a kind of temperature waveform as a function of the angular position is recorded (Figure 6.11b). By processing these waveforms the direction of the flow is found. When flow velocity increases, the heater's power has to be increased to keep its temperature constant, so the average heater's power gives information on the flow rate.

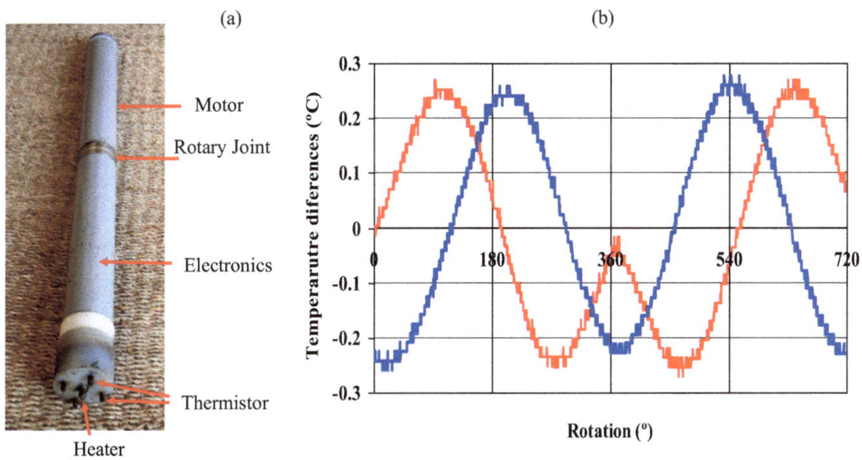

Fig. 6.11: (a) Rotary Device Probe. (b) Temperature waveform.

6.5.2.3.3 Groundwater Acoustic Doppler Velocimeter (GADV)

Type 1

There is a U.S. patent describing a method based on the Doppler shift of an acoustic wave (Yankielun, 1998). The sound source and sensors provided as examples are piezoelectric transducers working at 4 and 10 MHz. Acoustic pulses are generated by a central sound source and detected by sound sensors positioned at a short distance from it. If four sensors are used, they can be positioned exactly north, east, south and west of the sound source. The ensemble is lowered in a screened borehole and placed below the water table. When the water moves, frequencies in the sensors will differ from the source frequency due to the Doppler effect. The patent claims that frequency differences are mathematically related to the water velocity by

$$V = \frac{c\Delta f}{2f_S} \tag{6.4}$$

where V is the water velocity, f_s the sound source frequency, Δf the difference in frequency and c the speed of sound in water. The direction and magnitude of water velocity is obtained by vector addition of the north-south and the east-west components.

Differences in frequencies are obtained by pulse counters which count the number of pulses detected during a predefined counting interval. Some examples are given in the patent. One of them shows that an instrument having a 10 MHz sound source needs 12 h to measure a water velocity of 2.3×10^{-6} m/s with good resolution. This time could be unsuitable in some applications. For example, if the relation of groundwater dynamics to the tidal level has to be studied, such counting times are undesirable, and perhaps other methods should be used.

Type 2

Another GADV, quite different from the above, is described by Wilson et al. (2001). This flowmeter does not measure fluid velocity directly, but tracks the velocity of suspended particles in the water column. It is similar to the Acoustic Doppler Velocimeter described in Section (5.4.6). It is approximately 1.2 m long with a 0.075 m external diameter. It can be deployed in wells with an inner diameter of 8.75×10^{-2} m.

The probe consists of one centrally mounted acoustic emitter and three receivers positioned on radial arms. A guard cage protects the probe from potential damage during field work. The maximum sample volume of the probe is 1.6×10^{-9} m^3 and the frequency of measurement is 25 samples per second. The instrument is deployed with a cable that provides the power and communication capability. It uses a flux-gate magnetometer as a compass. Borehole flows from 1×10^{-4} to 2.4 m/s are measured accurately.

6.5.3 Seepage Meters

As stated above, groundwater discharge in coastal zones may be estimated by collecting the water coming out of the sea bottom in a funnel buried in the sediment. The water captured by the funnel is conducted through a tube where a flowmeter, which is part of the seepage meter, measures the flow as a function of time (Fig. 6.12).

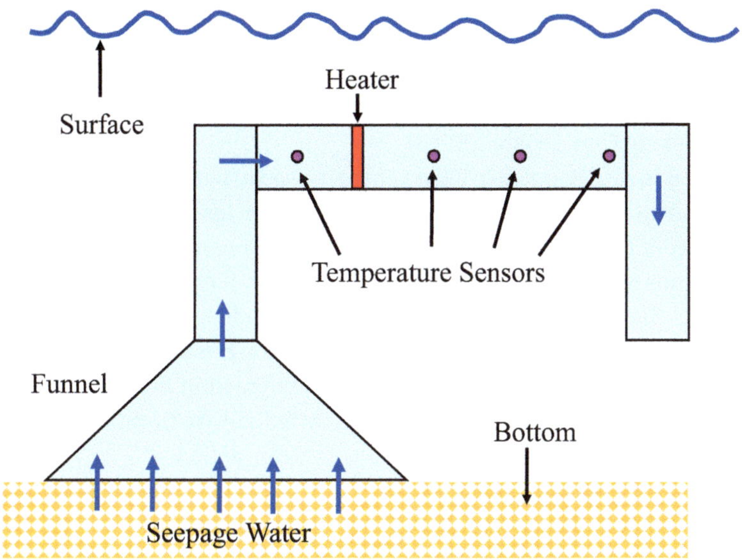

Fig. 6.12: Thermal seepage meter. A funnel buried in the sediment collects the seepage water from the bottom discharge. Heater and sensors measure the flow.

A thermal method for measuring the flow velocity in the tube (V_t) consists in producing a pulse heat in the water inside the tube and measuring the temperature at 0.05, 0.1 and 0.15 m downstream the heater, and at 0.05 m upstream it (Taniguchi & Fukuo, 1993; Taniguchi et al., 2007). The downstream sensors are measuring the ambient temperature plus the temperature due to the injected heat. The upstream temperature has to be subtracted from the downstream ones to exclude the natural changes in water temperature. The heat pulse is applied for two seconds every five minutes and the time when the temperature measured by the thermistors reaches a maximum (t_{max}) recorded. Since the relationship between log V_t and log t_{max} is almost linear, the calibration curves allow V_t to be estimated from t_{max}. Then the groundwater velocity (V_g) is calculated by multiplying the velocity V_t by the ratio of the cross-sectional areas of the tube and the funnel. In this kind of application the flowmeter requires measuring water flow velocities in the range from 4.5×10^{-4} m/s to 0.01 m/s.

An ultrasonic seepage meter has been patented that differs from the former one, basically in the way that flow velocity is measured in the tube. It uses two piezoelectric transducers that continually generate burst of ultrasonic signals from one end of the tube to the other. The speed of the water affects the speed of the ultrasonic signal. The measurement of the ultrasonic signal speed provides information to get the speed of water in the tube. It is claimed that speeds of water as low as 10^{-7} m/s are measured.

6.5.4 Flow in Remediation Works Where Velocities Are Very Low

6.5.4.1 In Situ Permeable Flow Sensor (ISPFS)

The instrument called In Situ Permeable Flow Sensor is used to measure the direction and magnitude of the three-dimensional groundwater flow vector in unconsolidated, saturated porous media (Ballard, 1994). The sensor is permanently buried in direct contact with the porous media and measures, approximately, the average velocity in a cubic meter around the sensor. It is claimed that the instrument is able to measure flow velocities in the range from 5×10^{-8} to 1×10^{-5} m/s. Being in direct contact with the formation the ISPFS avoids all problems related to the interaction between aquifer flow and observation wells (James et al., 2006; Lengricht & Kai-Uwe Graw, 2002), and there is no need to know the hydraulic conductivity of the formation.

The ISPFS consists of a rod of low thermal conductivity, 0.75 m long and 0.05 m in diameter, containing a heater and surrounded by an array of 30 thermistors capable of measuring within 0.01 °C (Fig. 6.13). When the rod is buried in some unconsolidated formation where groundwater velocity is to be measured, a thermal perturbation is applied to the media by the heater. The power applied is between 60 and 120 W, which increases heater temperature in about 20 to 25 °C over the temperature of the formation. Power should be as much as possible to get maximum sensibility, but not so much to produce natural convection of water.

If there are no asymmetries in the construction of the instrument, the heat flux leaving the rod should be uniform over the surface of the cylinder. In such a case, the distribution of temperature on the surface of the cylinder would vary as a function of the direction and magnitude of the groundwater velocity. Comparing the measured temperature distribution with theoretical temperature maps obtained from thermal equations, the three-dimensional vector velocity is estimated. The theory assumes a long cylinder, buried in perfect contact with an infinite saturated porous media whose thermal and hydraulic properties are homogeneous and isotropic. Buoyancy effects due to changes in water density are neglected.

Fig. 6.13: In Situ Permeable Flow Sensor. Photography courtesy of Stanford Ballard, Sandia National Laboratories, Albuquerque, New Mexico, USA.

In practice it is complex to fulfill the theoretical assumptions underlying the method; drilling techniques deform the microstructure of the ground, sometimes compacting it (percussion) and sometimes making it less consistent (rotation), which is likely to change the uniformity required by the theory. Nor is it easy to achieve close contact with the saturated soil, because during the installation air may be trapped near the walls of the sensor. Although the air may be displaced by water, the porosity near the sensor could become anisotropic. Therefore, in order to succeed in measuring with the In Situ Permeable Flow Sensor a correct installation is a key topic.

6.5.4.2 Conductivity Flowmeters

6.5.4.2.1 Point Velocity Probe

This method was developed to measure groundwater velocity in unconsolidated non cohesive media such as sand (Labaky et al., 2007; Devlin, 2002; Devlin et al., 2009). It consists in measuring the velocity of a tracer on the surface of a cylinder (apparent velocity Va), which allows groundwater velocity to be evaluated far form the cylinder ($V\infty$). The tracer must have electrical conductivity other than the conductivity of groundwater; it may be deionized water or water with a given saline concentration.

The device used to implement this method consists of a cylinder with a supply of tracer solution connected to an injection port which releases the saline tracer. The injector is constructed drilling a hole on the cylinder and placing a diffuser stone through which pulses of the tracer are released (Fig. 6.14).

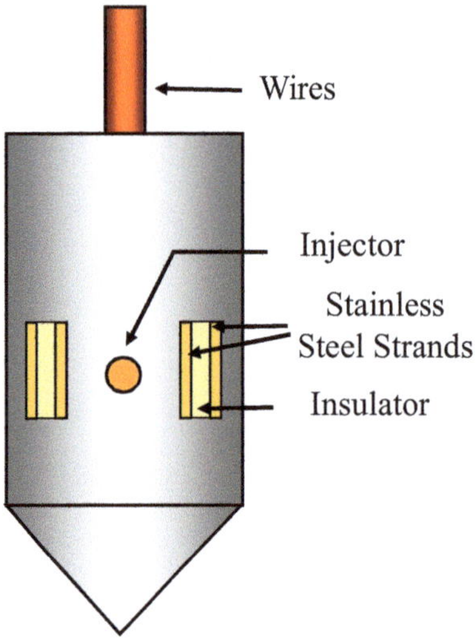

Fig. 6.14: Point Velocity Probe. This instrument is used in unconsolidated non cohesive media such as sand.

Two or more sensors are placed on the same horizontal plane than the injector to measure water electrical conductivity. Each conductivity sensor consists of two stainless steel strands.

An alternating current is applied to the strands, which work as a conductivity cell sensor (Section (4.10.1)), and voltage is measured on a constant resistor connected in series (Section (4.3)). When groundwater is over the sensors, they measure a background resistance attributable to the aquifer water. As the saline tracer passes over the sensor, the electrical conductivity between both strands decreases (the opposite happens if deionized water is used). The conductivity as a function of time is recorded at each sensor and these curves are used to estimate *Va* by fitting them to a solution of the advection-dispersion equation. Next, the angle α between the aquifer velocity direction and the injection port radius is estimated, and the velocity $V\infty$ calculated. The inaccuracy in the estimation of α can introduce errors in the estimation of $V\infty$. According to Labaky et al. (2007), laboratory test demonstrated that the method works between 5.8×10^{-7} and 1.13×10^{-5} m/s with errors within ± 15% and the direction is estimated with an error of about ± 8°.

6.5.4.2.2 Advection – Dispersion Velocity Meter (ADVM)

This method was developed to assess the safety of geological disposal of radioactive wastes. It is implemented by means of a probe which is introduced in a single well (Kawanishi et al., 1999). This probe consists of a screened cylindrical frame 0.06 m in diameter and 0.02 m in height filled with 1 mm diameter glass beads. The probe has a central common electrode and 12 equally spaced electrodes located around the central electrode on a circle 0.03 m in diameter (Fig. 6.15). The central electrode is a pipe through which the tracer solution (distilled water) is introduced. Distilled water has an electrical conductivity lower than that of the groundwater being measured.

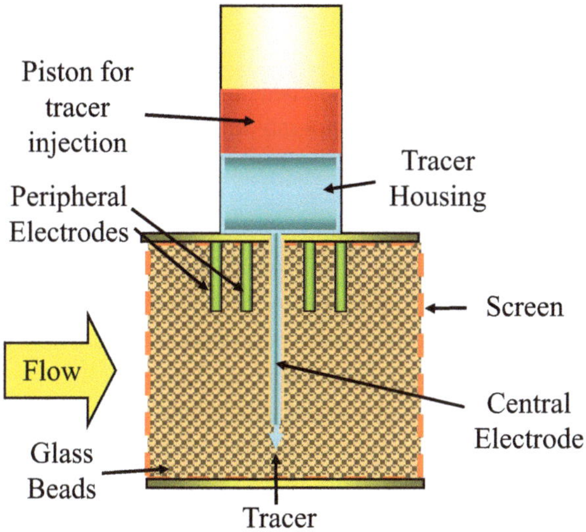

Fig. 6.15: Advection – Dispersion Velocity Meter.

The probe is lowered into the well where groundwater velocity is to be measured. Below it a rubber packer is placed to isolate the measuring section. Then the section is filled with a porous material such as sand to create a continuity of the porous medium inside the well, and an upper packer installed to isolate the section from vertical water displacements.

When groundwater enters the electrodes section and displaces the distilled water, the conductivity between the central electrode and the surrounding electrodes changes. The tracer injected through the central electrode will produce a migrating plume. When the plume is in the middle of the distance between the central electrode and the surrounding electrodes, the conductivity will be a minimum (or the electrical potential a maximum) (Kawanishi et al., 1999).

The velocity and direction of groundwater flow is evaluated by measuring the time at which the conductivity reaches a minimum (Δt). Theoretically, the velocity within the electrode section (V_0) is related to this time by

$$V_0 = \frac{\Delta x}{2\,\Delta t} \tag{6.5}$$

where Δx is the distance between the central electrode and the surrounding electrodes. The average flow inside the glass beads is different from the real flow velocity (V_r) due to the screen, glass beads and surrounding sand. Then a correction factor (β) has to be found empirically by a calibration process,

$$V_0 = \beta\, V_r \tag{6.6}$$

A measuring range from 10^{-9} to 10^{-5} m/s is reported in the literature (Kawanishi et al., 1999).

6.6 Discussion

Continuous research work is made having as a goal the measurement of groundwater flow in a single well by means of a simple, easy to operate and low cost measuring device. Some comparative field tests (Wilson et al., 2001) and many field and laboratory work seem to demonstrate that still more effort has to be dedicated to this objective before having a mature and reliable technology.

Wilson et. al. (2001) explain the importance of the search for a solution to the measurement of the horizontal component of groundwater flow velocity from inside a well. Their work compares the flowmeters previously described as HHPF, CB and GADV Type 2, with the technique called hydrophysical logging (Wilson et al., 2001). Unfortunately, because there are no standard methods for field use that allows the actual groundwater velocity to be known, it was not possible to determine which of the tested instrument gave a more accurately measurement, but some comparative results were found. Among the conclusions of this work, the following ones should be stressed:

1. "None of the tools consistently provided repeatable measurements of velocity and direction."
2. "A comparison of the measurements made in each well indicated that the three tested flowmeters rarely measured similar velocities and flow directions."
3. "The velocities estimated with the hydrophysical logging were typically very low and were most comparable to the velocities measured with the HHPF."
4. "The CB and GADV Type 2 measured similar velocities; however, the two tools seldom measured a similar flow direction."

6.6.1 Summary of Direct Methods Characteristics

Table 6.1 summarizes in each column the main characteristics of the direct methods described above: maximum and minimum velocities (Vmax and Vmin), the environment surrounding the probe (Environ), the time needed to perform one measure (Time) and the approximate volume sampled by the method (Volume). Values of volume and time are estimated from different literature sources that not always are coincident.

Table 6.1: Direct methods characteristics

Variable Method	V max (m/s)	V min (m/s)	Environ	Time (h)	Volume (mm³)
USGS	1	10^{-3}	water	0.25	10^{6}
CB	3×10^{-2}	unclear	water	0.5 *(3)	1
GLV	10^{-4}	10^{-7}	water	Not reported *(4)	10^{-4}
HHPF	3.5×10^{-4}	7×10^{-7}	s.p.m.	0.5	10^{6}
RDP	$> 10^{-3}$	$< 10^{-4}$	Water	0.25	10^{6}
GADV 1	unclear	unclear	Water	12	10^{6}
GADV 2	2.4	10^{-4}	Water	0.16 *(2)	10^{6}
ISPFS	10^{-5}	5×10^{-8}	s.p.m.	*(1)	10^{9}
PVP	1.13×10^{-5}	5.8×10^{-7}	s.p.m.	Velocity dependent	10^{6}
ADVM	10^{-5}	10^{-9}	water	Velocity dependent	10^{6}

*(1) Initially it requires 20 h to reach thermal equilibrium. *(2) After Wilson et al. (2001). *(3) and *(4) require to reach laminar flow.
References: USGS = U.S. Geological Survey Vertical flowmeter; CB = Colloidal Borescope; GLV = Groundwater Laser Velocimeter; HHPF = Horizontal Heat Pulse Flowmeter; RDP = Rotary Device Probe; GADV1 = Acoustic Doppler Velocimeter (Type 1); GADV2 = Acoustic Doppler Velocimeter (Type 2); ISPFS = In Situ Permeable Flow Sensor; PVP = Point Velocity Probe; ADVM = Advection–Dispersion Velocity Meter; s.p.m. = saturated porous medium.

6.6.2 Limitation of Direct Methods

It was underlined in Chapter 1 (Section (1.1)) that a desirable condition of any measuring method is to disturb as little as possible the phenomenon being measured. Among the direct methods there are two groups with different measuring requirements: one group requires placing the instrument inside an observation well, and the other needs to install the instrument in close contact with the aquifer soil. In both cases there are some conditions that hinder the fulfillment of the assumptions made in the theory of the method, i.e. that the flow passing the instrument is the actual flow in the aquifer.

Among the problems observed in the first group, the most remarkable were: the presence of the observation well in itself modifies the actual groundwater flow, and it is not yet sufficiently understood how filter slots and the small-scale soil composition around the borehole interact to produce a representative sample of the groundwater flow of the aquifer inside the well (Lengricht & Kai-Uwe Graw, 2002).

Most of the time, flow inside the well follows an eddy flow pattern and seldom a rectilinear one. The positive and negative accelerations that water suffers in passing the screen slots contribute to the turbulence inside the screen (Lengricht & Kai-Uwe Graw, 2002; James et al., 2006). Flow at the well center was estimated to be one order of magnitude less than just inside the slots. Then calibration of the flowmeters under laboratory conditions is recommended (James et al., 2006) to estimate the flow in the surrounding formation from measurements made within the screened well.

The most important problem that affects the second group is to satisfy the premise of intimate contact of the instrument's probe with the saturated porous media required by the theory. It is not possible to introduce the instrument in the place where measurements are to be carried out without some degree of alteration of the porous matrix during the installation. Drilling procedures modify the characteristic of the soil where probes are placed, in some cases consolidating the soil and in others increasing the porosity. This change in the porosity with respect of the original soil also perturbs the flow around the instruments clearly.

References

Alliance for coastal technologies. (2007). *Workshop proceedings report. Wave sensors technology.* Indexing number ACT 07-03. St. Petersburg, Florida, March 7-9, 2007.

Ballard, S. (1994). The in situ permeable flow sensor: a ground-water flow velocity meter. *Ground Water, 34*, 231-240.

Barstow, S. F., et al. (2003). *In*: Hauser, D., Kahma, K. K., Krogstad, H. E., Lehner, S., Monbaliu, J., & L.Wyatt, L. R. (Eds.), *Measuring and Analysing the Directional Spectrum of Ocean Waves, COST 714 Working Group* 3, Helsinki 18.10.2003.

Clayson, C. H. (1989). *Survey of instrumental methods for the determination of the high frequency wave spectrum.* Report no. 267, Institute of Oceanographic Sciences, Deacon Laboratory.

Datawell (2009). *Datawell Waverider Reference Manual WR-SG, DWR-MkIII, DWR-G.* Datawell BV Oceanographic Instruments, The Netherlands. Available at: http://cdip.ucsd.edu/documents/index/gauge_docs/mk3.pdf

Devlin, J. F. (2002). *Groundwater velocity probe.* U.S. patent no. 6,393,925 BI.

Devlin, J. F., Tsoflias, G., McGlashan, M., & Shilling, P. (2009). An inexpensive multilevel array of sensors for direct water velocity measurements. *Ground Water Monitoring and Remediation*, 29(2), 73-77.

Doong, D., Lee, B., & Kao, C. (2011). *Wave Measurements Using GPS Velocity Signals.* Sensors 2011, 11, 1043-1058; doi: 10.3390/s110101043. Available at: www.mdpi.com/journal/sensors

Foster, J. W., & Fryde, L. J. (1990). *Groundwater azimuth detection.* U.S. patent no. 4,963,019.

Guaraglia D., Pousa, J., Kruse, E., & Mayosky, M. (2009). A rotary thermal probe for measuring groundwater velocity. *Instrumentation Science and Technology, 37*, 1-16.

GWPD 14. (2010). *Measuring continuous water levels by use of a float-activated recorder.* Groundwater Technical Procedures of the U.S. Geological Survey. Version: 2010.1 Available at: http://pubs.usgs.gov/tm/1a1/pdf/GWPD14.pdf

Hess, A. E. (1986). Identifying hydraulically conductive fractures with a slow-velocity borehole flowmeter. *Canadian Geotechnical Journal, 23*, 69-78.

James, S. C., Beauheim, R. L., Mandell, W. A., Pedler, W. H., & Jepsen, R. A. (2006). Simulation to verify horizontal flow measurements from a borehole flowmeter. *Ground water, 3*, 394-405.

Kawanishi, M., Miyakawa, K., & Hirata, Y. *Development and application of groundwater flow meter in fractured rocks: measurement of velocity and direction of groundwater flow in single well.*

International Atomic Energy Agency SM-357/80 (1999). Available at: http://www-pub.iaea.org/MTCD/publications/PDF/csp_006c/PDF-Files/paper-80.pdf
Kearl, P. M. (1997). Observation of particle movement in a monitoring well using the colloidal borescope. *Journal of Hydrology, 200*, 323-344.
Kerfoot, W., & Skinner, S. (1983). *Groundwater flowmeter*. U.S. patent no. 4,391,137.
Labaky, W., Devlin, J. F., & Gilham, R. W. (2007). Probe for measuring groundwater velocity at the centimeter scale. *Environmental Science & Technology, 41*(24), 8453-8458.
Lawrence, J., Holmes, B., Bryden, I., Magagna, D., Torre-Enciso, Y., Rousset, J. M., Smith, H., Paul, M., Margheritini, L., & Cândido, J. (2012). *D2.1 Wave instrumentation database*. Marine Renewables Infrastructure Network (Marinet).
Lengricht, J., & Kai-Uwe Graw. (2002). Interaction of aquifer flow and observation wells detected by particle image velocimetry. *Proceedings 13th Congress APD-IAHR, 1*, 560-565. Singapore 2002. Available at: http://www.uni-leipzig.de/~grw/lit/texte_100/122_2002/122_2002_well_apd_iahr.pdf
Melville, J.G., Molz, F. J., & Guven, O. (1985). Laboratory investigation and analysis of a ground-water flowmeter. *Ground Water, 23*(4), 486-495.
Moltz, F. J., & Young, S. C. (1993). Development and application of borehole flowmeters for environmental assessment. *The Log Analyst, January-February*, 13-23.
Momii, J. K., Jinno, K., & Hirano, F. (1993). Laboratory studies on a new laser Doppler velocimeter system for horizontal groundwater velocity measurements in a borehole. *Water Resources Research, 29*(2), 283-291.
Mulsow, C., Schulze, M., & Westfeld, P. (2006). *An optical triangulation method for height measurements on instationary water surfaces*. Institute for Photogrammetry and Remote Sensing, Dresden Technical University, Germany, IAPRS Vol. XXXVI, Part 5. Available at: http://www.isprs.org/proceedings/XXXVI/part5/paper/MULS_657.pdf
NDBC. (1996). Nondirectional and Directional Wave Data Analysis Procedures. Technical Document 96-01. Stennis Space Center, National Data Buoy Center, National Oceanic and Atmospheric Administration, U.S. Department of Commerce.
Payne, G. (2008). *Guidance for the experimental tank testing of wave energy converters*. The University of Edinburgh. Available at: http://www.supergen-marine.org.uk/drupal/files/reports/WEC_tank_testing.pdf
Payne, G. S., Richon, J. B., Ingram, D., & Spinneken, J. (2009). Development and preliminary assessment of an optical wave gauge. *Proceedings of the 8th European Wave and Tidal Energy Conference, Uppsala, Sweden*. Available at: http://www.see.ed.ac.uk/~shs/Wave%20Energy/EWTEC%202009/EWTEC%202009%20(D)/papers/261.pdf
RBR Limited. (2011). *WG-55 Wave Gauge User manual*. Ottawa, Ontario, Canada. Available at: http://www.rbr-global.com
Sears, F. W., & Zemansky, M. W. (1964). *University Physics*. Reading, MA, USA: Addison-Wesley Publishing Company, Inc.
Taniguchi, M., & Fukuo, Y. (1993). Continuous measurements of groundwater seepage using an automatic seepage meter, *Ground Water, 31*(4), 675-679.
Taniguchi, M., Ishitobi, T., Burnett, W., Wattayakorn, G. (2007). Evaluating ground water-sea water interactions via resistivity and seepage meters. *Ground Water, 45*(6), 729-735.
Teng, C., Bouchard, R., Riley, R., Mettlach, T., Dinoso, R., & Chaffin, J. (2009). NDBC's Digital Directional Wave Module. OCEANS 2009, MTS/IEEE Biloxi - Marine Technology for Our Future: Global and Local Challenges.
Van Rijn, L. C., Grasmeijer, B. T., & Ruessink, B. G. (2000). *Coasted 3D measurement errors of instruments for velocity, wave height, sand concentration and bed levels in field conditions*. Department of Physical Geography; University of Utrecht WL, Delft Hydraulics.
Wilson, J., Mandell, A. W., Paillet, F., Bayless, E., Hanson, R., Kearl, P., Kerfoot, W., Newhouse, M., & Pedler, W. (2001). An evaluation of borehole flowmeters used to measure horizontal ground-water flow in limestones of Indiana, Kentucky, and Tennessee. *Water-Resources Investigations Report 01-4139*. U.S. Department of the Interior, U.S. Geological Survey.
Yankielun, N. E. (1998). *Doppler velocimeter for monitoring groundwater flow*. U.S. patent no. 5,796,679.

7 Wind, Rain and Solar Radiation Measurements

7.1 Preliminary Discussion

Before addressing the technological aspects of the subjects to be considered in this chapter, it is relevant to discuss some more general aspects. Because the title of our book refers to modern instrumentation, we have been tempted to obviate some topics such as mechanical anemometers or tipping bucket rain gauge. But upon reflection, the decision to include mechanical equipment with some level of detail prevailed. The reasons behind this decision were mainly three: 1) It is not clear that reliable existing methods will be soon replaced by new ones; 2) In general, postgraduate students are not yet familiar with the use of these instruments and how mechanical information is converted into electrical information; 3) We think that descriptions result necessary to advice readers about instrument limitations and operational problems.

7.2 Wind Measurements

7.2.1 Introduction

Wind velocity is a vector which is among the most traditionally measured parameters in meteorology. There are some anemometers that calculate the resulting vector by adding two horizontal wind velocity components measured in two perpendicular directions. It is also common practice to measure the average horizontal wind in polar coordinates, namely, wind speed (the magnitude of the wind vector) and wind direction (the orientation of the wind vector) (US EPA, 2000.). Incidentally, in meteorology wind direction is defined as the direction from which wind blows, and it is measured in degrees clockwise from true north.

When information on the variability of the wind is required, the peak gust and the standard deviations of wind speed and direction are considered (WMO, 2008). Instruments used for in situ measuring of wind are called anemometers, and in general may be grouped into mechanical and non-mechanical sensors.

In order to average natural turbulent wind fluctuations it is required to extend the averaging during several minutes. Usually, the wind average period for forecasting purposes is 10 minutes, and for climatological statistics is 60 minutes. When it is desired to know peak gusts, a running average filter with a certain integration period T must be used. This period is defined as the **gust duration**, and the maximum observed wind speed over a period of time (such as one hour) is defined as the **peak gust**. For example, if a running filter of 30 s is used to average the output of an anemometer for an observation time of one hour, and the maximum value registered is 100 km/h, it is said that in that particular hour there was a peak gust of 100 km/h with 30 s of duration.

As it happens with other instruments, the time constant could be used to characterize anemometers (Section (2.4.5)). For a first-order system it is the time required for a device to reach about 63 per cent of a step-function change. One limitation of the time constant is that for rotating anemometers it was found in wind tunnel tests that it depends on wind speed (the variable one wants to measure). Then another particular constant not dependent on wind is defined for rotating anemometers: **the response length or distance constant** (l_0). It is the length of fluid flow (the length of the cylinder of air) which has to pass through the anemometer before it has attained the 63 per cent of its final response to a step-function change of the input speed (it is usually measured in meters). The cup anemometer could be considered a first-order filter whose time constant τ is given by

$$\tau = \frac{l_0}{V} \tag{7.1}$$

where V is the mean wind speed. It should be noted that l_0 is independent of the wind speed (Kristensen & Hansen, 2002).

It was also found that the natural frequency ω_0 of the vane system is proportional to V, and then a **distance constant** (l_v) for the vane is also defined as (Kristensen, 1993)

$$l_v = \frac{V}{\omega_0} \tag{7.2}$$

These two constants help compare anemometers performance and fit the most adequate anemometer for the application needs.

7.2.2 Mechanical Anemometers

Current mechanical anemometers usually have two parts: the actual mechanical part that produces a mechanical response to wind excitation and the transducers to convert mechanical variables into electrical ones. The first usually consists of a rotating device to measure wind speed and a vane to measure wind direction; and the second is composed of sensors and circuits which provide electrical outputs from the mechanical inputs.

With respect to the mechanical parts, there are basically two aerodynamic solutions: one is the **cup and vane anemometer**, and the other is the **aerovane**.

Regarding the sensors and circuits devoted to transform a mechanical rotation into an electrical signal, there are several solutions already explained in Section (4.14). Every manufacturer uses his own sensors according to his experience and the anemometer application. For example, when electrical energy is not available and batteries and solar panels are not of great help, as could be in boreal or austral winters, it could be useful to use anemometers that produce their own electrical energy from the wind. This solution requires paying a price because to produce electrical energy it is required to take mechanical energy from the wind. These sensors impose an electrical load on the mechanical device which tends to brake the rotor; as a result, low wind velocities cannot be measured. On the contrary, when it is desired to measure low velocity and electrical energy is not a constraint, sensors with the minimum inertia and frictional torque should be selected.

7.2.2.1 Aerovane

The anemometer known as aerovane consists of a body which aligns itself to the main wind direction due to a vane or tail (Fig. 7.1). The wind speed is measured by a propeller which may have two or more blades. The propeller is placed on the front of the body and its transference has a linear behavior in a certain operating range; it has a starting threshold below which the force produced by the wind is not enough to make the propeller rotate, and extends up to some maximum speed where the rotation can no longer follow the increasing wind speed. The transference of the propeller depends on the angle between the plane containing the propeller and the wind direction; the propeller axis must be always parallel to the wind direction for the calibration to be valid.

Fig. 7.1: Aerovane anemometer. (Photograph courtesy of Ricardo Zuazquita, Naval Hydrographic Service of Argentina).

Some particularly designed propellers have a variation of the rotational speed which is nearly a cosine function of the angle between the plane containing the propeller and the wind direction; it is said that these propellers have a cosine response. Two such propellers orthogonally mounted, and whose axes are in a plane parallel to the ground, may be used to determine the vector components of the horizontal wind. If a third propeller is added vertically, the vertical component of the wind can also be measured. Because the propeller response deviates from the cosine response for angles between 80° and 90°, it is recommended by some government agency (US EPA, 2000) that users of vertical propeller anemometers should consult with the manufacturer on proper handling of the data.

The starting threshold of propellers depends on the mechanical design. When light weight materials such as molded plastic are employed, starting thresholds below 0.5 m/s are achieved. More robust anemometers made of metallic parts may have thresholds between 0.5 m/s and 1 m/s; accuracy is typically about ± 0.3 m/s. Once again the selection of the adequate anemometer depends on the use. Probably for polar regions with high winds and where frequent maintenance is not possible, a robust metallic propeller should be selected.

Because of their relatively quick response times, these sensors (aerovanes) are also suitable for use in determining the standard deviation of the along-wind-speed fluctuations. Some sensors have distance constants of about one meter.

7.2.2.2 Cups and Vane

A cup anemometer is shown in Figure 7.2. It consists of three or four hemispherical or cone-shaped cups mounted symmetrically around a vertical axis of rotation. The three cup anemometer is recommended by some government agencies (US EPA, 2000). These sensors are omnidirectional, which means that the angular speed of the device is independent of the wind direction. They average the wind speed and direction in space, which is a useful characteristic because turbulent fluctuations can be considered, to some extent, as a spatial variable rather than a temporal fluctuation ([Kristensen, 1993). Cup anemometers are robust and easy to operate. The only precaution during installation is to make sure that the rotation axis is in the vertical direction. As with all mechanical anemometers they have an initial speed threshold and the transference has a useful linear range beyond which it loses sensitivity. When cup sensors are combined with a wind vane, they provide information about wind speed and direction.

There has been a considerable interest in understanding the dynamics of the cup anemometer (Kristensen, 1993), particularly in knowing its response when exposed to a turbulent wind. It has been found that cup anemometers 'overspeed', because they respond more quickly to an increase in the wind speed than to a decrease. For example, a robust cup anemometer for use in Antarctica (Fig. 7.3) was tested by the authors and it was found that it has a time constant of 0.5 s with a given increasing wind step, whereas for the same decreasing step the time constant was 4.5 s. Thus, due to this "asymmetric" response, it is expected that the measured mean wind speed of a turbulent wind will be higher than the real one.

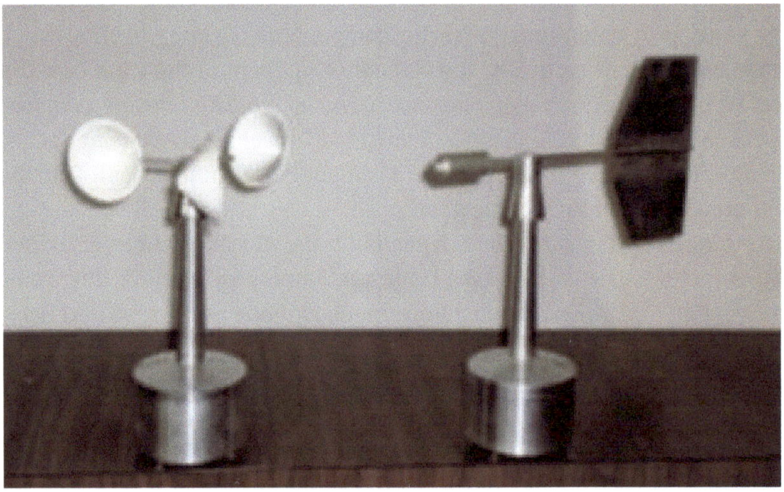

Fig. 7.2: (Left) Cup anemometer. (Right) Wind vane. (Photograph courtesy of Ricardo Zuazquita, Naval Hydrographic Service of Argentina).

The wind vane and the cup speed sensor are spatial filters of the wind speed and direction. When an anemometer is designed, it is important to match the distance constant of the vane to the distance constant of the cup. In this way both speed and direction would "filter" the information in a similar way, or in other words, both constitutive parts of the anemometer "average" approximately the "same wind" (Kristensen, 1993).

Fig. 7.3: Robust cup anemometer for use in Antarctica made of metallic parts. (Photograph courtesy of Ricardo Zuazquita, Naval Hydrographic Service of Argentina).

7.2.2.3 Electrical Information Proportional to Wind Speed and Direction

As described in Sections (4.14.1) and (4.14.2), there are several operating principles to convert rotor and vane mechanical information into electrical information. According to Section (2.2), it is convenient that the anemometer's rotors (cups and propellers) produce an electrical signal proportional to their rotation rate (linear transference) and it would be suitable that this signal be easy to process and store. Beyond a starting threshold, rotors have a range where the relationship between wind speed and rotation rate is linear. Thus, an electrical device with linear electrical output as a function of rotation is a suitable solution. There is always a starting threshold due to shaft friction which could be increased by the transduction method used to obtain the electrical signal.

As stated above, those anemometers that generate electrical energy by means of a rotating shaft produce an electrical braking torque which poses some limitations in terms of achieving low thresholds and quick response times. Then these methods are not the most adequate when the mechanical energy available from the wind is low, because they tend to stop the shaft rotation. This effect is observed as an increase in the threshold of the transference, i.e. as an increase in the inability of the device to rotate when the wind velocity is low.

Optical devices where a light beam is interrupted by a rotating slotted disk are frequently used in applications where low thresholds are required because their small braking torque is due only to the shaft friction. As the cup or propeller rotates the slotted disk chops the light beam generating pulses on the photo detector. The pulsed output frequency from this type of transducer is proportional to the rotation rate. Pulses are easy to handle by digital systems since they are simply counted over a time window. The number of slots used in anemometers may be about 100 to eliminate signal fluctuations which may arise with low wind speeds (US EPA, 2000).

Some devices for measuring wind direction with a vane can also be found in Chapter 4. Potentiometers are a low-power consumption classical solution, but they suffer from maintenance problems associated with the friction caused by the wiper on the potentiometer track. Also, this friction drops the sensitivity of the vane to respond to low velocity winds. When electrical energy to supply instruments is not a constraint and the electronic circuits have some processing capability such as a microprocessor, optical encoded discs are a good choice because they do not have parts to wear out (Section (4.14.2)).

7.2.3 Sonic Anemometers

The first sonic anemometers were developed about 1950 for special purposes (Kaimal & Businger, 1963). The initial application was further extended and at present they are used as an alternative to the traditional mechanical anemometers. Sonic anemometers consist of an array of acoustic transducers and a microprocessor-based electronic circuit. The array of transducers determines the capability of the instrument to measure wind velocity in one, two, or three axes. Most of the anemometer models use three orthogonal axes, each axis having two transducers. In the example shown in Figure 7.4 the transducer supports in the x and z directions are both in the same vertical plane, whereas the transducer support in the y direction lies in a vertical plane perpendicular to the first.

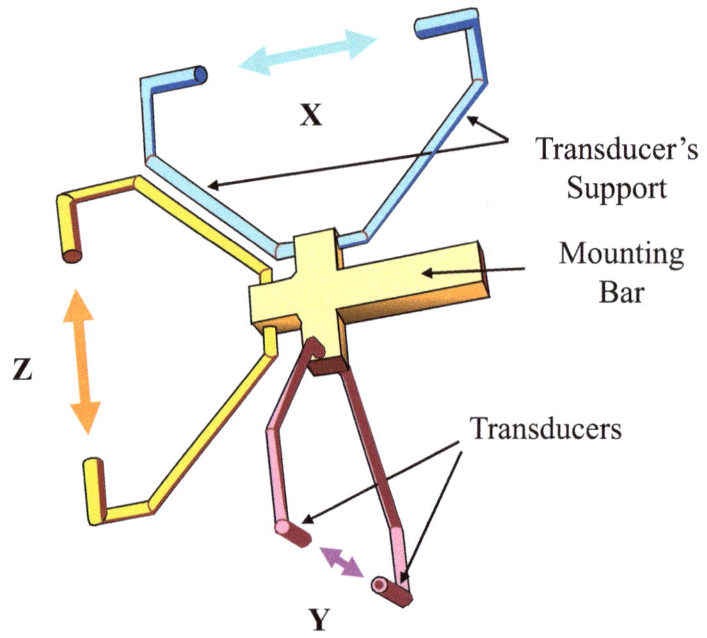

Fig. 7.4: The mounting bar holds the three transducer's supports. Each support keeps two faced acoustic transducers. The mounting bar is horizontal, Z and X are in a vertical plane, and Y is in other vertical plane, perpendicular to the first. The instrument's axis measuring the vertical component of wind is Z.

Under high speed horizontal wind conditions, the array that has three orthogonal axes produces some flow distortion because of the turbulence created by the transducer supports. In such cases, three non-orthogonal axes may be used where only the horizontal wind components are of interest (Fig. 7.5).

The ultrasonic transducers consist of piezoelectric crystals enclosed in metal housings; they are about 10 mm in diameter and 25 mm in length. Transducers are acoustically isolated from the housing and sealed to prevent contact with the outside environment.

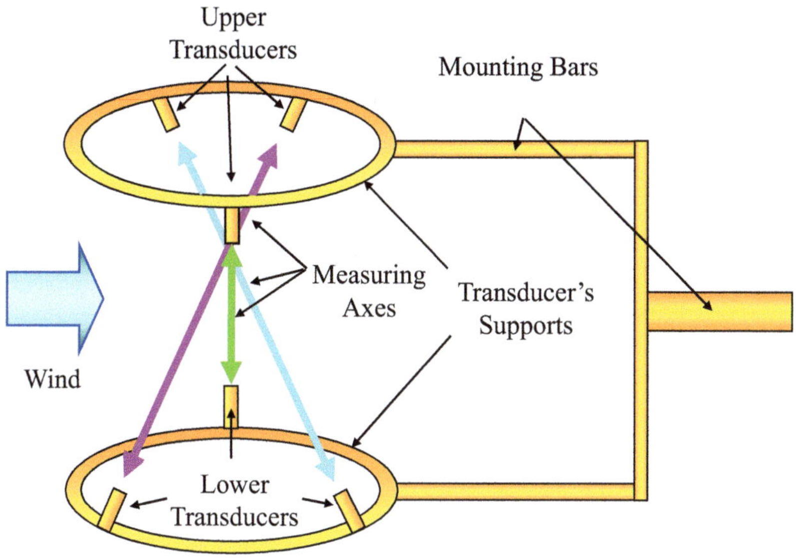

Fig. 7.5: The transducer's supports minimize the horizontal flow distortion. Each of the upper transducers faces one lower transducer.

By means of the electronic circuits, acoustic pulses are generated at one transducer and received by the opposite one. Figure 7.6 represents two transducers, A and B, separated by a distance L. The periods of time t_{AB} and t_{BA} are, respectively, the travel time required for the acoustic wave to travel the distance L from the transducer A to B and vice versa (Eqs. (7.3a) and (7.3b)). Based on the flying time of the pulses, the microprocessor calculates the wind velocity (Section (5.4.1)).

The system of equations (7.3a) and (7.3b) can be solved for the wind velocity component v (Eq. (7.3c)

$$t_{AB} = \frac{L}{c+v} \tag{7.3a}$$

$$t_{BA} = \frac{L}{c-v} \tag{7.3b}$$

$$v = \frac{L}{2}\left(\frac{1}{t_{AB}} - \frac{1}{t_{BA}}\right) \tag{7.3c}$$

Note that v is the velocity component of the flow velocity (**V**) in the direction of each pair of transducers ($v = V\cos\theta$), θ being the angle that **V** forms with the axis of the two faced transducers, and c the speed of sound in air (which depends on the air temperature). These calculations are repeated for each of the orthogonal axes.

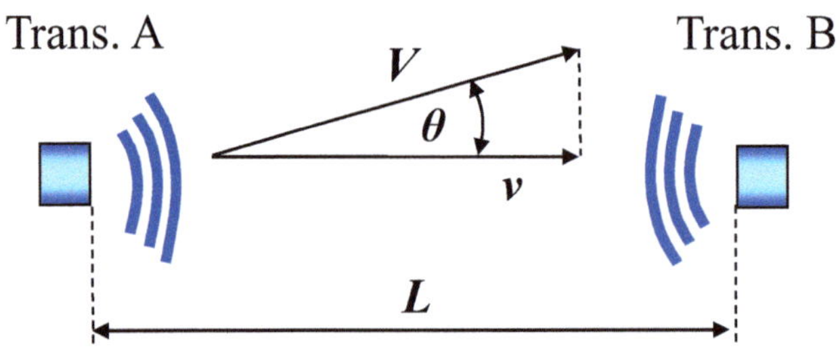

Fig. 7.6: Two faced transducers forming a measuring axis. A pulse is sent from A and received at B and vice-versa. The times of flight in both directions are measured. L is the distance between transducers and v is the component of the wind velocity **V** along the axis direction. θ is the angle between wind velocity and the axis direction.

The sampling rate is generally user selectable and the instrument can record as many as 10 samples per second. In turn, each one of these ten samples represents the average of several measurements (for example, 10 measurements).

This kind of anemometer may record each single component of wind velocity and does not require moving parts. Because each axis measures one component of the wind velocity vector along known direction axes, the composition of the total wind vector is straightforward. The maximum wind velocity may be between 15 and 60 m/s (54 and 216 km/h), depending on the array sensor geometry. Accuracy is typically ± 0.1 m/s.

7.2.4 Comparisons of Anemometer Measurements

A pair of examples will be analyzed to show that in some applications even different operating principles could give similar results, whereas the same principles could result in similar, but not identical measurements, in other applications. It should be noted that in the first case comparisons are made between averaged measurements, whereas in the second, dynamic behaviors are compared.

7.2.4.1 Comparison of an Aerovane and a Sonic Anemometer

Aerovane and sonic anemometers were compared for a particular aeronautical application (Fisch, 2010); the average and maximum wind speeds were analyzed. The aim of this study was to find the relations between both measuring systems (aerovane and sonic anemometers) in order to substitute the mechanical instrument by a sonic one. The aerovane had been working for years and historical data recorded with this technology had to be related to the new one acquired by the sonic anemometer. By means of this study, the values which relate both data sets were found.

The comparative research consisted in analyzing data acquired during 6 days during the high wind season of a rocket launch center. The analysis considered both the average and maximum wind speeds for one and ten-minute time intervals.

The results showed that there was no significant difference between the results obtained from both instruments and for both time intervals (one and ten-minute time intervals). The data analysis of the two instrument time series showed that sonic measurements were a little higher than those from the aerovane; the differences for the averaged wind speed values were around 0.5 m/s. For maximum wind speeds, differences increase to a value around 1.0 m/s. Therefore, the researchers considered that to join the past data set acquired with the aerovane and the future data set to be acquired with the sonic anemometer, they only needed to add the above differences to the old data set.

This is an interesting result which shows that **for some applications** anemometers with different technologies may give comparable results.

7.2.4.2 Comparison Among Different Cup Anemometers

Wind power plants feasibility studies require knowledge of wind energy. An investigation of the characteristics of five commercial cup anemometers being used in wind energy has been carried out under wind tunnel and laboratory conditions (Pedersen et al., 2006). All anemometers had three conical plastic cups but differ in their mechanical design (cup size, shaft, bearing and body). They were analyzed to get their torque and angular response characteristics, and friction. The representative parameters of such devices were fitted to two different cup anemometer numerical models, which were used to simulate time responses under free field operational conditions.

It was observed from this study that even for similar anemometers, the details on the design of the cup, rotor, shaft, bearings and body have their influence on the angular and dynamic characteristics of the devices.

This work also shows 30-second time traces of the five cup anemometers compared to one propeller anemometer, all of them working at 8 m/s wind speed. The graphs permit visualizing the magnitude of the so called "inherent turbulence", which characterizes most cup anemometers. The figures show that the variations of the cup anemometer readings are significantly higher than those of the propeller; the propeller readings look like a low pass filtered version of the cup readings.

These results show that propeller anemometers tend to spatially and temporally filter wind speed fluctuations, thus giving average values of wind speed.

7.3 Rain Gauges

7.3.1 Introduction

Rain gauges are instruments used to gather rain precipitation. Also known as pluviometers, they are basically composed of two parts: a rain collector (usually a funnel) and a rain meter. The second may be as simple as a transparent graduated container (Fig. 7.7) that measures the volume of water precipitated in a given time period, for example 12 hours.

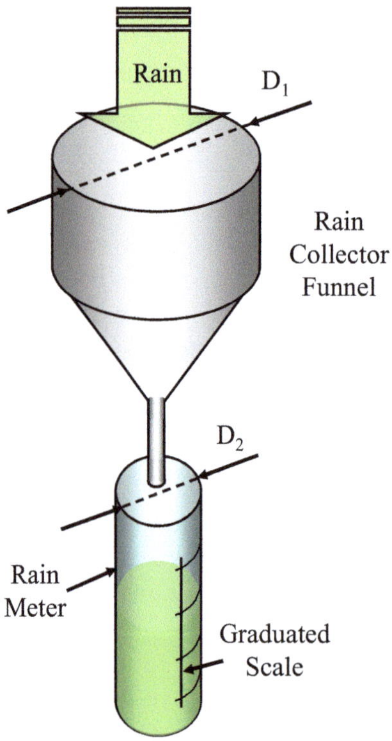

Fig. 7.7: A simple rain gauge or pluviometer

The rain volume measured in the container is referred to the area in which it was collected; then rain precipitation is generally expressed in millimeters. For example, if the graduated scale shows a height of h_2 mm that correspond to a collected volume of rain V_2, then the rain fallen (h_1) in mm is given by

$$h_1\,(\text{mm}) = \left(\frac{D_2}{D_1}\right)^2 h_2 \tag{7.4}$$

where D_1 and D_2 are the diameters of the funnel and the graduated container, respectively.

This simple description corresponds to a manual rain gauge. This elemental way of evaluating the fallen rain requires that a person pour off the device each time it reads the scale and enter the measured value in a spreadsheet at fixed time intervals, usually from once to three times a day. It should be noted that the data gathered in this way is an average value of the precipitation over the period between two readings. This average may be very useful for agriculture purposes but could not be good enough for other applications, e.g. for predicting floods in a city for evacuation alerts.

7.3.2 Tipping Bucket Rain Gauge

A more automatic way of measuring the rain precipitation is achieved by means of the tipping bucket rain gauge. It is also composed of a funnel that collects the rainfall and a different rain meter. Figure 7.8 shows a photograph of the internal parts of a typical tipping bucket rain gauge where the tipping bucket can be observed. The bucket is separated at the center forming two symmetrically divided containers and it can rotate round a pivot between two rest positions.

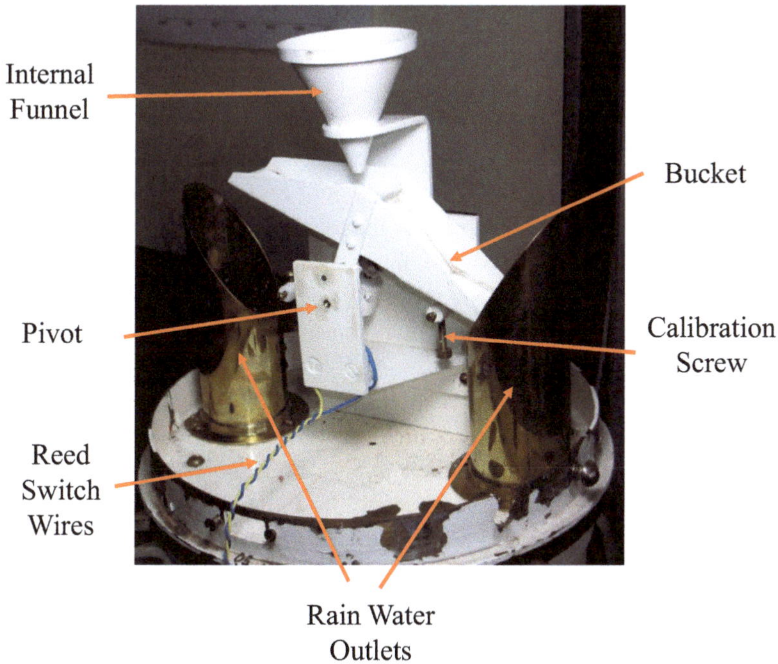

Fig. 7.8: Tipping bucket rain gauge.

The external rain collector funnel discharges into the internal funnel, which in turn fills one of the bucket's containers. As it receives water the center of mass of the whole bucket changes; and it is forced to tilt towards the opposite rest position. A magnet attached to the bucket then activates a reed switch (Section (3.7.19)) when passing close to it (Fig. 7.9).

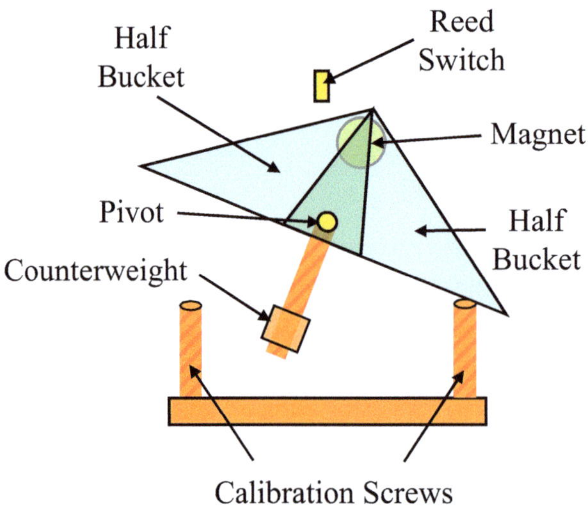

Fig. 7.9: Schematic of the bucket with the sensor (magnet and reed switch). Calibration is performed by adjusting the counterweight and the calibration screws.

The system is calibrated by means of a counterweight and a couple of calibration screws. Calibration is performed in such a way that each time the switch is activated a precise amount of water is discharged into the water outlets. Both outlets discharge in a common container placed below the rain meter. This container accumulates all the fallen rain, which permits the number of switch activations to be verified to see if they agree with the gathered volume.

Modern rain gauges have an electronic circuit with a precise clock. Thus, the amount of switch activations in a period of time is counted and recorded in memory. Generally, this period may be user selected; if a short period is selected, say ten minutes, it is possible to register the rain rate of strong short time precipitations.

In some applications such as flood warning in cities even ten minutes may be too much time, because to perceive a tendency at least three points are needed, which implies waiting at least half an hour. Then, for these kinds of applications it is required to register and transmit the instant at which the bucket is tipped. The instrument shown in Figure 7.8 was designed to evaluate the performance of the discharge ducts of a city during strong rain events. For this reason it has both possibilities: counting the dumps on a fixed time and registering the precise time when the bucket rotates. For flood warning this information has to be transmitted in real time. Figure 7.10 shows a precipitation of about 10 mm in 15 minutes. In order to get a discretized temporal axis, dumps were post processed and integrated over a period of one minute. After a delay of 15 minutes the water arrives to a discharge channel and its level begins to rise; 20 minutes later it reaches about 2 m.

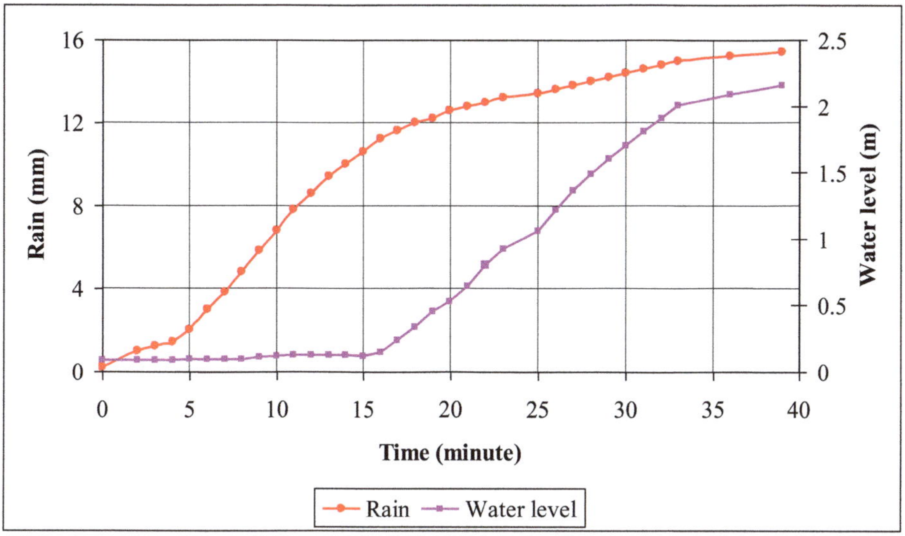

Fig. 7.10: Rain precipitation of about 10 mm in 15 minutes. The water arrives at a discharge channel whose level rises 2 m in 20 minutes.

The resolution of a tipping bucket is given by the volume of water that makes the bucket tip up. This volume is referred to the catchment area of the funnel and expressed in mm. The World Meteorological Organization (WMO) recommends rain tipping gauges with a catchment area of 200 cm^2 and a resolution of 0.1 mm.

Because the rain gauge meter performance is based on a seesaw mechanism, it has to be installed leveled to work correctly. Some models provide leveling screws and a bulls-eye level for easy and precise adjustment in the field. For use in cold weather some models are provided with a heater to prevent the seesaw mechanism from freezing. A proper functioning of the device requires a periodical maintenance, birds perching and insects could clog the funnel. Figure 7.11 shows a top view of the internal funnel after a storm; insects have obstructed water flow in spite of the screen that covered the funnel. Students should be aware that these problems could occur and try to prevent them from happening.

Fig. 7.11: Internal funnel obstructed by insects. It causes the overflow of the funnel and the distortion of measurements.

7.3.3 Rain Gauges Without Moving Parts

7.3.3.1 Siphon Rain Gauge

Most of the earth's surface is covered with water, and most of the rain information collected at sea in the past was due to ships. Rain gathered by ships suffers some problems due to the air flow pattern over the ship's superstructure: the roll, pitch and translation motion; the sea spray that is caught by the rain gauge, etc. It was found that placing rain gauges in meteorological buoys anchored at fixed positions in the ocean was a good solution to collect more and better data at sea. It is obvious that tipping bucket rain gauges cannot be used for this purpose due to buoy's motion; a particular rain gauge design with no moving parts is required for buoys.

There are some possible designs. One of them has the following components: a catchment funnel, a measuring device and a siphon to discharge the accumulated water (Fig. 7.12). The precipitation caught by the funnel is conducted to the measuring tube where the height of the water column is measured at fixed periods of time. The precipitation volume is referred to the funnel catchment area and expressed in millimeters.

The measuring tube and the left vertical tube of the siphon are communicated at the bottom, so water has the same level in both. When the water reaches the siphon's upper bend level, a dumping triggers and the measuring tube is discharged through the siphon up to the zero level.

The height of water in the measuring tube may be measured in different ways. A capacitive measuring device is frequently used (Case & Michelena, 1981; Holmes & Michelena, 1983). In this case there is a central electrode coated with an insulating material. The electrode is the inner conductor of a cylindrical capacitor, the insulting material is the dielectric, and water works as the external conductor of the capacitor (Section (3.7.8)). Because in cylindrical capacitors capacitance is directly related to the capacitor's length, measuring the capacitance the height of water is known. For this purpose, a suitable electronic circuit is used to measure the capacitance and record the data.

Controlled heaters allow operation at very low temperatures and the collected snow is melted and measured as rain precipitation. The siphon process that empties the measuring tube, usually, lasts less than a minute.

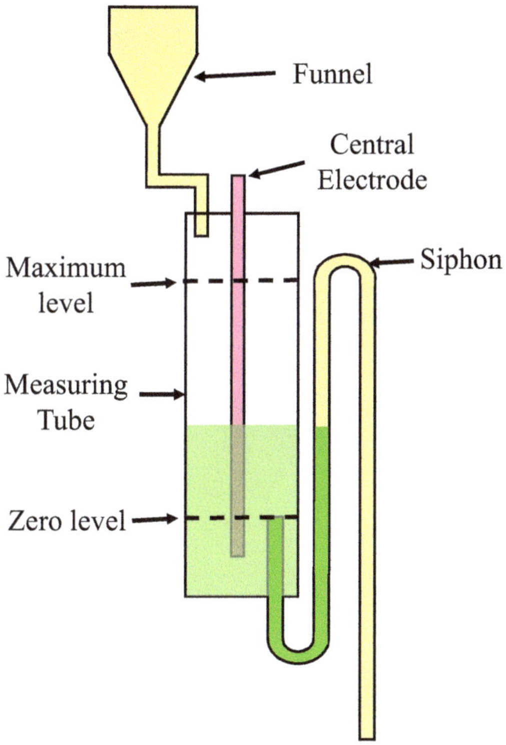

Fig. 7.12: Siphon rain gauge.

7.3.3.2 Optical Rain Gauges

This technology is based on the alterations produced by raindrops in the propagation of a plane optical wave between a transmitter and a receiver. A raindrop illuminated by a plane wave creates a scattered field that can be considered as spherical waves emitted from the center of the raindrop. An interference pattern created by the incident plane wave and the scattered spherical wave, which depends on the raindrop size, appears at the receiver plane. Interferences are perceived on the receiving plane as shadows, which cause fluctuations in the received light intensity. These signals are known as optical scintillation.

Scintillations due to falling raindrops along the beam path allow rainfall rates to be estimated. The signature of the raindrop-induced scintillation is acquired on board the instrument, and its characteristics used to estimate rain parameters. This technology has been granted numerous international patents.

In early studies He-Ne lasers were used as transmitters (Ting-i Wang et al., 1979), photo detectors as receivers, and a distance of several meters between both was adopted. Present instruments (Fig. 7.13) based on this principle use a modulated near-infrared LED source as transmitter, a photodiode as a receiver, and the distance between source and receiver is about half a meter or even less (Ting-i Wang and J. D. Crosby).

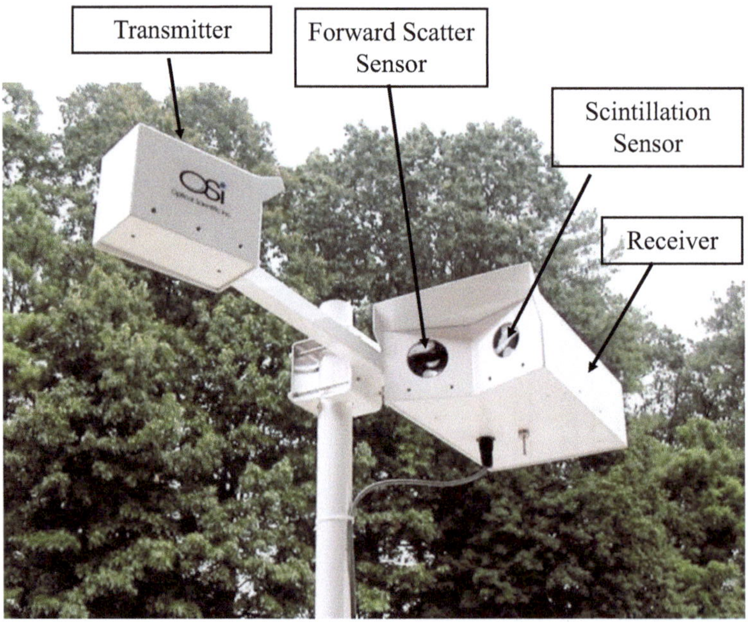

Fig. 7.13: Optical Rain Gauge with scintillation and forward scatter sensors. (Photograph courtesy of Optical Scientific, Inc. Gaithersburg, MD, USA).

The horizontal beam makes the instrument sensitive to vertically falling raindrops, but not to horizontal (wind blown) ones. The height of the horizontal slot aperture of the receiver makes the sensor sensitive to precipitation particles but not to sand and dust. The short distance between transmitter and receiver permits both elements to be mounted on the same base, facilitating accurate alignment and easy transport and installation. Because moving parts are not needed, as is the case of the tipping bucket rain gauge, the optical rain gauge is suitable for deployment at ocean platforms such as ships and meteorological buoys.

Variants of the Optical Rain Gauges (ORG) that can differentiate between rain and snow are known as Optical Weather Identifiers (OWI). Some versions of the OWI (Fig. 7.13) combine the scintillation technology with an off-axis sensor that measures the optical forward scattering, which is sensitive to fog and drizzle. In order to discriminate ice particles from water droplets, some models add an acoustic sensor. All the information gathered by the three sensors is processed by a particular algorithm; field results show that the combination of the three technologies onboard a single instrument discriminates ice pellets and hail stones quite well (OSR 0303, no date). Various models of ORG and OWI have been used in airports, roads and hydrology applications.

7.3.3.3 Weighing Gauge

The principle of the weighing gauge consists in weighing the precipitation caught by a bucket. It is able to determine the precipitation of rain, snow or hail. For this purpose it uses a high-precision load cell to weigh the bucket contents and a temperature sensor for compensating the temperature changes in the load cell. The rainfall rate is obtained as the difference in rainwater accumulated over a given time interval. The bucket has to be emptied after a given period of precipitation to prevent it for overflowing. It can be emptied by hand of with the help of a pump. There are reports that mention some attempts to automatically drain the bucket but with limited success (Nystuen, 1999).

For temperature below zero, it is recommended to use antifreeze liquid because if the collected precipitation freezes completely, the collecting bucket could suffer severe deformations.

7.3.3.4 Disdrometers

They measure the drop size distribution and the rain rate. There are different operation principles but the most long-established disdrometers are based on the generation of an electric signal by every rain drop that hits the surface of a sensor.

One of these instruments is known as the **Joss-Waldvogel disdrometer (JWD)** (Fig. 7.14). It transforms the vertical momentum of impacting raindrops on a sensor into electrical pulses whose amplitudes (U_L) are given by

$$U_L = k\,D^n \tag{7.5}$$

where D is the drop diameter, and k and n are constants.

For a given particular design the sensor signal is amplified and the drop diameters in the range from 0.3 to 5.4 mm produce pulse amplitudes from 0.3 to 10 V, respectively; for this range $k \approx 0.02586$ and $n \approx 3.7$ (a few changes in the design would produce different constants). The manufacturer provides the calibration table for 127 drop size intervals, but in order to reduce the amount of data a pulse height analyzer classifies pulse amplitudes into 20 size classes; the instrument records the number of counts on each class for a 30 s period.

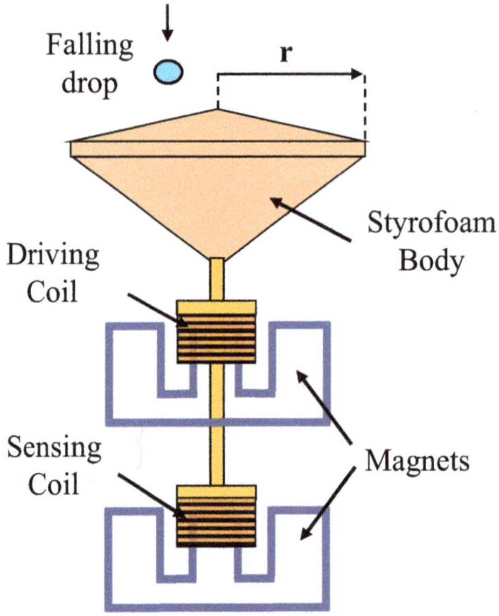

Fig. 7.14: Joss-Waldvogel disdrometer.

The sensor consists of two coils fixed to a conical Styrofoam body which receive the raindrop impact; one coil senses the drop impact and the other resets the initial position. The sensing and driving coils are placed in magnetic fields produced by two magnets. The kinetic energy of the drops produces a displacement of the conical body which moves de coils. On the moving sensing coil a voltage is induced according to Faraday's law (Section (3.7.12)). This signal is amplified and injected into the driving coil such that a force that opposes the displacement is generated. In this way the excursion of the sensor is reduced. Thus, it takes a short time to return to the original initial position being ready to measure a new impact. For a better understanding of the working principle see Sections (4.4.2), (4.4.3) and (4.13.2). The upper cone makes the raindrops slide so that the droplet impact area stays without water. If water remains on the top it will reduce the impact of the next drops on the sensor, modifying the sensor signal (Stijn de Jong, 2010).

Once the relation between the sensor output signal and the drop diameter (D) is known by means of a calibration process, the volume of fallen water (V) can be calculated:

$$V = \sum_{all\ drops} \frac{\pi D^3}{6} \tag{7.6a}$$

Because the radius of the sensor (r) is known, V can be referred to the surface of the sensor to express the rain in millimeters of height (h_1):

$$h_1\ (mm) = \sum_{all\ drops} \frac{D^3}{6\, r^2} \tag{7.6b}$$

The amount of energy of a drop hitting the sensor depends on its mass and velocity. An assumption of the method is that the drops that hit the sensor have reached the terminal velocity (velocity remains constant). It is the velocity at which the resultant of upward and downward forces on the drop is null. Drops are not perfect spheres and the drop size is expressed with an equivalent diameter; it is the diameter of the drop if it would have been a perfect sphere.

Some precautions should be taken about the ambient acoustic noise of the place where the instruments will be installed. It has been reported (Tokay et al., 2003) that this type of disdrometer is sensitive to background audio noise. While a noise level of 50 dB or less had little effect on signals corresponding to drops, a noise level of 55 dB reduces the detected number of those diameters ranging from 0.3 to 0.4 mm. When noise increased to 70 dB, the detection of drops of 0.3–0.8 mm in diameter was almost completely suppressed.

7.3.3.5 Piezoelectric Disdrometer

The conical cap that prevents the puddle effect of the raindrops remains similar to the previously described one, but the generation of an electrical signal proportional to the raindrops energy is based on a piezoelectric disc (Stijn de Jong, 2010). A schematic of the sensor is shown in Figure 7.15.

The operating principle and equations are similar, but the way in which the impact of the drops is measured changes. This disdrometer was an attempt to make these instruments cheaper but this design still needs further improvement.

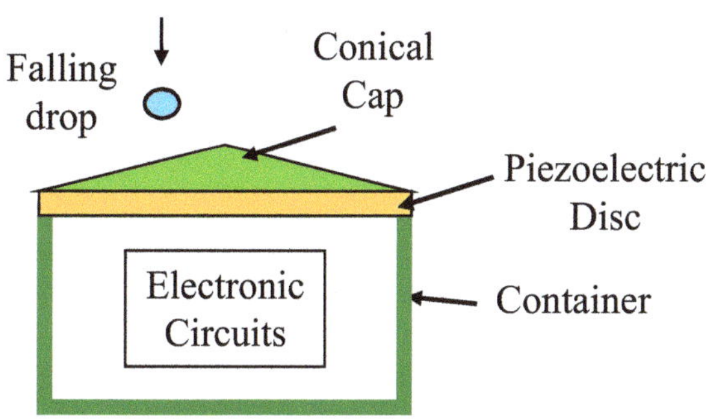

Fig. 7.15: Piezoelectric disdrometer

7.3.3.6 Acoustic Rain Gauge (ARG)

When rainfall hits the surface of a body of water a particular underwater sound is generated; analyzing this sound a quantitatively measure of the rain can be obtained. The rainfall-generated acoustic underwater signals are louder than other sound sources such as breaking waves, biology, etc., and can be distinguished from them. The sound signature for different drop sizes has been studied (Ma & Nystuen, 2005), thus the drop size can be estimated from the sound signatures. Estimates of drop size distributions can then be used to estimate the rainfall rate.

The frequency range of rainfall induced underwater sounds is from 1 to 50 kHz, the dominant sources of sound from raindrops being tiny bubbles trapped underwater during the splash. Large raindrops produce larger bubbles with a lower frequency sound (2-10 kHz). The sound produced by large raindrops is highly correlated with the rainfall rate and can be used to detect and measure it.

The ARG used by Ma and Nystuen (2005) consisted of a hydrophone, amplifiers, band pass filters, CPU, a memory card, a data output and a battery pack. The signal is processed on board the instrument and the power spectra of several time series are averaged. The spectra are evaluated by means of special algorithms to detect the acoustic signature of the rainfall.

Results of several months of field work were compared with other rain gauges with good agreement. It has also been established that the acoustic signal from rain is wind speed dependent; at low frequencies the signal from rainfall is polluted by breaking waves sound. It is expected that this kind of rain gauge will be useful in ocean meteorological buoys where long time unattended instrumentation is needed.

7.3.4 Comparative Analysis of Rain Measuring Systems

An interesting work (Tokay et al., 2003) compares four impact Joss–Waldvogel disdrometers (JWD) and 27 tipping bucket (TB) rain gauges operated at 11 different sites during 44 days. Disdrometers were collocated with one or two rain gauges at all sites; collocated refers to gauges less than 2 m apart.

Analysis of data measured with two TB of the same model collocated at eight different sites, showed that the total mean percent errors were 1.3, 3.0, 3.1, 3.9, 4.2, 8.2, 13.8 and 16 %. Results from other field experiments, presented in the same work, show total mean percent errors of 0.8, 2.6 and 6.3 %. For those not already familiar with rain measurements, these results could be indicative of the errors that they could expect when measuring with instruments of the same manufacturer and model. Therefore, if instruments with different specifications or from different manufacturers are used in the study, perhaps the measurement spread could increase. Also, it is worth mentioning that out of the 27 TB deployed, some suffered problems, few of them due to vandalism, others to known causes and several to unknown failures.

This indicates that if it is desired to avoid loosing valuable information, installing a redundant amount of instruments is advisable.

Tipping buckets (TB) rain gauges and Joss-Waldvogel disdrometers (JWD) were collocated at four sites. In this case, instruments with different operating principles were compared. In two of the four installations, disdrometers registers showed a lack of small drop data. It was attributed to the noise level generated by the drops hitting the surface of the metal roof at these sites. In another installation lack of particular sized drops were suspected to be due to the noise of a diesel generator used to power the equipments. For the total period of the research, in two of the four installations the disdrometers measure in excess of the TB with mean percent errors of 18.2 and 8.7 %. In the other two sites they measured 18.6 and 5.8% below. As expected, when two different principles are used the spread in results increases. A valuable conclusion of this work is that the site to install the JWD requires being far from any sound background noise source because it reduces the JWD ability to detect small drops effectively.

We consider it necessary to mention some experiences reported by the authors (Tokay et al., 2003) because they are very enlightening for future users of these instruments and they can help to prevent foreseeable failures. Furthermore, we strongly agree with them.

Urban areas should be avoided to prevent vandalism.

Collocation of multiple gauges at a short distance is required.

Disdrometers should be collocated with at least two gauges.

Gauges should be calibrated prior to and at the end of each field campaign.

Gauges and disdrometers should be visited weekly for short field campaigns and biweekly for annual or longer operations.

7.4 Instruments for Measuring Solar Radiation

There are several kinds of instruments to measure direct and indirect radiation coming from the sun. They have different purposes and their geometries and optical properties are suited for these specific purposes, but all of them have some common features in their operating principles. Also, they have a common component called "radiation detector". Therefore, it has been considered that the simplest way to explain these instruments begins with the description of the detectors.

7.4.1 Radiation Detectors

The name of "radiation detectors" is given to sensors (Section (2.1)) that convert the radiation signal into an electrical signal. There are two methods to get the electric output in radiation detectors. In one of them, the radiation is first converted to a differential temperature; in the other, radiation is directly converted to an electric parameter.

For instruments measuring radiation over a wide spectral range (pyrheliometers, pyranometers and pyrradiometers), transducers that convert radiation into temperature are frequently used (IAMAP, 1986). Most of the standard or reference instruments are of the thermal type. Instead, when measurements in particular spectral bands are required (sunphotometers), transducers with photoelectric outputs are preferred (IAMAP, 1986).

7.4.1.1 Radiation to Temperature Conversion

The incidence of solar radiation on a thermal resistor produces a temperature difference along the heat flow direction that is directly proportional to the received radiant energy. Thus, measuring the temperature difference along the resistor allows knowing the solar radiation. As stated in Section (4.7.4.2), thermocouples are well suited for differential temperature measurements; also, resistance thermometers and other temperature sensors could be used (Section (4.7)) to measure temperature on thermal resistors.

Thermal resistors for this application should absorb all the incident radiation arriving at them. With this purpose they are coated with special paints whose absorption coefficients are very close to 1 (they have very little reflectance). Furthermore, these resistors should not irradiate; for this reason they are of low thermal resistance so that the resulting temperature difference is low, thus reducing infrared losses (Fig. 7.16).

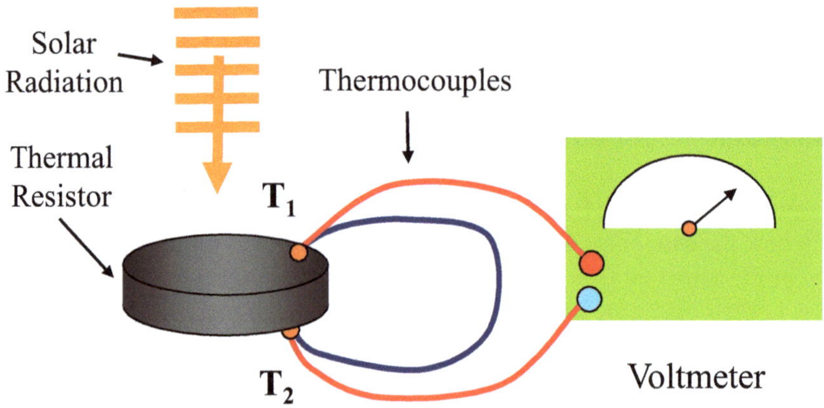

Fig. 7.16: Radiation to temperature conversion with a thermal resistor

With the purpose of getting an electrical output, the difference in temperature is measured by thermocouples which transform thermal energy into electrical energy. Only two thermocouple junctions are schematically depicted in the figure, but in a real device several junctions are placed on both measuring points summing up their voltages. This multiple junction arrangement, called a **thermopile**, increases signal level, decreasing noise influence. The junctions placed on the face receiving the radiation (temperature T_1) are called hot junctions, while those placed on the shaded face (temperature T_2) are called cold junctions. In general, cold junctions are on a metal body called a heat-sink which is at a constant reference temperature.

In this detector, the signal input suffers two changes, in the first step the incident light is converted into temperature, and in the second, temperature is converted to electricity.

7.4.1.2 Radiation to Electric Energy Conversion

In order to convert radiation directly into electric current, sensors based on the photoelectric effect are used; among them, photo-resistors and photodiodes are frequently utilized.

Photo-resistors are also known as light dependant resistors (LDR). When light with enough energy impinges on the semiconductor material of which the LDR is composed, photons produce free electrons. Thus, these special resistors exhibit a decrease in its resistance when illuminated.

Photodiodes are semiconductor devices which contain an interface in their crystal structure called a p–n junction. Light impinging on the junction generates electron–hole pairs, which gives origin to a photocurrent approximately proportional to the light intensity. Photodiodes can be operated also in a photovoltaic mode, generating a voltage when illuminated, but in this case light – voltage transference is nonlinear.

In LDR and photodiodes, radiation is directly converted to an electric parameter, without going through the conversion of radiation into temperature, as it was in the previous case.

7.4.2 Instruments for Total Radiation Measurements

Instruments to measure total radiation are those measuring the complete spectrum of the sun, the most known are pyrheliometers, pyrometers and albedometers.

7.4.2.1 Pyrheliometers

A pyrheliometer measures the direct component of solar radiation at normal incidence. The field of view of pyrheliometers is small to minimize the registration of circumsolar radiation; it frequently is between 6 and 10 degrees. For this reason, the radiation detector is placed at the lower end of a tube which restricts the view angle of the instrument by means of the field stop and the aperture stop (Fig. 7.17). There are different types of pyrheliometers based on similar ideas to measure the direct solar radiation.

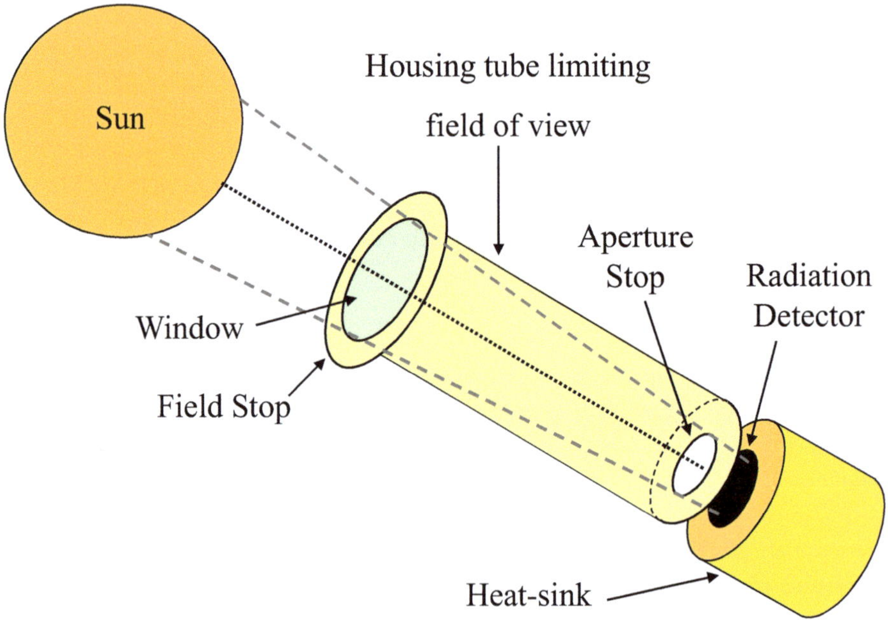

Fig. 7.17: A pyrheliometer

7.4.2.1.1 Reference Pyrheliometers

They are not found working as field instruments, nor are they of recent design, but because their design did not significantly change, the description of their operating principle helps to understand modern instruments.

1.- The Angstrom Compensation Pyrheliometer is a reference standard instrument. It has two strips of manganin (a resistive material whose resistivity is quite constant with temperature). These strips are coated with a black absorber and one of them is directly exposed to solar radiation while the other is protected from sunlight. The strip in the shadow is electrically heated until the temperatures of both strips equal.

When both strips are at the same temperature it can be said that the sun energy arriving at the detector per unit time equals the electric power provided by the current (i) to the manganin electric resistor, so Eq. (7.7) can be used to calculate the solar radiation (S). Remember that power dissipated on a resistor is proportional to the square of the current through it (Section (3.7.7)).

$$S = k\,i^2 \tag{7.7}$$

where k is a constant that accounts for the geometrical and optical constructive factors of the pyrheliometer as well as for the resistor value. Temperature difference is measured by differential thermocouples attached to both strips.

2.- The standard instrument of the Smithsonian Institution (USA) consists of two blackened cavities surrounded by a special chamber through which water flows at a constant rate. One of the cavities is exposed to solar radiation while the other is shaded and heated through electrical resistors (IAMAP, 1986). A null difference in cavities' temperatures is achieved by adjusting the electrical input to the resistive heaters of the shuttered cavity. Again the equality between solar and electrical energy is used to show up the solar radiation. Temperature difference is also measured by means of thermocouples.

In both standard instruments, obtaining a true irradiance value from the measured electrical power is a difficult task. It requires the accurate knowledge of the electrical power through the resistor, the aperture area (through which the radiation enters the tube), the amount of heat losses through the metallic wires of the resistor and sensor, the absorption coefficients of the cavity receiver as a function of the spectral range of the incident solar radiation, etc. (IAMAP, 1986).

Briefly, measuring solar radiation with this kind of instruments implies comparing the heat produced by the radiated energy with the heat generated by a known electric energy source. It is assumed that all the received energy is converted into heat, and that no heat loss exists. Because these conditions are difficult to achieve, a number of correction terms has to be employed in the calculations to account for deviations from the ideal values.

Pyrheliometers are usually supplied with mechanical adjustments to allow the movement of the housing tube in elevation and azimuth to point the instrument to the center of the sun. These adjustments may be done by hand or with some automatic device.

7.4.2.1.2 Modern Pyrheliometers

The operating principle of modern pyrheliometers is similar to the one described above, thus the following explanations are based on Figures 7.16 and 7.17.

The window of modern instruments (Fig. 7.17) is an optical filter that defines its spectral range. Generally they are chosen so that approximately 98 % of the solar radiation spectrum is transmitted through the window and impinges on the detector.

As stated in Section (7.4.1.1), the detector consists of a thermal resistor and thermocouples. The top surface of the thermal resistor is black painted; this paint has a porous finish that helps to absorb more than 97 % of the incident radiation. The paint is selected to have a constant absorption as a function of the radiation wavelength. Some manufacturers state that absorption is practically constant within 2 % on the spectral measuring range (Kipp & Zonen B.V., 2008). The characteristics of this paint have to be very stable over long periods of time because the transference of the instrument strongly depends on it.

The radiation entering the window arrives at the thermal resistor warming its surface. Heat flows through the resistor to the heat-sink which may be the pyrheliometer body (Fig. 7.17). The temperature difference along the resistor is measured by a thermopile composed of several thermocouple junctions. The thermopile generates a voltage directly proportional to the temperature and thus to the radiation. For this reason, the ideal transference of the instrument is a constant value called the sensitivity.

The constructive characteristics of the thermal resistor and the thermopile are somehow variables; therefore, each pyrheliometer has to be individually calibrated.

Direct solar radiation (E_{DC}) expressed in [W/m²] is calculated by

$$E_{DC} = \frac{V_{\text{out}}}{S} \tag{7.8}$$

where V_{out} is the output voltage of the thermopile [V] and S is the sensitivity of the instrument [V m²/W].

Even when instruments are designed for outdoor use and are almost hermetically sealed, sometimes, due to a difference between inner and external pressure, some moisture can enter the instrument housing. Condensation of the moisture inside the window decreases the radiation arriving at the sensor. This problem is solved using a desiccant (silica gel) inside the instrument. Also dust deposited outside the window would result in radiation underestimation. In order to keep accurate measurements the desiccant must be changed periodically and the window cleaned frequently.

Ultraviolet radiation downgrades the black paint on the sensor surface; therefore, for assuring the measurements quality, a yearly or every two-year recalibration is required. Recalibration is performed by comparison with a reference pyrheliometer.

In order to always point to the center of the sun, pyrheliometers can be mounted on an automatic sun follower called a tracker; it may be either of clock-driven or computer-controlled type. Some clock driven sun trackers have a GPS integrated receiver to automatically configure location and time data. Because the time is periodically checked by the GPS receiver any potential clock drift is corrected.

Other sun trackers are based in photo detector devices that sense the intensity of the sun on a surface perpendicular to the housing tube; they are called active sun trackers. The active trackers keep the instrument pointing to the maximum sun intensity by means of some electronic and software working on a two degrees of freedom mechanical motorized device.

7.4.2.2 Pyranometer

Ideally, a pyranometer measures radiation in an elevation angle ranging from horizon to horizon and over an azimuth range of 360 degrees. In other words, it is an instrument for the measurement of all hemispherical solar radiation, direct and diffused (IAMAP, 1986; Kipp & Zonen B.V., 2013). In this instrument the sensor is horizontally placed so that the radiation arriving to it comes from the sun and the sky.

One model of pyranometer has a radiation sensor based on a thermal resistor, as that described for pyrheliometers (Fig. 7.18a). When radiation impinges on the sensor the developed heat flows through the sensor to the pyranometer housing which works as a heat-sink; the temperature difference on the resistor is measured by means of a thermopile. The temperature on the thermal resistor is affected by wind and rain; also, there are some long-wave radiation losses. Therefore, in this design, the sensor is protected by two glass domes to decrease the influence of the above factors.

Other pyranometers have a different shape for the radiation to thermal converter device; they have a disc divide into 12 equal circular sectors which are alternatively painted black and white (Fig. 7.18b). A thermopile formed by six pairs of junctions is place on the disk in such a way that hot junctions are on the circular sectors painted black and cold junctions on those painted white. Because the black and white surfaces have the same absorption factor in the infrared spectral region, the instrument is sensitive only to short-wave radiation (the signal we want to measure), and only one protective glass dome is required.

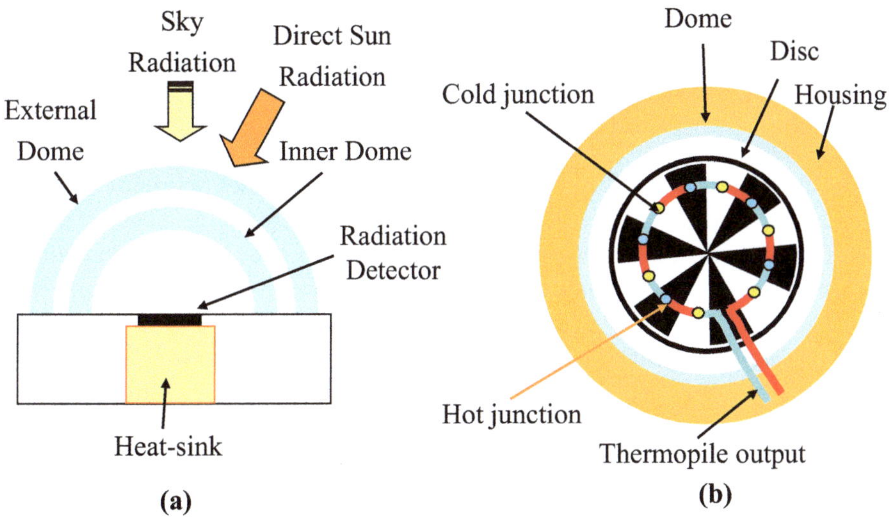

Fig. 7.18: Two models of pyranometes. (a) Lateral cross section view of a model which uses a sensor similar to that of a pyrheliometer and requires two domes. (b) Top view of the model which uses a disc divided into 12 circular sectors; it requires only one dome. In both models temperature is measured by a thermopile.

The glass domes of pyranometers work also as an optical filter with an approximately flat transference for the wave-length radiation range from 300 to 3000 nm.

In order to collect radiation from the entire hemisphere of the sky, pyranometers have to be installed at a height of about 2 m above ground level. No obstructions should be above the horizon and no object (buildings, trees, etc.) should project their shadows on the instruments at any time of the day. The instrument must be mounted leveled on a platform free from vibrations.

Some operating problems that could lead to underestimation of the signals are dust or moisture on the external face of the pyranometer's dome. Also, the condensation of moisture inside the dome decreases the signal. Dust should be removed frequently with a brush and dew or frost eliminated by means of a fan producing a current of air over the dome. In general, an internal recipient with desiccant prevents moisture condensation inside the dome.

Calibration of pyranometers should be performed yearly and the recommended way is to place a secondary standard pyranometer alongside the operational instrument being calibrated (IAMAP, 1986). At least 100 samples from both instruments should be taken in diverse radiation conditions (clear and cloudy skies). The ratio between the two measures for each sky condition is calculated and averaged. This average together with the calibration factor of the secondary standard is used to estimate the new calibration factor of the operational instrument.

7.4.2.3 Albedometer

Albedo is the fraction of solar energy reflected from a celestial body that does not emit light by itself. It has been established that it takes values between 0 and 1, 0 being the albedo of a body which does not reflect any light and 1 the albedo of a body that reflects the entire incident light. For example, a surface on the Earth that reflects back into space 40 % of the incident energy is said to have an albedo 0.4 (http://www.astromia.com).

Water is highly absorbent of sunlight, so it has a low albedo, but most sunlight impinging on ice covered with snow is reflected, so it has a high albedo. As an example, the albedo is about 0.15 for grass and 0.5 for dry sand (Kipp & Zonen B.V., 2013).

A way of measuring the albedo is by using two pyranometers. A pyranometer is placed as usual, and a second one is mounted in an inverted position, pointing towards the floor. The upper pyranometer measures global solar radiation and the lower measures solar radiation reflected from the surface below. The albedo is calculated from the irradiance (W/m^2) measured by both instruments. For this application, pyranometers should be at a minimum height of 1.5 m above ground level to minimize shadow effects and to attain spatial averaging (Kipp & Zonen B.V., 2013).

7.4.3 Spectral Measurements

7.4.3.1 Pyrradiometers and Pyrgeometers

These instruments are designed for measuring long-wave irradiance on a plane surface coming from the hemisphere above the instrument. Pyrradiometers are sensitive to radiation of wavelengths from 0.3 to 60 μm whereas pyrgeometers are sensitive to radiation from 2 to 60 μm (IAMAP, 1986).

The operational principle is similar to the instruments previously described, where a thermopile on a thermal resistor is used to transform radiation into voltage. They can be installed upwards to measure downward radiation from the sky or downwards to measure radiation from the ground. In the second case instrument should be mounted 1 to 2 m above a uniform surface covered by short grass. Care should be taken to avoid interference of the mounting pole on the instruments.

7.4.3.2 Filtered Pyrheliometers and Pyranometers

Pyrheliometers and pyranometers measure all the solar radiation wavelength range. Then, if it is desired to measure only in some spectral band, it is possible to interpose special optical filters before the radiation sensor. Due to the non ideal transference of the filters, it is somewhat complicated to evaluate the solar radiation in the desired band with pyrheliometers. The problem increases in the case of the pyranometers because the transmission of a filtering dome depends on the direction of the incoming radiation, making the interpretation of such measurements somewhat questionable (IAMAP, 1986).

7.4.3.3 Sunphotometer

A sunphotometer is an instrument to measure only some spectral bands of the solar radiation. It uses a silicon photo-electric sensor as those described above in Section (7.4.1.2). The sensors themselves have a filtering behavior and optical filters are added to select specific narrow bands. The bandwidth of the filters is 5 nm for high-accuracy instruments. The WMO recommends a minimum of three wavelength intervals centered at 368, 500 and 778 nm to measure aerosol optical depth. A sunphotometer for additionally measuring ozone, water vapor and nitrous oxide would need to measure in twelve spectral bands. Silicon detectors are sensitive to temperature changes. Therefore it is required to measure the temperature of the detector to apply a correction factor or to control the temperature of the detector by heating it.

References

Case, B., & Michelena, E. (1981). *Rain gauge for NOAA data buoys*. IEEE/MTS Symposium, Oceans 1981, Marine Technical Society, Boston, MA, 463-467.

Fisch, G. (2010). Comparisons between aerovane and sonic anemometer wind measurements at Alcântara Launch Center. *Journal of Aerospace Technology and Management*, *2*(1), 105-110.

Holmes, J., & Michelena, E. (1983). *Design and testing of a new rain gauge for NDBC meteorological data buoys*. Fifth Symposium on Meteorological Observations and Instrumentation, Toronto, ON, Canada, American Meteorological Society, 34-37.

IAMAP (International Association of Meteorology and Atmospheric Physics - Radiation Commission). (1986). Revised Instruction Manual on Radiation Instruments and Measurements. Claus Frohlich & Julius London (Eds)., WCRP Publications Series No. 7, WMO/TD - No. 149.

Kaimal, J.C. & Businger, J. A. (1963). A Continuous Wave Sonic Anemometer – Thermometer. *Journal of Applied Meteorology*, *2*, 156-164.

Kipp & Zonen B.V. (2008). *Instruction Manual. CHP Pyrheliometer*. Delft, The Netherlands.

Kipp & Zonen B.V. (2013). *Instruction Manual. CMP series Pyranometer, CMA series Albedometer*. Delft, The Netherlands.

Kristensen, L. (1993). *The Cup Anemometer*. Risø–R–615(EN). Risø National Laboratory, Roskilde, Denmark.

Kristensen, L., & Hansen, O. F. (2002). *Distance Constant of the Risø Cup Anemometer*. Risø–R–1320(EN). Risø National Laboratory, Roskilde, Denmark: Pitney Bowes Management Services.

Ma, B. B., & Nystuen, J. A. (2005). Passive Acoustic Detection and Measurement of Rainfall at Sea. *Journal of Atmospheric and Oceanic Technology*, *22*, 1225-1248.

Nystuen, J. (1999). Relative Performance of Automatic Rain Gauges under Different Rainfall Conditions. *American Meteorological Society*, *16*, 1025-1043.

OSR 0303 (Optical Scientific Report 0303). *WIVIS with HIPS sensor performance*. Optical Scientific, Inc., Gaithersburg, MD 20878-4008 USA.

Pedersen, T. F., Dahlberg, J. A., & Busche, P. (2006). *ACCUWIND - Classification of Five Cup Anemometers According to IEC61400-12-1*. Risø-R-1556 (EN). Risø National Laboratory, Roskilde, Denmark.

Stijn de Jong. (2010). *Low cost disdrometer. Improved design and testing in an urban environment.* Master Thesis report, Delft University of Technology.

Ting-i Wang, Earnshaw, K. B., & Lawrence, R.S. (1979). Path-Averaged Measurements of Rain Rate and Raindrop Size Distribution Using a Fast-Response Optical Sensor. *Journal of Applied Meteorology*, *18*, 654-660.

Ting-i Wang, & Crosby, J. D. Taking Rain Gauges to Sea. Global Climate Monitoring Gets New Tool: Rain-Induced Optical Scintillation for Measuring Precipitation at Sea. Reprinted from *Sea Technology* magazine. Optical Scientific, Inc. www.opticalscientific.com.

Tokay, A., Wolff, D. B., Wolff, K. R., & Bashor, P. (2003). Rain Gauge and Disdrometer Measurements during the Keys Area Microphysics Project (KAMP). *Journal of Atmospheric and Oceanic Technology*, *20*, 1460-1477.

US EPA (United States Environmental Protection Agency). (2000). *Meteorological Monitoring Guidance for Regulatory Modeling Applications*. EPA-454/R-99-005, Office of Air Quality Planning and Standards.

WMO (World Meteorological Organization). (2008). *Guide to Meteorological Instruments and Methods of Observation*, WMO-No. 8. Seventh edition.

http://www.astromia.com/glosario/albedo.htm

8 Ground-Based Remote Sensing Systems

A traditional approach to introduce remote sensing systems is to group them into two categories, namely passive and active systems. In passive remote sensing the equipment used to acquire the information from the target (object under study) gathers the data from the **natural energy** emitted and/or reflected by the target. The measuring system described in Sections (4.7.6) and (4.7.7) can be considered a kind of passive remote system. In active remote sensing, instead, the measuring system **emits energy** that "illuminates" the target, and receives part of **its own energy** reflected or scattered by the target (Lillesand et al., 2004). In both cases the energy arriving from the target is analyzed to draw information about the characteristics of the target (velocity, position, size, temperature, aerosol, etc.).

Remote sensing can be performed in different media: solid, fluid or vacuum. Acoustic systems used for measuring flow, as those presented in Section (5.4.4.), could be considered remote sensing systems in a fluid (water) where the sensing wave is a sound wave. Other systems use electromagnetic waves in air, as in the case of radar, or in a physical solid medium such as in fiber optic or soil; an example of the last method is called **Distributed Temperature Sensing** (**DTS**) and will be described below.

In this chapter we will pay attention to active remote sensing systems, where energy is irradiated from the measuring systems to the target, the propagating wave being either electromagnetic or acoustic.

8.1 Distributed Temperature Sensing (DTS)

8.1.1 Introduction

In the near past, thermal hydrological processes have been studied using point measurements, as can be obtained from thermocouples, thermistors and Integrated Circuit sensors (I.C.). In the last few years, a very interesting system with unique characteristics for measuring temperature in environmental sciences and industrial applications has emerged (Selker et al., 2006). It permits the temperature of thousands of points to be simultaneously monitored, which could reveal interactions and processes yet unknown.

This technology is based on the use of fiber optics (f.o.) as sensors and allows environmental processes to be continuously studied over a much greater spatial scale than has been done until now. The measuring of temperature and pressure in laboratory conditions by means of f.o. is not new, but the methods to obtain the information from f.o. sensors and the portability of the instruments have evolved in the last years. These advances have made it possible to use these new tools in field conditions (Selker et al., 2006).

In order to introduce the measuring principles that have been reported as most adequate for environmental applications, it is necessary to explain what f.o. are and how they are employed by the different methods for measuring temperature.

8.1.2 Optical Fibers

Optical fibers are flexible filaments of circular section made of glass or plastic which are barely thicker than a hair. They have an internal transparent core surrounded by a layer of material of lower refractive index. This difference in indexes causes total internal reflection of the light, making the fiber behave as a waveguide.

Because light is confined to the core, which permits light to be transmitted between the two ends of the fiber, they are commonly called "light pipes". The external layer, also called cladding material, is surrounded by a protective coating made of a polymer (Fig. 8.1).

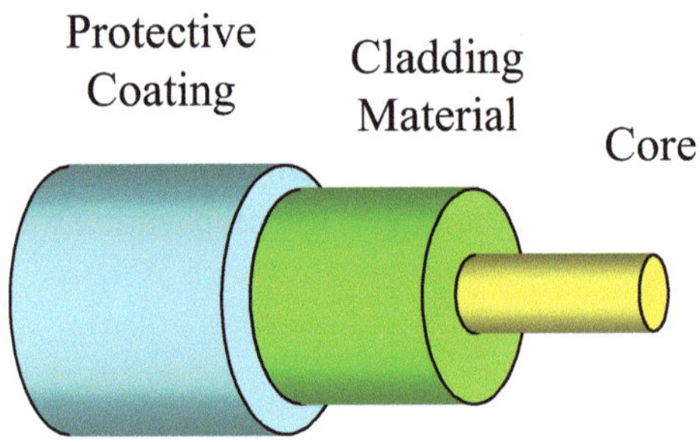

Fig. 8.1: Fiber optic components.

Fiber optics are capable of transmitting high bandwidths over long distances with low losses, and is not interfered by electromagnetic noise. For these reasons they are replacing copper wires in local area data transmission networks and are of great importance in transoceanic communications.

8.1.3 Bragg Grating

Usually, in a f.o. used for communications, it is desired that the light passes through the fiber without internal scattering; the core is thus made as uniform as possible along its length. But when the f.o. is used as a sensor, it may be desired to introduce some periodical variation of the refractive index of the core along the length of the fiber to obtain a Bragg grating. It has been mentioned in Section (3.5.1) that the Bragg law is valid for electromagnetic waves, as is the case of light propagating in f.o.

The Bragg scattering law gives the conditions for the backscattered light to interfere constructively and permits the f.o. to be used as a sensor. Therefore, the refractive index of the core is increased by design at regular intervals *d* (Fig. 8.2), where *d* is an integer number (*n*) of half-wavelengths of the light (λ) that is desired to be backscattered, i.e.,

$$d = n\frac{\lambda}{2} \qquad n = 1, 2, 3, \ldots \tag{8.1}$$

By measuring the wavelength of the scatters it is possible to know *d* from the end of the fiber. Then, if a certain phenomenon is modifying the grating it could be detected just by measuring the scattered wavelength. For example, if the fiber core length is modified due to a mechanical stress, or a thermal expansion or contraction, the distance *d* will vary, and this variation will be evident by a change in the wavelength of the scattered signal.

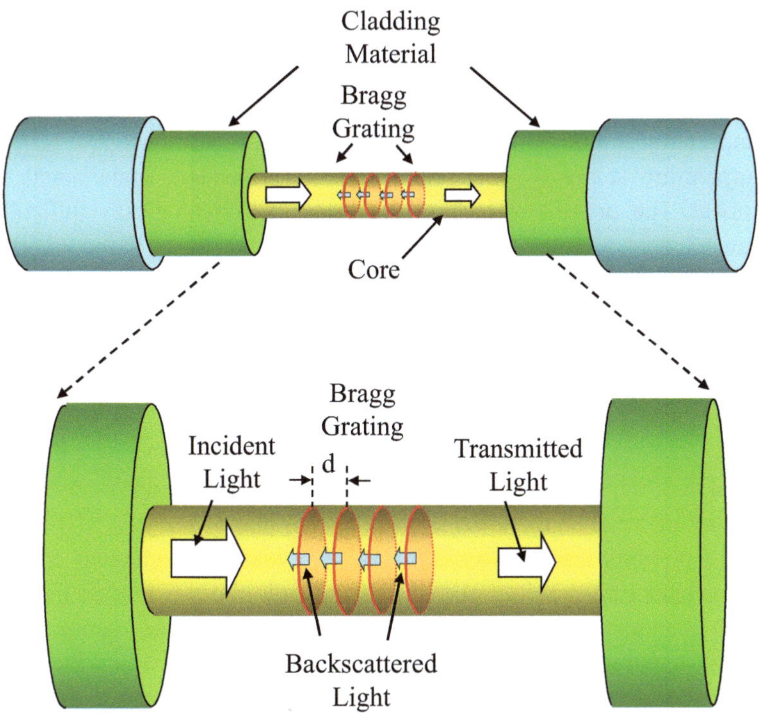

Fig. 8.2: A Bragg grating is "impressed" on the fiber core by deliberately increasing the refractive index along its length. Light of a particular wavelength is backscattered by the grating.

Thus, by means of a calibration process, the function (f) that relates temperature to wavelength may be established (Eq. (8.2)). Then we could know the temperature in as many points of the fiber as **different gratings** we could "impress" on the fiber. Usually, the process of impressing the gratings is called etching.

$$T = f(d) = f(\lambda) \tag{8.2}$$

This means that we could lay a fiber of several kilometers in length along a river or a beach to simultaneously know the temperature at many points by measuring and recording just at one end of the fiber. This constitutes a huge advantage over the process of reading and recording each one of the points individually. Many gratings can be distributed along a fiber; distances between gratings may be very close (less than 1 mm) or very far (more than 10 km). The distance from the end of the fiber to the grating is known because it is defined during the manufacturing process. The distance to the spatial point where the scattered light comes from is thus known, and so is the place where temperature is measured.

Therefore, by etching each grating with different distances d suited for specific ranges of propagating wavelengths, and placing them in well defined points on the fiber, the temperature of each specific point can be identified by the range of the scattered frequency. Thus, it is possible to know the places the scatter comes from, obtaining a graph of the temperature (by means of Eq. (8.2)) as a function of length. Hence, the temperature as a function of time and distance can be known.

When temperature at many points along the f.o. is needed, some practical limitations appear. The process of etching the grids is technologically difficult, resulting in an expensive method. But following the same ideas based on the scattering of light in f.o. we will discuss other methods that do no require modifying the fiber, as the etching of grating does.

8.1.4 Scattering in F.O.

Some conceptual insight into the scattering processes in f.o. is required to understand how some methods measure temperature by means other than gratings.

Scattering of photons inside f.o. can be produced by local changes in the characteristics of light transmission. The analysis of how the transmitted light is affected by these fiber modifications permits the f.o. to be employed as a sensor.

What it is not apparent are the internal physical processes by which the light interacts with the f.o. material producing the light scattering. Therefore to describe how f.o. work as sensors we need to introduce this subject first.

There are different types of scattering processes present in f.o., some of them known as Rayleigh, Brillouin and Raman scattering. Taking into account the energy and frequency involved in the scattering processes, two types of scattering are recognized: elastic and inelastic. In elastic scattering the scattered photons have the same energy and frequency as the incident photons; Rayleigh scattering is elastic and is due to the fluctuations in density of the matter that produces the scatters. This kind of scattering is related to the way the molecules are organized in the f.o. material and is not of interest in the use of f.o. as a measuring device.

Conversely, in inelastic scattering, photons lose or gain energy and undergo a frequency change. A loss of energy corresponds to a down shift in frequency, and an increase in energy, to an up shift in frequency; the first is called Stokes-type scattering and the second anti-Stokes scattering.

Raman and Brillouin are inelastic scatterings and are of interest in measuring temperature and other parameters by means of f.o. Therefore, we will try a simple description of both scatterings just to gain some insight into the measuring process.

The scattering of an X-ray beam was explained in Section (3.5.1), and this explanation is valid for electromagnetic and acoustic waves. The explanation of the scattering processes in f.o. materials has two possible approaches, the classical and the quantum physics ones.

8.1.4.1 Brillouin Scattering

Electromagnetic forces due to the passage of intense laser light through f.o. produce molecular thermal agitation that sets off density changes in the fiber. These density changes propagate as acoustic waves or phonons (Selker et al., 2006). The change in density produces a change in the light refractive index that travels as an acoustic wave in the fiber. As a result, the light scattered by these moving inhomogeneities will experience a Doppler shift, and this shift can be measured from one of the ends of the f.o. as is done in the case of the method using Bragg gratings.

Because the velocity of the acoustic wave in the f.o. depends on the temperature and strain of the medium, a system devoted to the analysis of the Brillouin scattering in f.o. results in an adequate tool to perform strain and temperature measurements. From a quantum physics point of view, thermal motions of atoms in a material such as f.o. create acoustic vibrations, which lead to density variations that can be treated as a density wave or vibrational quanta (phonon). Thus, Brillouin scattering can be studied as an interaction between an electromagnetic wave and a density wave (photon-phonon scattering).

It has been found that Doppler frequency shifts of the electromagnetic wave in f.o. are linearly related to the temperature of the f.o. (Eq. (8.3)) (Minardo, 2003). Thus the f.o. becomes a temperature sensor.

$$\Delta f = K\,\Delta T \tag{8.3}$$

Summarizing and simplifying, the Brillouin scattering in f.o. is due to variations of the refractive index which travel as sound waves and scatter the light wave producing a Doppler shift on it. The velocity of the acoustic waves in the f.o. is a function of temperature so changes in temperature produce changes in the frequency shift of the scattered light. Relating frequency shift and temperature, the second may be inferred from the first.

8.1.4.2 Raman Scattering

The Raman scattering in f.o. may be thought of as the redirection of energy when the propagating light wave finds an obstacle or inhomogeneity in the material (Hahn, 2007).

Individual atoms and molecules constituting the f.o. material have specific vibrational and rotational modes (or levels) with their own vibrational and rotational frequencies (v_{vib} and v_{rot}).

Light is an electromagnetic wave and its associated electric field perturbs the electrons in the molecules of the f.o. material with the same frequency of the light source that illuminates the fiber (v_s). Most of the scattered light is of this frequency (v_s) Nevertheless, a small amount of energy is scattered at different (lower and upper) frequencies due to some complex interactions at the atomic and molecular levels between the electromagnetic wave and the material structure. The difference (or shift) in the scattered frequency is related to the vibrational and rotational frequencies of the molecular structure. As the spacing between rotational levels is small, many molecules can be excited to higher rotational levels by the incoming light. Therefore, changes in rotational modes produce dim spectral lines on both sides of the line corresponding to the exciting light (v_s). Vibrational levels, on the other hand, are generally so widely spaced that only a few molecules are in higher vibrational levels at ordinary temperatures, and are capable of contributing with additional energy to a scattering process. Consequently, according to **Stokes law** of fluorescence, Raman lines due to changes in vibrational states are generally located towards frequencies lower than that of the exciting light (v_s), i.e., towards longer wavelengths (called *Stokes* lines). However, dim lines are also observed towards the region of higher frequencies with respect to that of the exciting light (v_s), i.e., towards shorter wavelengths. Since these lines have a frequency higher (shorter wavelength) than that of the exciting light (v_s), they represent a violation of Stokes law, and are thus called *anti-Stokes* lines (Jenkins & White, 1957).

The intensities of both Stokes $P_S(z, t)$ and anti-Stokes, $P_{aS}(z, t)$ Raman scattering are functions of the light intensity, but the anti-Stokes scattering is also very sensitive to temperature. Then the ratio $R(z, t)$ (Eq. (8.4)) is independent of the intensity and depends exponentially on the fiber temperature. Therefore, measuring $R(z, t)$ allows the temperature $T(z, t)$ to be calculated (Eq. (8.4)),

$$R(z,t)=\frac{P_S(z,t)}{P_{aS}(z,t)}$$
$$T(z,t)=\frac{\gamma}{C-\ln(R(z,t))+\Delta\alpha\, z} \tag{8.4}$$

where γ, Δα and C are calibration constants, and z distance along the f.o. (Suarez et al., 2011).

The accuracy of the temperature measurements depends on the total number of scattered photons observed; this amount increases with the time and spatial integrations of the measurements, and the proximity of the measured point to the light source (Selker et al., 2006).

8.1.5 DTS Technology in Environmental Sciences and Hydraulics

The Doppler shift of the scattered light for silica optical fibers is about 13 THz in Raman scattering, and on the order of some GHz in Brillouin scattering (Minardo, 2003). The energy associated with lattice vibrations (Raman scattering) is much greater than acoustic wave energy generated in spontaneous Brillouin scattering. In spite of these differences, both kinds of scatters have been used for DTS in hydraulics applications, but is seems that some instrument manufacturers prefer the Raman scattering for their equipment (Bao et al., 1993; Selker et al., 2006).

A simplified schematic of DTS equipment which uses Raman scattering is depicted in Figure 8.3. It consists of a laser source to excite the f.o., a detector to convert light into an electrical signal and demodulate the frequency shift, and a computer. Stokes and anti-Stokes intensity components of the scattered signal are separated and used in the computer to calculate temperature.

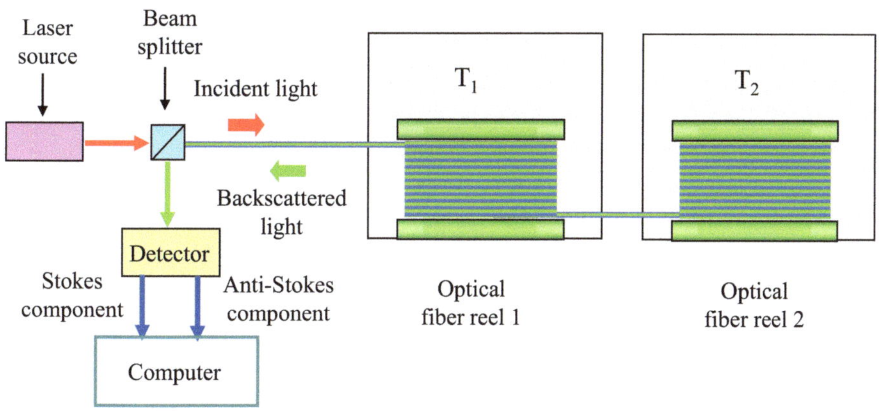

Fig. 8.3: Simplified DTS equipment measuring on an optical fiber coiled on two separate reels which are at different temperatures T_1 and T_2.

It has been mentioned that in the fiber sensors with Bragg gratings it is possible to know the distance the scatter comes from because the length from the grating to the end of the f.o. is fixed during the manufacturing process, and each grating is identified by the range of wavelengths that it can scatter. In other words, each grating is associated with a wavelength range which in turn is related to a position on the f.o. Unfortunately, this method cannot be used with f.o. using Raman and Brillouin scatterings.

With the aim of associating the scatter with the distance it comes from in DTS it is required to send light pulses and measure the pulse round-trip travel time. This method is also known as optical time domain reflectometry (OTDR) and allows distances to be estimated; in particular, it is used in DTS to relate temperatures as a function of position along the optical fiber.

As in acoustics Doppler Current Profilers (Section (5.4.4)) and other methods where pulses are launched to obtain information about the characteristics of the medium where the wave propagates, the length of the pulse also defines the sampling resolution in space for DTS. Resolution increases as the length of the optical pulse decreases. As in the acoustic profilers, the measuring range in the fiber could be thought of as divided into a number of measuring cells over which the scatters are spatially integrated. As will be shown later, this concept is also used in radars.

Stokes and anti-Stokes scattering have different extinction coefficients, so connectors, cables, and splices have a priori unknown effects on Stokes and anti-Stokes backscatter (van de Giesen et al., 2012). For this reason, DTS techniques require carrying out an individual calibration of the instrument together with the fiber sensor (Selker et al., 2006). Manufacturers provide calibration routines that require placing sections of the cable in known-temperature baths to establish the constants of Eq. (8.4) that will be thereafter used in the measuring process.

In Figure 8.3 the fiber sensor is coiled on two separate reels that are at different temperatures T_1 and T_2. This set up is frequently used to calibrate DTS instruments and permits the spatial resolution of the ensemble to be evaluated (equipment and f.o. sensor) (http://www.sumitomoelectricusa.com; van de Giesen et al., 2012).

Some calibration algorithms assume that the f.o. parameters remain uniform over the entire length of the fiber, but more sophisticated calibration techniques and specifically designed algorithms for hydrological applications have been developed (Hausner et al., 2011; van de Giesen et al., 2012; Suarez et al., 2011). Unfortunately, temperature controlled conditions are not readily available in the field, so that field calibration is difficult.

8.1.6 Specifying a DTS Equipment

A typical way for specifying the measuring performance of DTS instruments is, for example: "a measurement with 1 m spatial resolution taken 1000 m from the instrument and integrated for 1 min will have a standard deviation on the order of 0.1 °C" (Selker et al., 2006).

As interesting as the specification in itself, which will surely change with technological evolution, is the way the instrument should be specified. These three parameters **(distance, spatial resolution and time resolution)** define the condition in which the measuring standard deviation was established. These parameters are strongly dependent one another so that the specification makes no sense if one of them is missing. Table 8.1 shows the specifications for commercial DTS equipment (http://www.sumitomoelectricusa.com). This is not equipment easily operated in the field, but some portable equipment is emerging.

Table 8.1: DTS specifications of commercial equipment

Measurable range (km)		**15**	
Minimum measuring time (s)		40	500
Temperature resolution in °C as a function of distance	1 km	± 0.5	± 0.2
	2 km	± 0.5	± 0.2
	3 km	± 0.7	± 0.2
	5 km	± 0.7	± 0.3
	10 km	± 1.5	± 0.5
	15 km	± 3.5	± 1.0
Sampling resolution (m)	1		
Spatial resolution (m)	2		

Manufacturers' specifications for DTS equipment are **based on particular recommended fiber-optic sensor**. Then if it is attempted to make measurements on a telecommunication infrastructure already installed, as was the case of seasonal temperature profiles along the bed of a lake (Selker et al., 2006), an in situ calibration has to be performed, and perhaps the performance of the instrument could be somewhat degraded.

The utilization of DTS systems in environmental application sciences has been reported in several scientific works in recent years. They were used to observe temperature distribution along a stream, air-snow interface temperature profile, air-water interface temperature in lakes, temperature profile with depth in a flooded decommissioned mine (Selker et al., 2006), groundwater-surface water studies, soil water content interactions, tidal estuary behavior (Hausner et al., 2011), storm water sewers inspection, dam surveillance (van de Giesen et al., 2012), etc. This technology seems to have a promising future in some long spatial scale field applications in environmental sciences.

8.2 Radar

The most well known active remote sensing system is the radar; the word radar is an acronym for **Radio Detecting and Ranging.** Many remote sensing systems used in environmental sciences are based on the working principle of the early radar used to detect flying objects. Because the concepts needed to understand the first systems are easier to explain the radar, an introduction to the early radar will initially be carried out. A traditional way in which radar operating principles are introduced is by means of an acoustic analogy that we will also follow here.

If a person shouts in a big empty room, after a certain delay, a reflection of the sound on the walls will arrive back to the person as an echo. This delay is the time required for the sound to travel from the speaker to the reflecting wall and back to the speaker. Measuring the delay and knowing the speed of sound, the distance to the wall may be estimated by the product of both quantities divided by two. The divisor takes into account that the sound must travel the distance twice, to go to the reflecting object and to return from it.

A similar phenomenon occurs with electromagnetic waves. Radar emits electromagnetic energy pulses that travel through space at a constant speed of about 300,000 km/s. If the electromagnetic wave finds an object that reflects or scatters it, part of the energy will be received back at the place of the original emission. This returned signal is called backscatter or echo as in the acoustic case. As before, the product of the propagation speed of the electromagnetic wave (c) by the time delay (t_{delay}) may be employed to estimate the range (R) from the emitter to the target:

$$R = \frac{c\, t_{delay}}{2} \tag{8.5}$$

A more detailed step-by-step description of a radar signal generation and reception follows as an example.

In radar for airport traffic control, the emitting and receiving antenna is located at the center of a dish that rotates at a rate of about one turn every ten seconds (Fig. 8.4).

The dish is located at the upper end of a tower. The radar transmits short duration high-power electromagnetic pulses that are radiated into space by the transmitting antenna; when this energy arrives at the targets they produce a diffuse reflection in a wide number of directions (scatters). Some of this energy is received back by the radar antenna and the flying time of the pulse in both directions is computed. This time is used to calculate the range (distance) where the scatter comes from.

The radar's antenna is designed to radiate and detect energy in one-directional beams. They have a characteristic beamwidth similar to that described in Section (2.3.2). Thus, during the dish rotation, when the antenna illuminates the target, the direction where the reflection comes from is well known and the direction of the target is recognized.

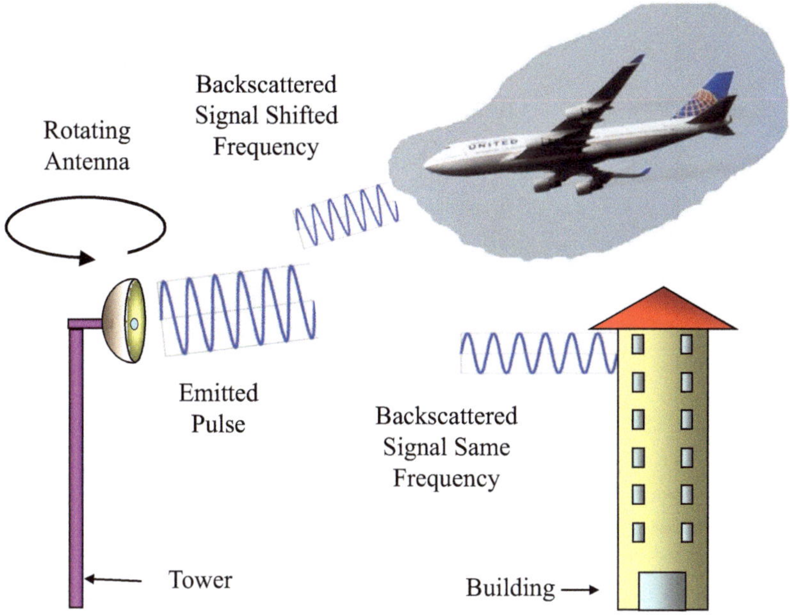

Fig. 8.4: The radar antenna rotates at the top of a tower. Fixed targets reflect the same emitted frequency. Moving targets reflect a frequency shifted with respect to the emitted one.

When the weak backscattered signals from the target arrive at the receiving antenna they are amplified and processed into the receiver. The receiver provides video output signals that can be displayed on the traditional plan position indicator (PPI) (Fig. 8.5). It consists of a circular display with concentric circles and with the rotating antenna represented at its center. Numbers on the circles are references that indicate the radial distances from the antenna measured in kilometers or miles.

As the radar antenna rotates, a radial trace sweeps the PPI display at a constant angular speed ω, synchronously with the antenna's beam. Thus, the trace indicates the pointing direction of the antenna's beam at each instant. When a target reflection is detected by the receiver a bright dot is drawn on the screen. The distance from the center of the screen to the dot represents the distance (d) from the radar to the target, thus the range and direction of the targets are shown; the PPI shows a picture of the targets present on the area covered by the radar beam.

If the target is fixed in space, as it could be the case of a building or a tree, the reflected frequency will be equal to the emitted one. But if the target is moving, there is a shift (Δf) of the backscattered frequency with respect to the emitted frequency (f_0) due to the Doppler effect (Section (3.3.1)). This shift is proportional to the product of the radial speed of the target (v) times the emitted frequency, and inversely proportional to the electromagnetic wave propagation speed (Eq. (8.6)). Therefore, the velocity of the target can also be known.

$$\Delta f = \frac{2\,v\,f_0}{c} \tag{8.6}$$

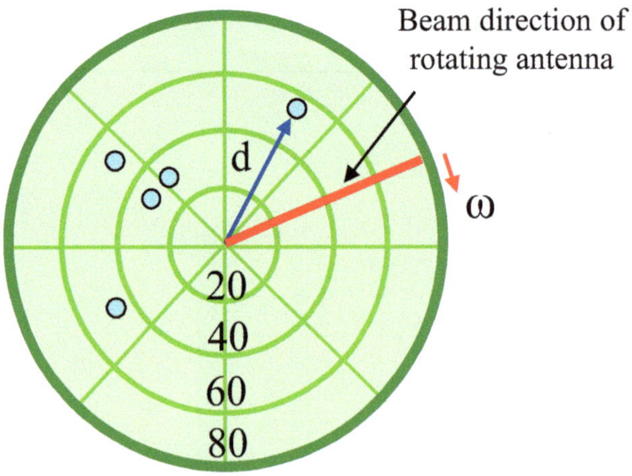

Fig. 8.5: A PPI display: the antenna beam direction is shown by the radial bar which rotates in synchronism with the antenna. Dots represent target positions. Numbers on the circles indicate their radial distances from the antenna.

A device called *duplexer* alternately switches the antenna between the transmitter and receiver, preventing the transmitted pulses from entering the receiver because their high energy could destroy the receiver. The duplexer requires a certain time to switch that is called the recovery time of the duplexer ($t_{recovery}$). Then those echo pulses coming from very close targets, which fall inside the transmitting pulse time (τ) plus the recovery time of the duplexer will not enter the receiver. This time in which the receiver cannot "see" the echo gives as a result a blind distance or minimum detectable range (R_{min}),

$$R_{min} = \frac{\left(\tau + t_{recovery}\right)c}{2} \tag{8.7}$$

For a typical transmitting pulse width of 1 μs (neglecting the recovery time that is characteristic of each particular radar), the corresponding minimum range is about

$$R_{min} = \frac{10^{-6} \times 3 \times 10^{8}}{2} \; \frac{\text{ms}}{\text{s}} = 150\ \text{m}$$

this means that this radar cannot detect targets that are closer than this distance.

Another important characteristic of radars is the range resolution. This is the ability of a radar system to distinguish between two targets at different ranges along the same direction. Figure 8.6 shows an example.

Neglecting radar and target particular characteristics, the range resolution depends basically on the width of the transmitted pulse (τ) and can be calculated from

$$R_{\text{resolution}} = \frac{\tau c}{2} \tag{8.8}$$

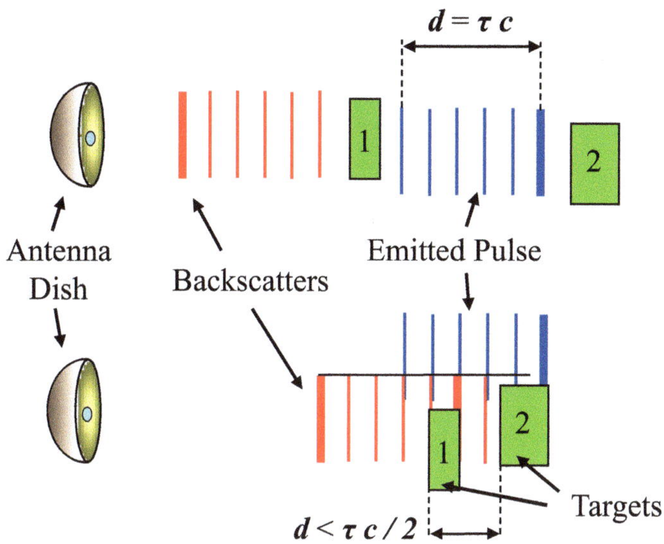

Fig. 8.6: Range resolution. The vertical parallel lines represent the radar wave fronts; the first wave front is shown thicker and a pulse is represented by six lines. In the upper part of the figure, the pulse is completely backscattered by the first target before the pulse arrives at the second target. In the lower figure the targets are closer to each other and the first target is scattering the fifth wave front while the second target is already scattering the second wave front.

The vertical parallel lines represent the radar wave fronts; the first wave front is shown thicker and a pulse is represented by six lines. In the upper part of the figure, the pulse is completely backscattered by the first target before the pulse arrives at the second target. In the lower figure the targets are closer to each other, and it shows that the instant at which the first target is scattering the fifth wave front of the pulse the second target is already scattering the second wave front. Thus, at the receiver the backscattered signals will be perceived as produced by the same object.

The range resolution and the beamwidth of the antenna define a volume that we will call the resolution cell. The shorter the pulse width τ and the narrower the beamwidth the smaller is the volume of the cell and the higher the ability to detect smaller targets (Fig. 8.7). Thus smaller cells result in higher spatial resolution.

We have mentioned that when a target is moving there is a shift of the backscattered frequency with respect to the emitted frequency due to the Doppler effect. Because the process of extracting the shift from the backscattered signal is a frequent procedure in remote sensing instrument receivers, a simple explanation of this process will be introduced.

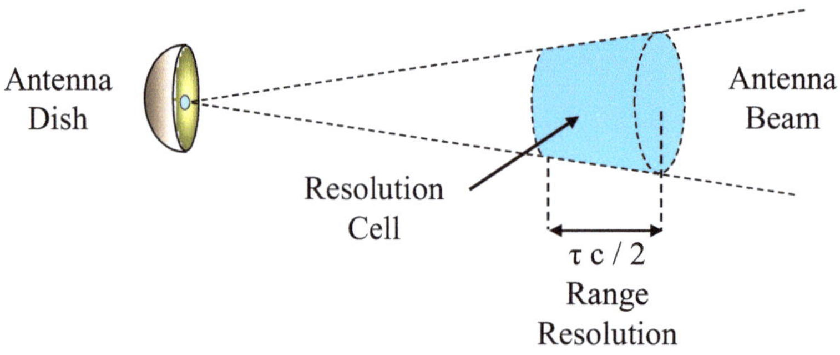

Fig. 8.7: The range resolution and the beamwidth of the antenna define a volume that we will call the resolution cell. The shorter the pulse width τ and the narrower the beamwidth the smaller is the volume of the cell and the higher the ability to detect smaller targets. Thus smaller cells result in higher spatial resolution.

Receivers used to get the frequency shift have some kind of mixer. A mixer is an analog multiplier or some kind of non-linear device that multiplies the local oscillator (LO) signal (which has the transmitter frequency) with the backscattered signal. Assuming that signals are sinusoidal, and the transmitted frequency is f_1 and the backscattered signal is $f_2 = f_1 + \Delta f$ the output $y(t)$ of the mixer is given by Eq. (8.9). As a result of the multiplying process, the difference and sum of the frequencies are obtained. Then a frequency filtering can be applied on the mixer output to keep the difference and reject the sum, thus a low-pass filtering of $y(t)$ gives a cosine signal with the frequency shift. A subsequent spectral analysis of the filter output will therefore account for the Doppler shift, and then for the velocity of the backscattering object. This kind of signal processing is called synchronous demodulation or homodyne demodulation.

$$\begin{aligned} y(t) &= \sin(2\pi f_1)\ \sin(2\pi f_2) = \frac{1}{2}\cos\left[2\pi(f_1 - f_2)\right] - \frac{1}{2}\cos\left[2\pi(f_1 + f_2)\right]; \\ y(t) &= \frac{1}{2}\cos\left[2\pi(\Delta f)\right] - \frac{1}{2}\cos\left[2\pi(2f_1 + \Delta f\right]; \\ y(t) &\Rightarrow \text{low pass filtered} \Rightarrow \hat{y}(t) = \frac{1}{2}\cos\left[2\pi(\Delta f)\right] \end{aligned} \tag{8.9}$$

So far we have used a well known type of radar to introduce some concepts. From now on we will use these concepts to describe how radar technology is applied to remotely estimate some parameters of interest in environmental sciences.

8.3 Upper-Air Remote Monitoring

One of the applications of remote sensing techniques in environmental sciences is the monitoring of upper-air meteorological parameters. Traditionally, the monitoring was done from fixed or moving instruments. Fixed instruments on meteorological towers measured parameters such as wind velocity, atmospheric pressure, temperature, solar irradiation, and humidity. Moving instruments allow measuring some meteorological parameters as a function of altitude; radiosondes with instruments on board a flying balloon measure and transmit the meteorological data to an earth-based station. In both cases it could be said that **measurements are made "in situ"** because instruments are in direct contact with the atmosphere. One important difference is that in the balloon case the wind is inferred from its trajectory as a function of time, and it is not a direct measurement made with a fixed anemometer as in the tower case. In the first case, it is said that it is a Lagrangian measurement of the wind because a particular parcel of air drives the balloon as it moves through the atmosphere as a function of time. In the second case it is an Eulerian measurement where wind velocity is measured on a specific volume in the atmosphere through which the wind flows as time passes.

In contrast with direct measurements, ground-based remote sensing systems provide means for the collection of upper-air meteorological data, **but no instrument is in contact with the measurand**. Unlike the case of in situ measurements the remote methods are based on **indirect measurements**; this means that the desired variables are derived from other variables that are measured directly. Active remote sensing systems for measuring atmospheric data are based on sound and electromagnetic wave propagation and on how the properties of the atmosphere modify the conditions for the propagation of waves.

This difference between direct and indirect methods has significant implications for the calibration and audit of upper-air measurement systems (US EPA, 2000), because a calibration of instruments (Section (2.2.3)) in the traditional way in which their measurements are compared to those of a standard instrument cannot be performed.

The best-known systems for remote sensing of the atmosphere are: Sound Detection and Ranging (SODAR), Radar Wind Profiler (RWP), Radio Acoustic Sounding System (RASS) and Light Detection and Ranging (LIDAR).

8.4 Wind Profilers

A wind profiler is a system capable of measuring the components of wind velocity in the atmosphere as a function of altitude. This is carried out by sending a sound or an electromagnetic wave towards the atmosphere. Part of the energy is scattered by the atmosphere, received back at the point of emission and analyzed by the system. Fluctuations in the refractive index of the atmosphere, produced by air density variations, are the cause of the wave energy scatters. Because wind carries these irregularities with a certain mean velocity, these turbulences become a tracer of the wind velocity. In other words, these turbulences behave as a moving target in the case of the previously described radar (Section (8.2)).

The backscatter signals are Doppler shifted due to the wind velocity, the shift being proportional to the average wind velocity in the direction of the propagating wave. Thus, the measurement of the Doppler shifts allows the mean wind velocity to be estimated. There are different technologies based on this operating principle. Some use sound waves and others electromagnetic waves to gather wind data; a third type uses both kind of waves and it is also able to estimate the atmospheric temperature profile.

8.5 SODAR

8.5.1 General Description

SODAR is an abbreviation for Sound Detection and Ranging; they are instruments used to measure upper-air wind in the atmosphere. They consist of acoustic transmitters and receivers located at the ground level and pointing to the sky. The transmitters send pulses towards the atmosphere, and the receivers "listen" to the signal coming back from it. When the same acoustic transducer is used for transmitting and receiving it is said that the emitter and the receiver are co-located, and the system is called mono-static; when they are separated the system is called bi-static.

The earth's boundary layer has inhomogeneities called **eddies** (Bahl et al., 2011), and these eddies are transported by the wind. When the instrument emits a short acoustic pulse upward, the wave traveling through the atmosphere finds these eddies and the acoustic energy is scattered in all directions. Because eddies move at the velocity imposed by the wind, the frequency of the scattered acoustic waves is different from the frequency of the emitted pulse due to the Doppler effect. The shift of the received signal is related to the radial wind speed (speed in the direction of the propagating wave). Therefore, by measuring the frequency shifts the radial wind speed may be estimated. With this information and the geometry of the transmitter and receiver beams, the horizontal and vertical wind speed components may be calculated (Section (5.4.4)).

The height where the scattering inhomogeneities are located may be calculated from the speed of sound and the delay in the returned signal with respect to the time the pulse was sent. Thus, a profile of the wind speed as a function of height can be obtained. The transducer beam length is discretized into equal length cells as explained in Section (5.4.4) and previously in this chapter (Section (8.2)). The minimum cell length gives the maximum vertical resolution of the system.

Most of the commercially available SODAR systems are of the mono-static type and are multi-axes systems; this means that they can send and receive signals along several directions. The ability to process signals from several radial directions permits the directional profile of wind speed to be derived (http://www.sodar.com).

Because SODAR transducers are pointing up, they require weatherproof housing. Early designs used specially arranged speakers to keep precipitation out; more recently arrays of several piezoelectric transducers have been successfully used to replace speakers. These arrays have the advantage of increasing the acoustic power simply by adding more elements (Section (4.13.4)). Nevertheless, the most remarkable advantage is the possibility of using the phased-array technology. This consists in combining the individual wave fronts of transducers to give a resulting wave front for the complete array (Section (3.5.3)).

The way in which the individual transducer's lobes are combined may steer the resulting sound beam (Fig. 8.8). The electronic circuits delay the phase of each transducer in such a way that the main lobe of the array may point at some desired angle. In this way the array works as a single transducer that can sequentially gather data along multiple axes. Thus, a single array can sweep the atmosphere to obtain data from multiple directions.

The hexagon of Figure 8.8 has 19 circles and represents the acoustic transducer array, each circle being one transducer. The wave fronts represent the acoustic beam sweep. It is worth noting that the array produces one wave front at a time (even when three wave fronts are drawn they are not produced simultaneously, but sequentially).

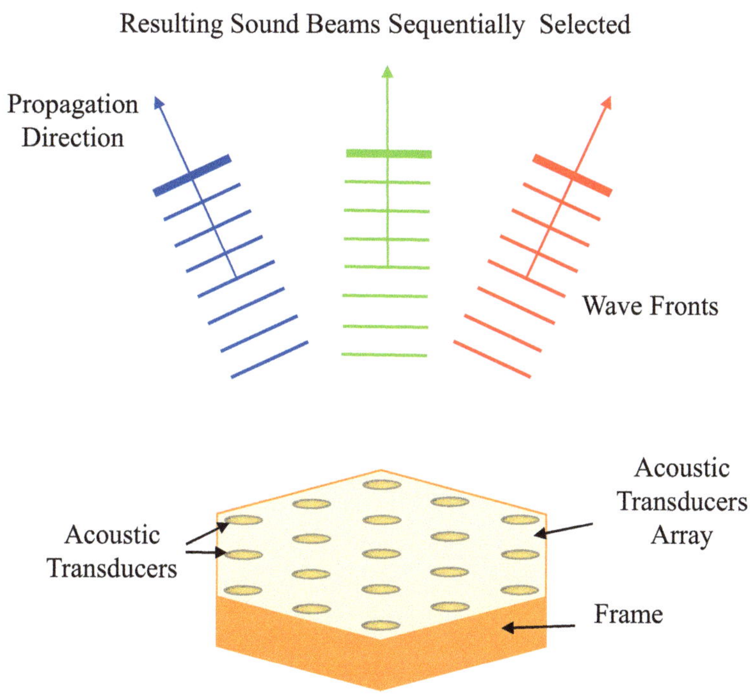

Fig. 8.8: Resulting sound beams selected one at a time.

Usually a multi-axes system consists of three to five beams. The array with five beams has a similar distribution to that shown in Figure 5.17 (Section (5.4.4.2)), in which one beam points vertically and the other four are tilted from the vertical, typically from 20 to 30 degrees. Once again, the beams are produced sequentially combining the individual lobes of each transducer by means of the beamforming process (Section (3.5.3)). Also, in the three-beam configurations, tilted lobes are placed along orthogonal directions on the horizontal plane (Fig. 8.9) (Bahl et al., 2011).

Other beam arrays use three titled beams horizontally offset by 120° (Bahl et al., 2011; Gustafsson, 2008). The beamwidth angle reported in modern literature is about 10 to 15°. The three emitting-receiving directions are switched to measure with one beam at a time (they cannot measure simultaneously in the three directions).

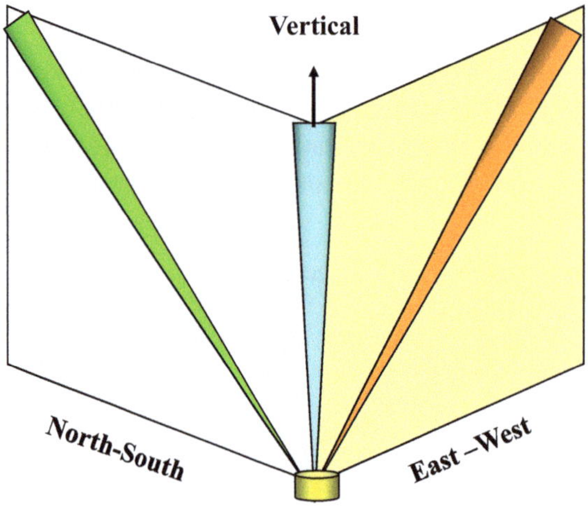

Fig. 8.9: Schematic of a three-beam configuration SODAR.

Generally, SODAR systems use the Fast Fourier Transform (FFT) to derive the signal Doppler shift. Signal averaging is used to improve signal detection, and can be performed either in the time domain or in the frequency domain (Bahl et al., 2011; Gustafsson, 2008). Due to the signal averaging, several minutes are needed to obtain a single piece of data.

It has been shown in Section (3.3.1) that the Doppler shift $\Delta f = f - f_0$ between the sent (f_0) and backscattered (f) frequencies is directly proportional to the velocity component of the inhomogeneities (v) along the wave propagation direction, and inversely proportional to the wave propagation velocity (c),

$$\Delta f \approx \frac{2\, v\, f_0}{c} \tag{8.10}$$

Because c varies with temperature (Eq. (8.11)) (Gustafsson, 2008) the actual temperature (T) in kelvins (K) is needed to correct the speed of sound,

$$c = \sqrt{401.8 \times T} \text{ m/s} \tag{8.11}$$

Before choosing a site for the installation of a SODAR system it is important to estimate the acoustic background noise level because noise may limit the maximum measuring height. Acoustic noise within the specific frequency band transmitted by the SODAR can seriously affect signal detection. Since the received signal power is inversely proportional to the square of the height it comes from, signals from greater heights are weaker and more prone to be affected by the background noise.

Another problem with a SODAR installation is that the beamwidth of the transducer (Section (2.3.2)) has a main lobe that covers the region where the desired signal comes from (upwards), but has also secondary lobes. Since these undesired lobes may point at an angle of a few degrees above the horizontal plane, the energy scattered or reflected by nearby objects such as trees, smokestacks, buildings and towers may reach the receiver. Although the secondary lobes have relative low sensitivity with respect to the main lobe, due to the great strength of the signals received from nearby objects noise may exceed the desired signal causing interference and detection problems. Thus SODAR have to be installed in open areas or a careful design should eliminate secondary-lobe effects.

For a mono-static SODAR system where the sound coming back to the receiver is only that scattered in the direction of 180 degrees, the backscatter-coefficient of the sound in the atmosphere is known as C_T (Jørgensen & Antoniou, 2002). For high wind speeds, when the turbulence can be high, C_T is low, backscattering less acoustic energy. Also, on cloudy days with small heat flux and temperature fluctuations in the atmosphere, C_T is low. Then in cloudy days with high wind speeds the signal arriving to the receiver may be so small that it could not be recovered from the background noise. Also, rain may limit SODAR operation capability (Jørgensen & Antoniou, 2002).

8.5.2 Some SODAR Characteristics

SODAR systems have maximum height ranges varying from a few hundred meters up to several hundred meters. As an example, some characteristics of two different SODAR systems found in the literature are shown in Table 8.2.

It is observed from Table 8.2 that System B with the higher altitude range has a lower spatial resolution (bigger height interval).

Table 8.2: Some Characteristics of two SODAR systems

Characteristic	System A	System B
Number of beams	3	3
Antenna beamwidth (°)	12	10
Operating frequency (Hz)	3144	2150
Pulse electrical power (W)	300	160
Horizontal wind speed range (m/s)	0-50	not available
Wind accuracy (m/s)	0.1	not available
Altitude range (m)	15-150	70-770
Height interval (m)	5	35
Averaging time for each data (minutes)	10	6

8.6 Radar Wind Profiler (RWP)

8.6.1 Introduction

The RWP is a radar system used to obtain the atmosphere's wind characteristics as a function of height. Its operation principle is based on the backscattering of high frequency electromagnetic waves due to fluctuations of the refractive index in the atmosphere caused by air turbulence.

8.6.2 Operating Principle

The operating principle of the RWP is similar to that described for the radar used to detect flying objects. Important differences are that the antenna of a RWP does not rotate and that it is installed fixed on the earth surface, its beam points vertically and pulses are emitted towards the sky. Another difference is that the antenna is composed of several smaller antennas and the set is called a phased array antenna in a way similar to the SODAR's transducer array. The signals of the individual antennas are controlled in phase and relative amplitude so that the resulting beam may be directed at different angles due to the constructive and destructive interferences of the individual wave fronts (Section (3.5.3)).

The distance and velocity of turbulent structures are estimated from the electromagnetic energy backscattered by the fluctuations of the refractive index of the atmosphere. Because these structures move with the wind speed, a profile of the average wind speed can be described. In order to gain some insight into the RWP characteristics, a brief description of a particular high frequency RWP follows.

A RWP for studying the atmosphere's boundary layer has a phased array antenna in the 915-MHz band of about 3 m in diameter and 1 m tall. The half-power beam width (Sections (2.3.2) and (2.4.4)) is 10° and five beams are sequentially switched at times *t*1 to *t*5 (Fig. 8.10) along five different directions. One of them is vertical and the other four are placed in two vertical mutually orthogonal planes, and their directions are about 15° from the zenith. Although once again the hexagon represents the antenna array and the circles represent the individual antennas, it has to be underlined that in this case they are electrical devices to produce propagating electromagnetic waves. The beamforming process produces the five sequential beams as a result.

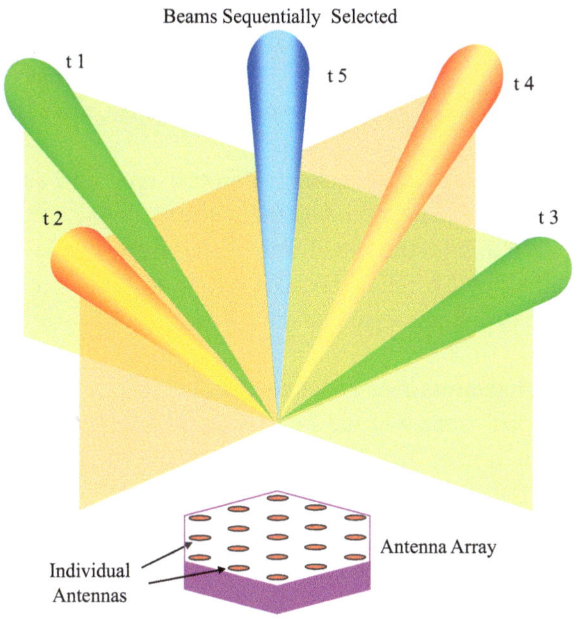

Fig. 8.10: An example of a RWP. The hexagon represents the antenna array and the circles the individual antennas. Each antenna is an electrical device which produces an electromagnetic wave. The beamforming of these individual waves produces the five sequential beams.

8.6.3 Height Coverage and Spatial Resolution

Turbulence varies with the time of the day and the height above ground level. Thus, the availability of RWP data depends on some parameters which are out of the system's user control. The maximum radar range depends on the radar frequency, transmitter power, transmitted pulse length and antenna size, which are under the control of the designer. Sometimes the designer makes some of these parameters selectable by the users, thus making instruments more adaptable to the weather conditions.

As the height in the atmosphere increases, turbulence increases in size (DeTect, 2009), thus the upper atmosphere lacks small scale turbulent structures. As explained in Section (3.5.1) turbulent structures with sizes half the radar wavelength are needed to get the biggest backscattered energy (Bragg law). Therefore, in the upper atmosphere high frequency radar energy (small wavelengths) propagates without suffering backscattering since it does not find turbulent structures that meet the Bragg law. For this reason, lower frequency radar (longer wavelength) has to be used at high altitude to meet turbulent structures that produce strong backscattering.

There are generally three different ranges of RWP systems; the RWP for use in the boundary layer has the shortest range and transmit in ultra high frequency (UHF) around 1 GHz. Since attenuation of the electromagnetic waves usually increases with frequency they have limited height coverage of about 2 or 3 km. The frequency has to be decreased to about 400 - 500 MHz to extend the useful operation range, up to the mid-troposphere (3 to 14 km) and full troposphere (20 km).

Based on radar characteristics (operating frequency, antenna type and size, transmitter power, etc) there are mathematical models that permit the maximum reachable measuring height of RWP to be estimated. These models depend also on atmosphere's temperature and humidity. Therefore, because atmospheric conditions change with time, it is not possible to define a fixed maximum height of coverage for the RWP.

Due to the above reasons, manufacturers usually define the maximum measuring height reachable during a certain percentage of the time. According to the World Meteorological Organization (Dibbern et al., 2003), the relative availability in % is defined as

$$\text{relative availability in \%} = \frac{\text{number of valid values}}{\text{number of possible values}} \times 100 \qquad (8.12)$$

Sometimes in defining the maximum measuring height reachable by a RWP it is recommended to use a relative data availability of 80% (DeTect, 2009).

There is also a minimum height below which the RWP cannot measure. It depends on the type and size of the antenna, the environmental electromagnetic noise in the installation site of the RWP, and the pulse length sent by the radar. This minimum height for boundary layer RWP is from 75 to 125 m and for tropospheric RWP, from 100 to 200 m (with availability greater than 80 %).

Similarly to the case of an Acoustic Doppler Current Profiler (Section (5.4.4)), the energy sent by the RWP antenna is a train of pulses (in the RWP, electromagnetic pulses) whose scatters contain the desired information about wind speed. Short pulses correspond to short distances along the beam direction, and then to high spatial resolution (Eq. (8.8)). As in all remote sensing systems there is a maximum spatial resolution (minimum cell length) that depends on the transmitted pulse length. Higher frequency RWP, as those used for boundary layer, have spatial cells between 50 and 200 m. Tropospheric radar have spatial cells between 250 and 1000 m.

For the same kind of profiler (same frequency, power and antenna), longer transmitted pulses can obtain data from higher tropospheric layers because the backscattered signals are longer and contain more information, which improves the possibility of extracting the signal from the noise. But at the same time, longer pulses will produce lower spatial resolution. With the purpose of solving this compromise, some profilers can work in two modes, one using short pulses for gathering high spatial resolution data but to a limited height; and other with a longer spatial range but with lower spatial resolution.

8.6.4 Averaging Time and Accuracy

The phenomenon being measured (wind) is variable in time and space. Since the antenna beams point to different spatial regions, they may be measuring in different regions of air turbulences giving different instantaneous wind speed values. Therefore, it is necessary to time average several measurements to have representative average values; typical averaging periods are between 15 and 60 minutes.

It has been determined by long-term comparisons with radiosondes that for a one hour averaging period, wind accuracy measured by RWP is on the order of ± 1 m/s in speed and ± 3° in direction (Coulter, 2005).

8.6.5 Installation Site Characteristics

Since RWP are very sensitive to electromagnetic noise they must be installed far from highways because cars' engine sparks could produce interferences. Also, they have to be far from power lines and radio frequency transmitters. The RWP beams have to be free from obstacles such as trees, towers, and bird and aircraft flyways. Other facilities that the site should have are electrical power, availability of data communications and security. In some places an approval for radar operations may be required by local authorities.

8.6.6 How to Specify a RWP

Specifying RWP is a complex task because their performances, such as height coverage, depend on the natural variability of atmospheric conditions and the installation site environment. Some manufacturers have prepared (DeTect, 2009) technical guides to facilitate user's introduction to the subject. These guides recommend collecting as much information as possible about the performance of commercially available wind profilers and to compare the information with users needs. As it happens with other instruments, customers sometimes require specifications that are beyond their real needs, which increase significantly the cost of the system.

8.6.7 Some Additional Considerations on SODAR and RWP Measurements

The wind data monitored by the above-mentioned measurement systems consist in volume averages rather than point measurements. Normally, radar wind profilers provide wind data averaged over 60 to 100 meter vertical intervals whereas SODAR data are typically averaged over intervals from 5 to 100 m.

The resulting data represent the average wind of the layer over which the winds are estimated. Besides their spatial averaging, data are also averaged in time. Averaging periods for wind data from SODAR and radar profilers are usually on the order of 15 to 60 minutes (US EPA, 2000).

It should be noted that systems provide measurements of mean wind speed, but variable wind information such as gusts are not available. This is because many samples in space and time are averaged to obtain one measured value.

8.7 RASS (Radio Acoustic Sounding System)

8.7.1 Introductory Explanation

This introduction presents the working principle of a Radio Acoustic Sounding System (RASS) as a general approach to the concepts involved in this measuring system. Following it, three examples with different techniques to implement the principle are described.

RASS (Radio Acoustic Sounding System) are used for remote sensing the profiles of vertical air temperature and wind speed and direction in the atmosphere (Hennemuth et al., 2012; Kartashov et al., 2008). The range of the profiles reported in the literature is variable because different electromagnetic and sound frequencies are used by different equipments; some reports mention ranges from 0.1 to 2.5 km above ground level whilst others from 1.5 up to 14 km (Chandrasekhar Sarma et al., 2008).

RASS combines the technologies developed for RWP and SODAR. Basically, it consists of an RWP array of antennas and a SODAR array of transducers emitting simultaneously upwards into the atmosphere. The operating principle behind temperature profiling is based on the effect of atmosphere temperature on the speed of sound in air; thus it results a quite indirect mean to sense the atmospheric temperature.

The principle can be explained as follows: Initially, the acoustic component of the RASS system emits an acoustic wave in the vertical direction which produces compression and rarefaction (expansion) of the atmosphere's air (Section (3.3)). These changes in the mechanical properties of the atmosphere alter the electromagnetic refractive index of the air. Note that the perturbation (sound) is a wave that travels in the atmosphere so the refractive index alteration also travels.

Simultaneously with the sound wave, electromagnetic waves are sent into the atmosphere impinging upon the altered properties of the atmosphere, producing the scattering of the electromagnetic waves. Because the perturbations are moving at the sound speed, the scattered energy will contain information on the sound speed.

The speed of the perturbations (sound) can be derived in some way from the backscattered electromagnetic signal. A profile of the speed of sound can thus be obtained, and because the sound speed changes with the atmospheric temperature the temperature profile can be inferred from the speed profile.

In order to obtain a signal easily recoverable from noise, a strong scatter is desired. There is a maximum of electromagnetic energy scattered by the altered mechanical properties of the atmosphere when the electromagnetic wavelength (λ_e) is twice the acoustic wavelength (λ_a) (Eq. (8.13)). When this condition is satisfied it is said that the electromagnetic and acoustic wavelengths are Bragg matched (Section (3.5.1)).

$$\lambda_e = 2\lambda_a \tag{8.13}$$

The propagation speed of the acoustic wave depends on the temperature and moisture of the atmosphere. By definition, the virtual temperature (T_v) expressed in kelvins (K) is $T_v = T\,(1 + 0.61\,r)$, where r is the mixing ratio of water vapor in the air (in kg/kg), and T the air temperature (K). The speed of sound in air (c_a) as a function of the virtual temperature is given by

$$c_a = 20.047\sqrt{T_v} \tag{8.14}$$

Because r is on the order of 10^{-3}, T_v is very close to T (T_v is at most 3 K higher than T), but with an approximate estimation of the humidity profile, corrections to get a more accurate temperature profile can be performed.

As underlined before, due to the use of the same antenna to transmit and receive, pulsed radar cannot "see" the echo during a certain time, which results in a blind distance or minimum detectable range.

8.7.2 RASS Examples

RASS theory and technology are still evolving (Hennemuth et al., 2012). Because of this, dissimilar practical configurations of the method have been implemented; all of them report good agreement between the temperature so estimated and the temperature measured by radiosondes. Some examples of this kind of systems will follow to clarify and materialize the previously described ideas and to show some conceptual differences in the implementations of RASS.

As already mentioned, the RASS has two components, the acoustic wave generator and the electromagnetic wave generator (Fig. 8.11). It has also been explained that wind profile can be estimated by both acoustic (SODAR) and electromagnetic (RWP) waves. For the sake of clarity, some explanations are repeated below when describing the practical implementations of the following examples. Also some details are included in an attempt to visualize the different approaches for solving the same problem.

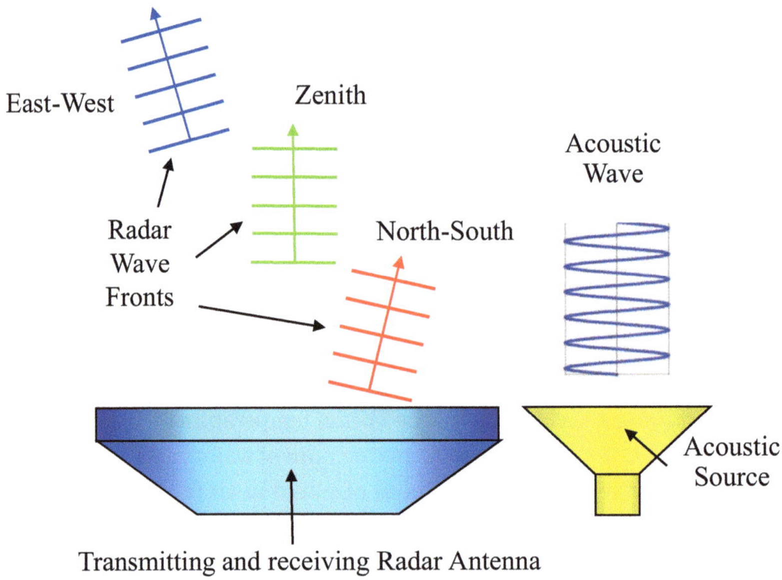

Fig. 8.11: Schematic of a RASS with its two components: the acoustic wave generator and the electromagnetic wave generator.

8.7.2.1 Example # 1

One kind of RASS is described by Pant et al. (2005); it works in a way similar to the general explanation previously described. The radar emits at 404 MHz and as in the RWP case it is capable of measuring horizontal and vertical wind velocities. The typical height coverage for wind is about 6–10 km and for temperature 2–3 km (depending on the weather conditions).

The wind profiler of the system works by emitting pulses of electromagnetic energy and switching the antenna between the emitter and the receiver by means of a duplexer (as in the RWP case). It uses an array of antennas which produce three sequential beams, one tilted along the east-west direction, another tilted along north-south direction, and the third looking at the zenith.

As explained in Section (8.6), **the wind profile** is obtained from the fluctuations of the refractive index caused by **natural turbulence**, which produces the backscattering of the radar waves. The air with different refractive indexes is carried by the wind at its mean velocity, giving origin to a Doppler shift of the backscattered signal. By measuring this shift the mean wind velocity is inferred. A spatial resolution of 300 m for wind measurements is achieved for this particular RASS.

In a period of about one hour at least ten sets of Doppler profiles corresponding to the three beams are available. Data obtained over the total observation period are passed through a process of consensus averaging which helps to eliminate, in part, the effects of transient interfering signals, outliers, etc.

Consensus averaging consists in determining if a certain percentage of the measured values agree within a certain range of differences (e.g., 1 m/s). Those values that meet the criteria are averaged to turn out the radial wind estimate.

The temperature profiler requires combining the radar and acoustic information. The radar operates at a wavelength λ_e and the acoustic wave has wavelength λ_a. The acoustic wave propagates vertically upward in the atmosphere creating regions of compressions and rarefactions which travel at the sound speed. Thus, the acoustic wave creates moving discontinuities of the electromagnetic refractive index in the atmosphere.

When the acoustic source is operated in such a way that $\lambda_a = \lambda_e /2$ it is said that the Bragg match condition is satisfied and the electromagnetic backscattered signal increases its intensity to the maximum. The radar antenna receives strong echoes from this artificially generated diffraction grating. This backscattered signal is Doppler shifted because the acoustic wave is propagating at the speed of sound. The Doppler shift is a measure of the local sound speed at the particular height. The height (range) where the reflected energy comes from is estimated from the electromagnetic pulse round-trip travel time as in Eq. (8.5).

The speed of sound depends on the temperature at the given height where the backscattered signal comes from. Thus knowledge of the Doppler shift as a function of height permits the atmosphere virtual temperature profile to be estimated. The local sound speed is related to the virtual temperature by Eq. (8.14).

Because the speed of sound varies with height due to changes in the atmospheric temperature, for fixed acoustic and electromagnetic frequencies the Bragg match condition would be valid only at some particular heights. In order to cover all the expected sound speeds in the atmospheric region of interest, a certain range of acoustic frequencies should be generated to match the Bragg condition. For the RASS being described, whose radar wind profiler operates at 404 MHz (~ 0.74 m), the range of acoustic frequency used is between 800 and 1000 Hz to cover the Bragg match up to heights of 3 km. The authors concluded that in practice, for the temperature expected in the place where the instrument was installed, a linear acoustic sweep interval of 40 Hz width, or a stepped approximation to such a sweep, could be used.

Going a little deeper, the Doppler shift is due not only to the speed of sound but also to the vertical component of the wind. Thus, the last must be subtracted to obtain accurate temperature measurements. If not corrected, a vertical wind speed of 0.3 m/s would result in a temperature error of about 0.5°C.

8.7.2.2 Example # 2

Another RASS system commercially available (Hennemuth et al., 2012) is quite different to the previous described. A SODAR is used as wind profiler and bi-static type radar (separated transmitting and receiving antennas) is used to measure sound speed (Fig. 8.12). Because the radar does not need to switch the antenna between transmitter and receiver, as was the case in the mono static pulsed radar, it can transmit and receive electromagnetic waves continuously. It consists of an electromagnetic transmitter at 1290 MHz and a corresponding receiver (other frequencies are also available); the transmitting and receiving antennas are separated by a distance usually between 4 and 6 m.

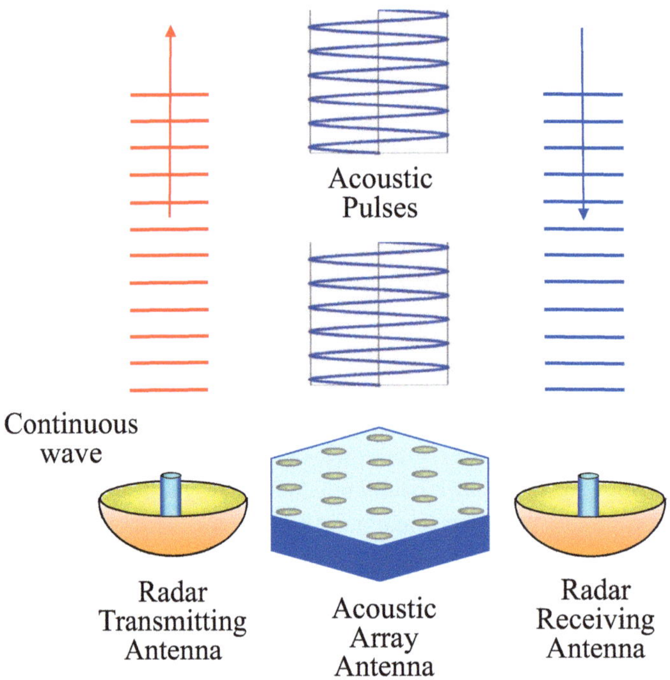

Fig. 8.12: Schematic of a RASS with an acoustic wave generator and bi-static type radar.

The SODAR component (wind profiler) of the RASS system transmits sound pulses in cycles of up to 5 beam directions and gets the Doppler shift of the acoustic signal to estimate the wind velocity (Section (8.5)). For temperature estimation the transmitting cycle of the SODAR is extended with an additional (sixth) sound pulse devoted to sound speed profiling. The Doppler shift of the electromagnetic wave due to the rarefactions produced by this extra acoustic pulse is used to estimate the speed of sound.

Two Doppler shifts are thus taken into account; one of them is the shift of the acoustic wave due to wind speed, and the other is the shift of the electromagnetic wave due to the travelling rarefaction produced by the sixth acoustic pulse. Once the acoustic and electromagnetic shifts are drawn from the respective signals, the same hardware and software can be used for processing the SODAR as well as the RASS echoes because the signal properties of both shifts are very similar (Hennemuth et al., 2012). With this system the temperature profile is known with a height resolution of 10 m, beginning at 35 m above ground level.

The use of bi-static radar permits the lower few hundred meters of the atmosphere to be sampled. It has been reported that this bi-static RASS system improves the overall quality of temperature profiles in the lowest 300 m. The disadvantage is that it requires a correction because the scattering angle is not exactly 180° and is a function of height.

8.7.2.3 Example # 3

Some characteristics of a commercial RASS system were evaluated by a US government agency (Coulter, 2005); they are summarized below. This RASS electromagnetic wave transmitting frequency is 915 MHz and its maximum range is from 3 to 5 km, depending on the atmospheric conditions. According to Coulter (2005), daily comparisons with data derived from radiosondes during more than one year gave the following quality figures:

Accuracy for wind speed: 1 m/s

Accuracy for radial wind components along the transmitter's pointing direction: 0.5 m/s

Accuracy for wind direction: 3°

Accuracy for virtual temperature: 0.5 K

This RASS uses a square antenna array of approximately 4 m side that transmits electromagnetic waves alternately along five pointing directions, one vertical, two in the north-south vertical plane and two in the east-west vertical plane. The distribution of the beams is similar to that shown in Figures 5.17 and 8.10; the beams are tilted about 14 degrees from the vertical. The system includes four fixed acoustic sources located at the corners of the antenna to estimate temperature profile (Fig. 8.13).

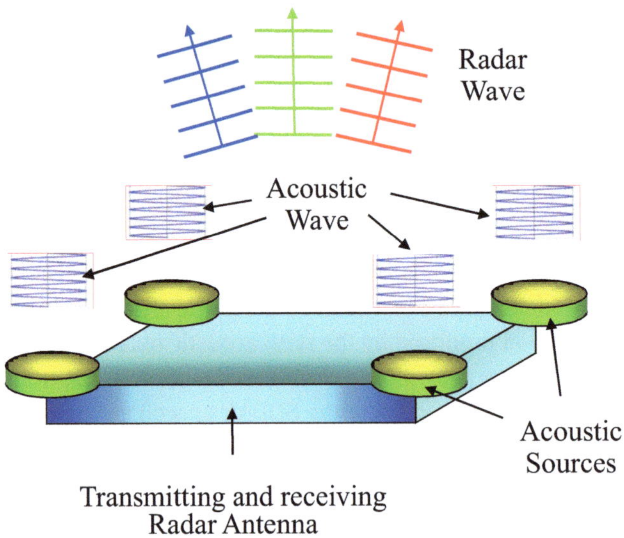

Fig. 8.13: Another type of RASS with a transmitting and receiving radar antenna and four fixed acoustic sources.

The radar transmits sequentially in each beam direction and receives backscattered energy from refractive index fluctuations that are moving with the mean wind speed. Radial components of motion along each pointing direction are determined. It takes 30 to 45 s to determine the radial components from a single pointing direction.

The system cycles through the five beams at low power, and then it cycles again at a high power (longer pulse length); the whole process is repeated every five minutes. About 12 estimates from each beam are saved for an averaging interval of one hour; these values are evaluated at the end of the period to determine the consensus-averaged radial components of motion. The radial values are then combined to produce the wind profile. The wind data reported by the system are: height, speed, direction, radial components, number of values in consensus, and signal to noise ratio.

When the beam of the radar is pointing vertically, an acoustic pulse of about 1-10 kHz is transmitted simultaneously into the atmosphere. Now changes in the index of refraction produced by the travelling sound wave are the source of the radar wave backscattering. The combination of the sound wave speed and the vertical wind velocity is estimated from the Doppler shift of the backscattered wave. Then it is sometimes necessary to compensate for vertical wind. Finally, the temperature profile is estimated from the sound speed as in the previous examples. The averaging time for this estimation is about 10 minutes. In normal operation, temperature profiles are determined during the first 10 minutes of every hour and the wind profile is averaged over the remaining 50 minutes.

8.7.3 Summary and General Considerations about RASS Systems

Wind speed in RASS systems is determined from the Doppler frequency shift of the waves, which are scattered due to **natural fluctuations in the refractive index** present in the atmosphere. The waves may be either acoustic (as in SODAR) or electromagnetic (as in RWP).

The virtual temperature is determined from changes in the speed of sound in the atmosphere. With the purpose of measuring the speed of sound a vertical sound wave is generated which **induces changes in the refractive index of the atmosphere.** The detection of the propagating velocity of this diffraction grating can be performed by means of the Doppler shift of the backscattered **electromagnetic wave.** The directly measured parameters have to be converted into the parameters of scientific interest: wind velocity and temperature profiles.

When analysing data from RASS systems, it should be taken into account that the wind speed is derived from backscattered signals coming from several beams spatially separated. To obtain the wind vector at a single height, the information of the cells of all the beams at this particular height are processed. The estimation of the wind assumes that the phenomenon is horizontally homogeneous, which could not be true. When the height increases, the separation between beams increases and the cells taken to calculate the wind are further apart.

Frequently Asked Questions

Coulter (2005) found that a very frequent question from users of RASS and balloon sounding system (or radiosonde) is: Why don't the profiler's values of winds and/or temperature agree with values from radiosondes?

This question should not surprise us because as explained in Section (7.2.4), even two anemometers installed close to each other, with the same operating principle but different rotation sensors (propeller and cup), measure different wind values; moreover, different cup anemometers (same rotation sensors) have different behaviors in turbulent wind (Section (7.2.4)).

These behaviors are due in part to differences in the transference of sensors; that is to say, both instruments respond in different ways to the same stimulus. But it is also because wind is not the same even in close proximity; in other words the stimulus changes spatially and instruments are "sensing different winds".

The answer to the frequent question follows the same reasoning as above. Actually, each instrument (RASS and radiosonde) "senses" the stimulus in a different way because they average the measurand with different time and spatial constants (different filters). The radiosonde measures with "traditional" instruments with high spatial resolution and fast time response; profilers, instead, average over greater volumes of atmosphere and average several measurements over a longer time. Also, it is true that they do not measure the "same" wind or temperature.

As introduced in Section (8.3), radiosondes perform spot measurements and travel with the mean wind in vertical and horizontal directions simultaneously; they do not measure in a vertical direction. The RASS measures over the same vertical column of the atmosphere and the resulting data is the outcome of values measured over a volume, which was previously defined as the resolution cell. This cell can range from tens to hundreds of meters in length, and have the diameter of the antenna beam. The RASS provide values averaged over a long time (usually 1 hour) and the radiosonde measure at one instant. Another difference is that the radiosonde measures the temperature directly and the RASS estimates the temperature from the speed of sound.

Summing up, the balloon sounding system and the RASS measure on different volumes of the atmosphere; that is to say, they measure different stimuli; the spatial and time scales of both measurements are quite different.

As a preliminary conclusion, it has to be accepted that understanding how instruments work allows us to recognize what is the measurand they are really measuring and why differences in measurements appear. These are some of the goals of this book.

8.8 LIDAR (Light Detection and Ranging)

In the same way as radio frequency waves are used by radar to study the atmosphere, the lidar uses laser pulses of very high frequencies (some in the frequencies of visible light) with the same purpose. Because of the small wavelengths (~ 0.4 - 2 μm) and high directivity of lasers, laser pulses are scattered by small dust particles (aerosols), particles much smaller than those “seen” by radar waves. The spatial resolution of the lidar is of a few meters, resulting much better than that of the radar. Often, the lidar wavelengths used for atmosphere studies range from the ultraviolet to the infrared.

Figure 8.14 is a schematic of the functioning principle of a lidar for the survey of atmospheric parameters. A laser is pointed vertically into the atmosphere and an optical amplifier system (such as a telescope) is pointed toward the laser beam. Aerosols, atmospheric gases and water drops, scatter the laser light along its path.

The optical signal captured by the telescope is filtered and focused on a photomultiplier that converts optical signals into electrical signals. The electrical signals, containing the information of the laser scattered energy, are digitized and processed by a computer, in a similar way as it is done in the radar system. In order to get Doppler shifts the backscattered signal is mixed with an optical local oscillator and then low-pass filtered in the same way as was explained for radars (Eq. (8.9)).

The frequency shift of the backscatters is a measure of the radial wind speed in the lidar line of sight. The distance (range) to the backscattering object is estimated from the pulse propagation speed (light speed) and the round trip flying time of the pulse.

As laser pulses propagate, they illuminate fractions of the atmosphere along the beam direction. Therefore, backscattered signals can be associated with given distances from the emitter; the time delays between pulses sent and received back permit the distance from the lidar to the analyzed zone of the atmosphere to be estimated.

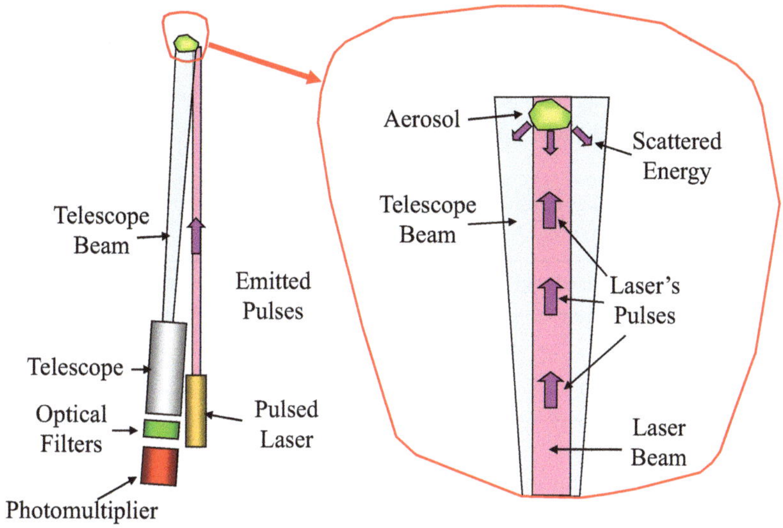

Fig. 8.14: (Left) The elements composing the lidar. (Right) How the laser signal is scattered and captured by the optical system.

Precision timing circuits segment the returned signals into a number of shorter time periods. These periods are linked to specified segments of the radial distance along the beam, called **range gates**. The concept of a range gate is similar to that described in radar systems, where it was called cell length (Section (8.2)). The backscattered signals within each range gate are used to calculate the average radial wind velocity for each gate. This process is repeated all along the beam.

The maximum lidar range depends on the pulse length, the power of the laser and the weather conditions. For example, for a commercial pulsed laser at 1.54 μm (near infrared), with a 50 m long pulse, a maximum height of 6,000 m can be reached under certain atmospheric conditions (http://www.leosphere.com). The minimum range depends also on the pulse length. The emitted laser pulse is so intense that at the first instants of the emission the fraction of the lidar pulse captured by the optics (telescope, filters, etc.) is more intense than the light backscattered by the atmosphere. Then there is a time window in which the lidar is blind, as it happens with radars, and this blind time fix the minimum measuring range. For the same lidar as above, a 50 m long pulse results in a minimum range of 100 m.

The laser sends to the sky high energy pulses of some hundreds of nanoseconds of duration (τ). This duration defines the spatial resolution of the measuring system (S_{res}). It is the ability of a lidar system to distinguish between two targets at different ranges along the same direction (Eq. (8.15)) (as explained for the radar case in Section (8.2), Eq. (8.8)).

$$S_{res} = \frac{\tau\, c}{2} \tag{8.15}$$

where c is the speed of light (3×10^8 m/s). Table 8.3 presents examples on how pulse duration, length of pulse and spatial resolution are related.

Table 8.3: The relation between pulse duration and length and spatial resolution

Pulse duration (ns)	Pulse length (m)	Spatial resolution (m)
100	30	15
800	240	120

The pulse repetition rate (*prr*) should be as high as possible in order to have a more frequent scanning of the atmosphere, i.e., to have more useful information, but it cannot exceed a maximum value to avoid ambiguity between returned signals (Eq. (8.15)). The time between pulses (1/*prr*) must be longer than the round trip time of flight of the pulse to the greatest height to be measured (h_{max}) (Eq. (8.16)) (Cariou & Boquet, 2011). Table 8.4 shows examples of *prr*.

$$prr_{\max} = \frac{c}{2h_{\max}} \tag{8.16}$$

Table 8.4: Examples of pulse repetition rate (*prr*)

Maximum height reached (m)	Maximum pulse repetition rate (kHz)
15,000	10
300	500

As stated above, the lidar measures the radial component of the wind or, in other words, the projection of the wind vector on the laser beam. Thus the schematic presented in Figure 8.14 would be able to measure only the vertical component of wind. Therefore, to provide the ability to measure the wind vector along other directions the beam is headed by means of a device called scanner. It moves the measuring array so as to direct the laser beam upwards along the lateral height of a cone around the zenithal direction (Fig. 8.15). The radial components are combined by means of simple trigonometric relations to get the wind components in the *x*, *y* and *z* directions. For these relations to be applied, some assumptions are made: (i) at a given altitude wind velocity is homogeneous in a horizontal plane; (ii) temporal variations in wind velocity are slow enough so as to consider that all measurements are simultaneously performed; and (iii) wind can be considered as constant in each one of the spatial cells into which the laser beam is divided.

The lidar beam may suffer attenuation due to three processes acting simultaneously in the atmosphere, namely energy absorption by molecules, molecular scattering and particle (aerosols) scattering. With intense fog or low clouds, the lidar cannot operate due to the high atmospheric attenuation

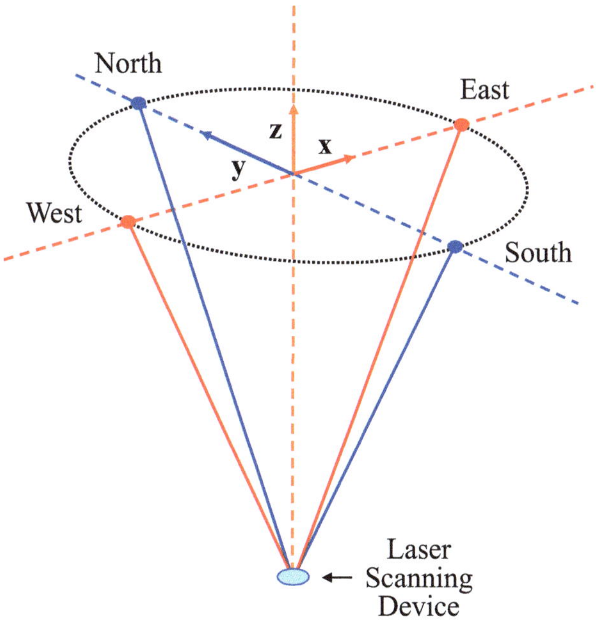

Fig. 8.15: The system shown in Figure 8.14 is rotated for the beam to describe an inverted cone with a vertical axis. The small dots represent the range gates (spatial cells) where the wind radial components are measured to be later converted into wind components along the *x*, *y* and *z* directions.

For the laser beam to be eye-safe it must be of wavelengths greater than 1.4 μm. The improvements of solid state laser technology makes the near infrared spectrum (1.4 μm - 2.2 μm) widely used for wind measuring lidar because they have good sensitivity to Doppler shift, good atmospheric transmission and are eye-safe (Cariou & Boquet, 2011; Rocadembosch et al., 2003).

Classic lidar systems emit very short high power laser pulses, but there are systems based on solid-state technology, known as microlidar, which are eye-safe and of low cost. They are of low power and modulate the optical carrier with pseudorandom sequences that permit the information from the noise to be extracted easier.

Lidar may have quite different specifications according to the desired application. For example, there some lidar used to evaluate installation sites for wind farms, which measure on a short range with great accuracy. Technical specifications of a commercial lidar for wind farm evaluations are presented in Table 8.5 (http://www.leosphere.com). Another lidar from the same reference, but for larger range applications, is presented in Table 8.6.

Table 8.5: A commercial lidar for wind farm evaluations

Range Min-Max	**40 to 200 m**
Data sampling rate	1 s
Number of measurement cells	12
Speed accuracy	0.1 m/s
Speed range	0 to 60 m/s
Direction accuracy	2°

Table 8.6: Another lidar, but for larger range applications

Maximum range	**15,000 m**
Wind measurement range on aerosol, depends on the range gate width and the accumulation time.	100 – 10,000 m 50-2,5000 m
Averaging time	1-10 s
Range gate width	20-50-100 m
Number of programmable gates	120-240
Radial wind speed accuracy	0.2 m/s below 1,500 m; 0.3 m/s above
Cloud detection	15,000 m
Azimuthal scanning	0 to 360°
Elevation angle	-10 to 190°
Angular resolution	0.5°
Maximum rotation speed	4°/s

Lidar are used to estimate the density of aerosols, the speed and direction of wind, the concentration of chemical species, temperature profiles and to measure height and thickness of clouds. When buying a lidar, it should be chosen according to the desired parameter to be studied (Rocadembosch et al., 2003). Also, lidar onboard ships and helicopters are used to measure coastal bathymetry; onboard planes, are used to get images of cultivated areas, canopy and forests. For more information on these applications more specifics books should be consulted (Campbell, 2008).

8.9 Weather Radar

The upper-air remote monitoring systems presented hitherto use fixed antennas placed at the ground level that are pointing to the sky. Instead, the weather radar is similar to the radar for airport traffic control, which has a rotating antenna for emitting and receiving electromagnetic waves (see Section (8.2)).

In order to understand the working conditions of the weather radar it is convenient to turn back briefly to the scattering of X rays by a crystal lattice (Section (3.5.1)). Recall that we had underlined that two matters were worth noting in the scattering of X rays by a crystal lattice, as well as in the scattering of electromagnetic and acoustic waves in general. First, the dimensions of the scattering sources are smaller than the wavelength of the incident wave and second, the distance between scattering centers is comparable to the incident wavelength. This is exactly what happens with weather radar systems used for detecting and quantifying rainfall. The dimensions of the scattering sources (raindrops) are smaller than the wavelength of the incident wave (on the order of some centimeters), and the distance between scattering centers (raindrops) is comparable to the incident wavelength. Therefore, because of the size of precipitation, most radar systems used to estimate rainfall are in the C-band microwave frequency, for example 5.6 GHz, which gives a wavelength of about 5 cm (Mushore, 2012). These radars have a spatial resolution of 2.5 km and provide a reflective map every 5 minutes.

The radiation backscattered to the radar antenna is proportional to the density of the water particles. In a volume of atmosphere illuminated by the radar, the power of these scatters is proportional to the sixth power of the particle diameter. Therefore, the radar reflectivity Z (or reflective factor) is (Mushore, 2012)

$$Z = \sum_i N_i D_i^6 \tag{8.17}$$

where N_i is the number of drops with a diameter D_i per unit volume of air.

The energy of the returned pulses depends on the amount of particles, particle size and shape, and particle state (solid or liquid):

(http://www.bom.gov.au/australia/radar/about/what_is_radar.shtml)

In order to scan a higher volume of the atmosphere, the elevation angle of the radar is not fixed as in the case of the traffic control radar, but the antenna is tilted after each rotation, thus it scans a different volume of the atmosphere each time (Fig. 8.16). This procedure is automatically repeated several times until the maximum tilt angle is reached. All the procedure takes a few minutes.

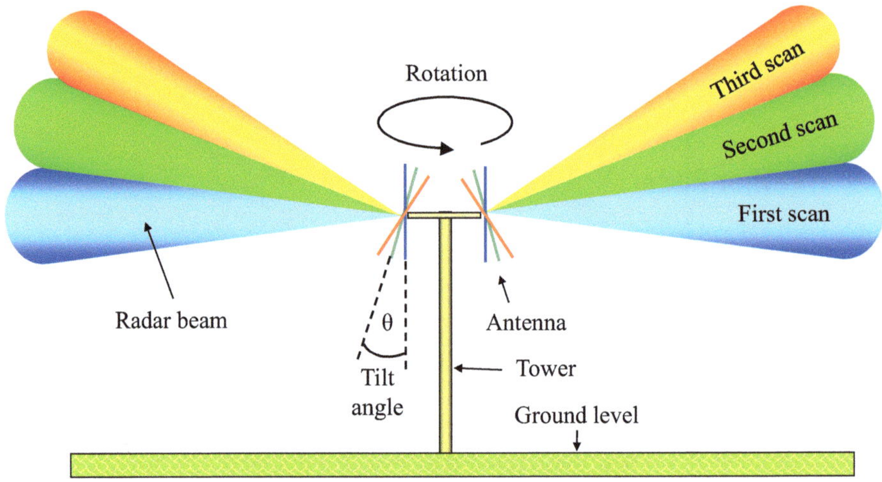

Fig. 8.16: The antenna rotates and the beam scans the atmosphere. After the first complete rotation the antenna is tilted and a second rotation begins. Each scan of the beam is shown with a different color.

The reflective factor Z (mm^6/m^3) must be converted into rain rate R (mm/h) using the reflectivity-rain rate (Z-R) relationship (Mushore, 2012):

$$Z = a\,R^b \tag{8.18}$$

where a and b are empirical coefficients. This means that a calibration has to be performed for each radar system in order to obtain such coefficients.

Rain radar data is generally presented to the users as an artificial colored map where each color corresponds to a different intensity of the precipitation. The map represents a given geographical area that depends on the radar range. The center of the map corresponds to the position of the radar. As shown in Section (8.2), this kind of map is called plan position indicator (PPI).

Due to the drag coefficient of air, falling-rain droplets are larger in the horizontal axis than in the vertical one. For this reason, in order to receive the maximum signal reflection from precipitation, traditionally C-band radar send horizontal polarized radar waves. In recent years, some radar systems are being modified to send alternatively horizontal and vertical polarized waves. This strategy allows more information about the shape and other precipitation's characteristics to be obtained. The new kind of radar is called C-band polarimetric radar or dual polarization radar.

Its operation is based on the fact that rain, solid precipitation, birds, insects, etc, produce different vertical and horizontal scatters. It is said that they have different reflective factors in both directions (Z_v and Z_h, respectively). For example, the ratio Z_v/Z_h is a good marker of the rain drop shape.

In spite of the capability of the meteorological radar to estimate precipitation in real time over large areas, measurements are affected by some sources of errors. One source of systematic errors is the reflectivity-rain rate relationship (*Z-R*) utilized to transform the radar reflectivity maps into precipitation maps (Piccolo & Chirico, 2005). For example, for the Royal Netherlands Meteorological Institution (KNMI) these coefficients are $a = 200$ and $b = 1.6$. But in a study performed for a particular area of the Netherlands it was found that the coefficient a ranged from 70 to 115, and the coefficient b from 1.1 to 1.7 (Mushore, 2012). Other used values for these coefficients are: $a = 300$ and $b = 1.4$ (Hunter, 2009). Therefore, adopting incorrect values for these coefficients could be an important source of error.

Improving the calibration of this relationship for a particular area can reduce these errors. This is done by correlating the radar data with a considerable amount of rain gauge data recorded simultaneously over a long period of time.

Another source of error is that if radar detects rain that does not reach the ground, then radar information does not agree with rain gauges installed at ground level. This phenomenon is called Variable Intensity Rain Gradient Aloft (VIRGA) and it happens when falling from a cloud, rain or hail evaporates before arriving at the ground (http://www.bom.gov.au/australia/radar/about/what_is_radar.shtml)

Because of the Earth's curvature and the decrease in the atmosphere refractive index (due to the decreasing density as height increases) the radar beam draws away from the Earth's surface. Thus, the beam of the radar does not detect rain in the low atmosphere when it is far from the radar. For example, radar that estimates 100 per cent of the actual rain-gauge amount at close ranges, detects only 25 per cent at a range of 100 km (World Meteorological Organization, 2008).

More recently, some errors were found to be introduced by the temporal and spatial scales that the radar utilizes in estimating the precipitation (Piccolo & Chirico, 2005). That is, the spatial resolution of the radar, the time interval between radar scans and the time performing each scan, play an important role in the results. In this research one C-band dual polarization radar with high spatial and temporal resolution was used. The range resolution was 75 m and a complete PPI could be acquired in just one minute (when a single elevation angle was used). In processing the data three different spatial scales (600 m, 1200 m and 2400 m) were used, which are close to the typical spatial resolutions of commercial operational radars.

This work showed that the error made in the estimation of the accumulated rainfall map can be significant for the sampling interval of 5-10 min usually employed in operational radars. In particular, for a 10 min sample interval with spatial scales between 600 m and 2400 m, the Normalized Standard Error ranges from 20 to 30 %, with error decreasing as the spatial scale increases.

Radars have the capability of measuring over inaccessible areas but are less suited to give accurate absolute rain-amount information. It is unlikely that radars will replace rain gauges because rain gauges provide additional information and are necessary to calibrate radar measurements. Radar is useful for the detection of potentially severe thunderstorms and tornadoes; also, digital processing of reflectivity can be used to detect hail. Nowadays computer systems can blend radar data from some radar with other type of information such as rain gauge data. The use of the Doppler effect on weather radars has led to the identification of some characteristic patterns or "signatures" of some meteorological phenomena such as a vertical column of rising rotating air (typically 2 to 10 km in diameter) called mesocyclone (World Meteorological Organization, 2008).

8.10 Measurement of Sea Surface Currents

Sea surface currents may be remotely estimated by means of high-frequency (HF) radar installed on the shore. The radar antenna emits electromagnetic pulses in the frequency range from 3 to 30 MHz, the corresponding electromagnetic wavelengths (100 to 10 m) being comparable to those of the mechanical surface ocean waves. For some wavelengths there exists a resonant backscattering of the electromagnetic wave emitted by the radar (Paduan & Rosenfeld, 1996). The scatters which arrive back to the shore are captured by means of antennas disposed in special arrays co-located with the emitting antenna The HF radar signal scattered by the sea waves contains information on the surface ocean currents.

Within the above-mentioned radar frequency band, sea waves will produce a Bragg scattering when the wavelength (λ_R) of the radar pulse is twice their wavelength (λ_S) (Section (3.5.1)). To see why this is so, consider the schematic of a HF radar (O) located at a height z over the sea shore (Fig. 8.17).

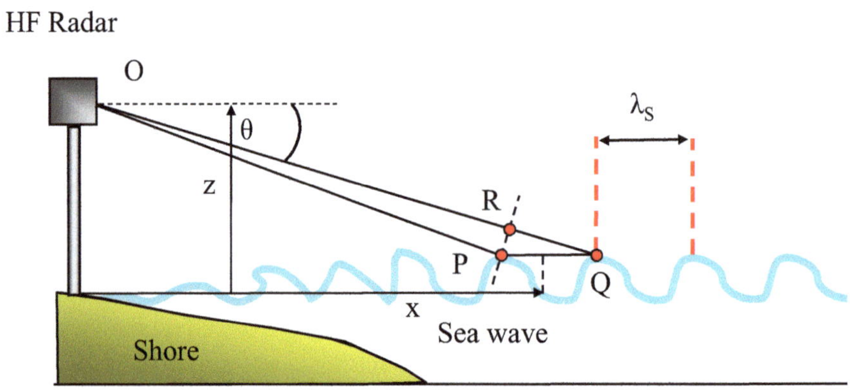

Fig. 8.17: Schematic of HF radar beams impinging sea waves.

A beam of rays strikes the crests of two consecutive waves, P and Q, located at a mean distance x from the shore. Let θ be the angle that the beam makes with the horizontal. The HF radar waves are backscattered by the sea wave crests P and Q and captured by the receiver antennas. A radar wave front drawn through P intersects the ray OQ at R (recall that a wave front is perpendicular to the direction of wave energy propagation). In the triangle PQR, the angle at Q is also θ.

Now, on coming back to the receiver antenna, the backscattered radar wave that struck the sea wave crest at Q will have traveled an additional distance $2\,RQ = 2\,PQ \cos\theta$ with respect to the radar wave that struck the sea wave crest at P. Since the angle q is very small, $\cos\theta \approx 1$, so $2\,RQ \approx 2\,PQ = 2\,\lambda_S$, where λ_S is the wavelength of the sea wave. If the backscattered radar waves are to arrive at the receiver antenna so as to interfere constructively, i.e. if they are to arrive in phase, the additional distance $2\,\lambda_S$ must be an integral multiple n of the radar wavelength, i.e.

$$2\,\lambda_S = n\,\lambda_R \tag{8.19}$$

where λ_R is the radar wavelength. Since the maximum energy of the backscattered wave is obtained for $n = 1$, it follows from Eq. (8.19) that this is equivalent to $\lambda_S = \lambda_R/2$. This is again the Bragg matching concept already presented in Section (3.5.1). In other words, for a radar frequency band from 3 to 30 MHz, the maximum radar wave energy arriving at the receiving antenna comes from those waves that have been backscattered by sea waves with a wavelength range from 50 to 5 m.

The Bragg scattering produces a very large energy peak at the receiver whose frequency can be determined accurately among the reflections from other surfaces (Paduan & Rosenfeld, 1996). The received signal, which carries the information on the sea waves, can thus be processed.

Because the target (sea waves) is moving at a certain velocity (v), the backscattered frequency is shifted with respect to the frequency sent by the radar due to the Doppler effect. The shift (Δf) of the backscattered frequency with respect to the radar-sent frequency (f_0) is proportional to the target velocity (v); they are related as in Eq. (3.8) (Doppler effect).

Two mechanisms are recognized in the wave displacement: the surface current speed (Vsc) and the phase speed of waves (u). Then Eq. (3.8) is presented slightly modified in Eq. (8.20), where c is the propagation speed of the electromagnetic wave.

$$v = Vsc + u = \frac{\Delta f\, c}{2 f_0} \tag{8.20}$$

According to the linear Airy water wave theory (or small-amplitude water wave theory), a sea wave of wavelength λ_S is considered to be a deep-water wave if it is in water of depth $h > \lambda_S/2$. The phase speed of deep-water gravity waves (u) is given by (Kinsman, 1965; Dean & Dalrymple, 1984; Sorensen, 1993) as

$$u = \sqrt{\frac{g\,\lambda_S}{2\pi}} \tag{8.21}$$

where g is the acceleration due to gravity. Therefore, since the emitting frequency of the radar (λ_R) is known because it is a characteristic of the radar defined when it is manufactured, and as $\lambda_S = \lambda_R/2$ due to Bragg law; u may be calculated from Eq. (8.21), and *Vsc* from Eq. (8.20). Thus, the surface current speed is obtained.

In order to get a reliable estimate of the spectral peak of the backscattered signal it is necessary to spatially average many observations over the sea surface and also over time. The range covered by this kind of radar is usually from 15 to 70 km, depending on the operating frequency, transmitted power and sea state conditions. Spatial resolution is about 3 km.

The calculated surface current speed is an average estimated over a depth on the order of 1 to 2 m (Paduan & Rosenfeld, 1996; Laws, 2001), depending on the radar operating frequency. High-frequency electromagnetic waves sample the sea current closer to the surface than low-frequency waves do.

The spectrum of sea echoes contains two well defined peaks shifted between about 0.2 and 0.6 Hz from the transmitted radar frequency. From these peaks surface sea currents are estimated (Laws, 2001).

The Bragg scattering occurs for waves traveling in the direction of the propagating electromagnetic wave (i.e. directly towards the radar or away from it), so in order to get a two-dimensional current map over a sea area it is necessary to measure with at least two radar (*A* and *B*) located at different places (Fig. 8.18); the information of both must be combined to estimate the vector current.

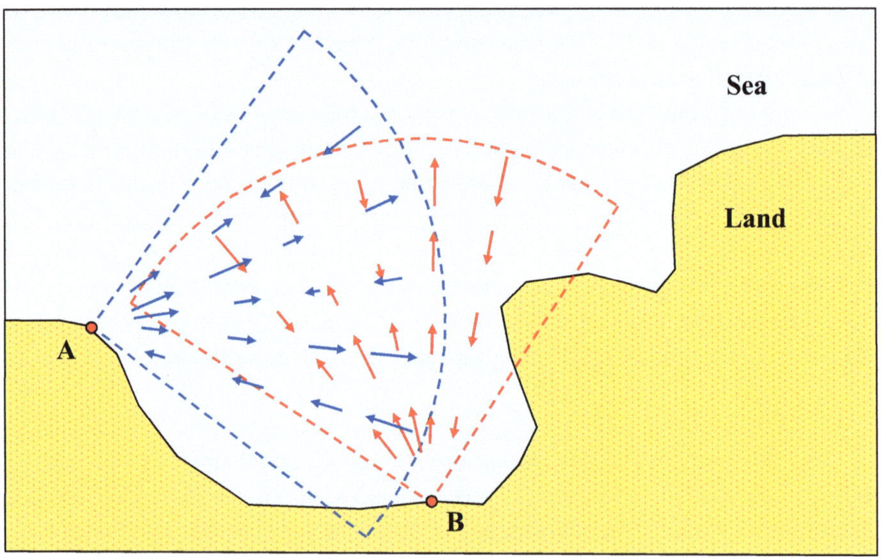

Fig. 8.18: The circles A and B indicate the shore installation of two radar (*A* and *B*). The dashed lines indicate the sea surface area covered by both radar systems. The arrows indicate currents measured by each of the radar. They must be composed to obtain the final current velocity field.

The circles A and B in Figure 8.18 indicate the places on the shore where radar are installed. The angular sectors (dashed lines) indicate the sea surface area covered by each of the radar; the arrows indicate currents measured by each of the radar. These measurements should be geometrically composed to obtain the final direction and amplitude of the average current in the area. The angular range covered by an array of radar antennas may be between 90 and 360° degrees. The angular resolution of an individual antenna may be from a few degrees to tens of degrees.

In general, the most frequent wavelengths of ocean waves, which produce the Bragg resonance, are from 5 to 50 m and the phase speeds in deep water range from about 3 to 10 m/s. For deep-water gravity waves, λ_S is given by

$$\lambda_S = \frac{g}{2\pi} T^2 \tag{8.22}$$

where T is the wave period. It follows from Eq. (8.22) that for the above sea wavelengths, the corresponding periods range from 5.7 to 1.8 s.

8.10.1 Example # 1

A real case of a radar installation is presented in Laws (2001) and will be described below. This system operates on four transmitting frequencies 4.8, 6.8, 13.4 and 21.8 MHz in a sequential mode, so that measurements on the four frequencies are done very fast. The transmitted pulse period is adjustable between 10 and 200 μs and establishes the range resolution in about 3 km. As explained before (Section (8.2)), there is a distance where the radar is blind. For this particular system the minimum distance from which information can be usable and currents can be measured is about 6 km from the radar's antenna.

In order to transmit with four frequencies, four power amplifiers with their corresponding antennas would be required. With the purpose of diminishing this number, the system was divided into a low band section operated at 4.8 and 6.8 MHz and a high-band section operated at 13.4 and 21.8 MHz, thus each one requires only one power amplifier. The power amplifier produced a signal pulse of about 50 W, but power can be increased to about 250 W.

Two separate half-wave vertical monopole antennas were used to irradiate the sea surface; each antenna has a resonant filter which permits them to be used to transmit two frequencies. The energy emitted by these antennas are backscattered by the sea waves and received by an array of antenna elements which works as a single directionally variable antenna (Section (3.5.3)). This array permits the direction where the signal comes from to be known. The elapsed time between the moment the pulse was sent and the instant the backscattered signal arrives allows the distance traveled by the signal to be determined.

The array consisted of eight loop antennas placed about 1 m above the ground and disposed aligned along the shoreline over a distance of about 50 meters, and separated by about 7 m (Fig. 8.19). The time delay between elements of this line of antennas is used to sequentially point the composed antenna at different directions as explained for the technique called beamforming (Section (3.5.3)). Recall that beamforming consists in shifting the phase of the signals coming from the individual antenna elements in such a way that when all the signals from the elements are summed, only those from a given direction will add in phase (Laws, 2001).

Each antenna has it own pre-amplifier and the signals from them are multiplexed and sent to the data processing unit. The radar system works approximately 12 minutes each hour to produce an hourly data file.

A boat with a transponder was used with the aim of calibrating the system. In this case a transponder is a device that emits an electromagnetic signal at a frequency similar to that of the radar. Thus, positioning the boat in known places and emitting signals which simulate those backscattered by the sea it is possible to calculate the amplification and phase shift needed for each antenna in order to properly calibrate the array with the intention that it effectively positions the boat. In other words, corrections were calculated so that the radar determination of the successive transponder positions agrees with the real positions. In the system presented herein the accuracy in the angle determination was estimated in ± 1.5°.

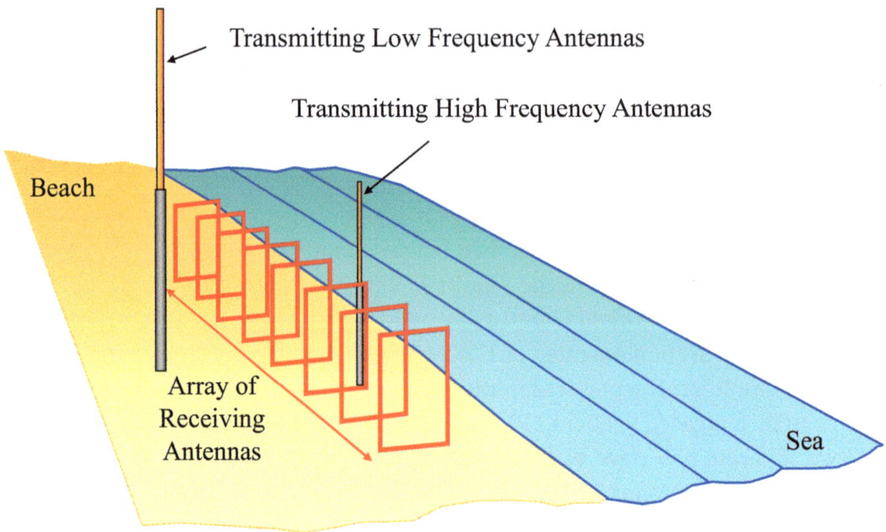

Fig. 8.19: Example of an array of transmitting and receiving antennas for measuring ocean currents (Laws, 2001).

8.10.2 Example # 2

Another radar system called Coastal Ocean Dynamics Applications Radar (CODAR) uses a more compact set of co-located antennas which sweeps the sea surface pointing at different directions. It requires much less space for installation and results easier to transport than the array of antennas previously discussed (Fig. 8.20). Two crossed looped antennas were mounted in orthogonal orientation with a single monopole antenna at the center (Paduan & Rosenfeld, 1996). CODAR operates at about 25 MHz so that the radar wavelength of 12 m resonates with ocean waves 6 m in length. The angular sector covered by the system is 120° and the range extends up to 15 km (Kovacevic et al., 2004). The system provides hourly radial velocities at 5° angular resolution and the data from two or more systems are combined to produce vector current maps with magnitude accuracy of 0.07 m/s and 10 degrees in direction.

Fig. 8.20: Pole with CODAR antennas (Courtesy of Andrea Mazzoldi CNR, Italy).

Some wide radar networks (more than 60 nested systems) to routinely monitor surface current in large areas have been reported (Garfield et al., 2011). In this case, CODAR systems of different characteristics were employed to cover a long coastal area of California, USA. Several 5 MHz systems that have an average range of approximately 180 km and spatial resolution of 6 km were nested with 12 and 25 MHz systems to cover ranges from 40 to 90 km with 1 or 3 km of spatial resolution.

In smaller areas where better spatial resolution is required, smaller networks of higher-frequency CODAR have been employed (Garfield et al., 2011). Four 42 MHz systems with a spatial resolution of 400 m are reported in the San Francisco Bay, California. This radar frequency produces a coherent Bragg scatter with waves of 3.5 m in length. These radar waves perceive the influence of currents to a water depth of 1.8 m and currents can be estimated with an accuracy of about 0.085 m/s. It is expected that with the addition of two CODAR of the same characteristics, the spatial resolution of this array will be increased to 200 m and the angular grid to one degree. It was also estimated that a 30 minute averaging time is adequate to predict currents in almost "real time". This information can be used to assist the racers of a boat competition with surface current information in planning the race strategy (Garfield et al., 2011).

An interesting finding followed the 9.0 magnitude earthquake off Sendai, Japan, on 11 March 2011. Tsunami signals were observed at five HF radar sites installed with the purpose of monitoring surface currents. Radar were installed on different continents and separated by a distance of 8,200 km. Authors report that the expected current flow and velocity oscillations due to the tsunami have been observed in the data from Hokkaido and California a short time before impacting on the shore (Lipa et al., 2011). They conclude that HF radar installations, in some locations where shallow-water bathymetry extends well offshore, could provide capability for the detection of approaching tsunamis.

In summary, the theory and technology of radar systems to monitor ocean dynamics from the shore have evolved quickly in the last 20 years. Radar systems were reduced in size and "in situ" data processing become feasible. At present it is possible to estimate ocean surface currents in wide areas with accuracy better than 0.08 m/s and with grid angular resolution of about one degree. Also, in some cases it seems possible to predict currents in a 30 minute averaging time which, for certain applications may be considered almost "real time". In some particular coast bathymetry it seems a useful tool in the early detection of tsunamis.

8.11 Ground Penetrating Radar (GPR)

It is a technique used to investigate the shallow subsurface of the Earth. It can provide information about the nature and depth of buried objects, and man-made or natural structures. Also, it has proved to be a useful tool in rescuing buried victims after building destructions (Cist, 2009). In addition, this technology has been used to find the ice thickness in the Polar Regions (http://www.sensoft.ca/FAQ.aspx). As in all of the radar already studied it works by emitting an electromagnetic wave and receiving its echoes. The information about the buried objects is derived from the echoes. Two antennas are used in this technique, one to emit and another to receive.

The GPR antennas are placed over the surface of the earth; one of them emits electromagnetic energy into the soil, exactly in the opposite sense to that of the radar used to profile the wind in the atmosphere (Section (8.6)). For this reason, this method uses principles analogous to those used in the seismic reflection method (Daniels, 2000).

The electromagnetic waves radiated into the subsurface travel at a speed (v), which is a function of the permittivity (ε) of the material. Then the propagation speed is different in materials with different electrical properties. The speed is inversely proportional to the square root of the permittivity times the permeability (μ) of the material (Skilling, 1974),

$$v = \frac{1}{\sqrt{\mu\varepsilon}} \tag{8.23}$$

Since the permittivity of earth materials is always greater than the permittivity of air, the speed of a wave in a material other than air is less than 3×10^8 m/s. The time interval required for a wave to travel from the transmitting antenna to the receiving antenna is called the **travel time.**

Let us consider the transmitting and receiving antennas placed as in Figure 8.21. In real instruments, both antennas are mounted on a trolley that permits them to be displaced on the Earth surface keeping their relative distance constant. When a pulse is radiated by the transmitting antenna (T_A), the receiving antenna (R_A) will receive first a pulse which traveled directly through the air (D). After some delay other pulses (R) will be received by the antenna; these correspond to waves reflected or scattered by subsurface objects. They travel through the material and arrive back at the surface. These pulses travel at a velocity determined by the permittivity (ε) of the soil material.

The reflections and scatterings of the single pulse sent from the transmitting antenna produce a sequence of pulses arriving at the receiving antenna that correspond to all the different echoes (travel paths). The recording of these pulses is called a **trace** and it is shown in Figure 8.22 for the example of Figure 8.21.

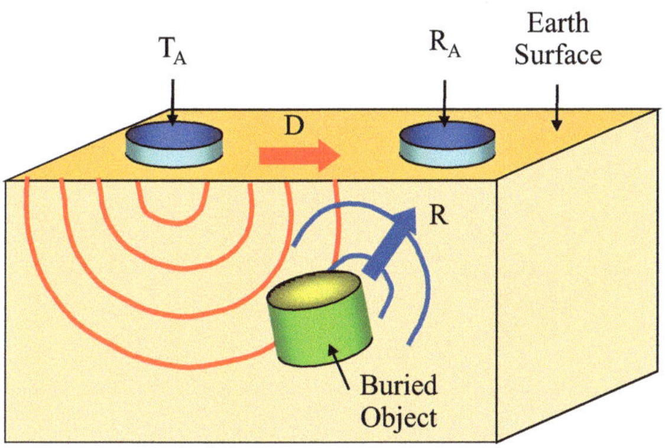

Fig. 8.21: A pulse is radiated by the transmitting antenna (T_A); the receiving antenna (R_A) will collect pulses (D) and (R). The first travels directly through the air; the second corresponds to waves reflected or scattered by subsurface objects.

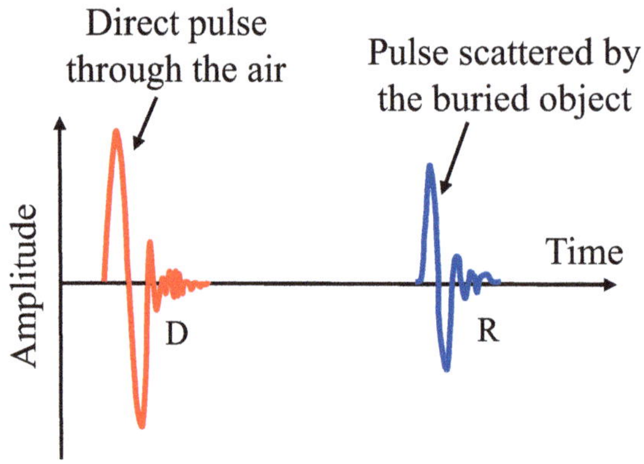

Fig. 8.22: Pulses arriving at the receiving antenna produce a time-dependent record called a trace.

The received pulses have different amplitudes and arrive with different delays; delays are greater for farther buried objects. If the propagation speed of the wave in the subsurface medium is known, delays can be used to estimate the depth of the scattering objects.

Assume that the subsurface material is homogeneous and that only one object of different permittivity is buried in it (Fig. 8.23). Displacing both antennas together on the surface from position P_1 to P_5 permits the object to be "viewed" from different directions. Because the electromagnetic wave travels downwards traces are usually drawn from above to below (Fig. 8.24a).

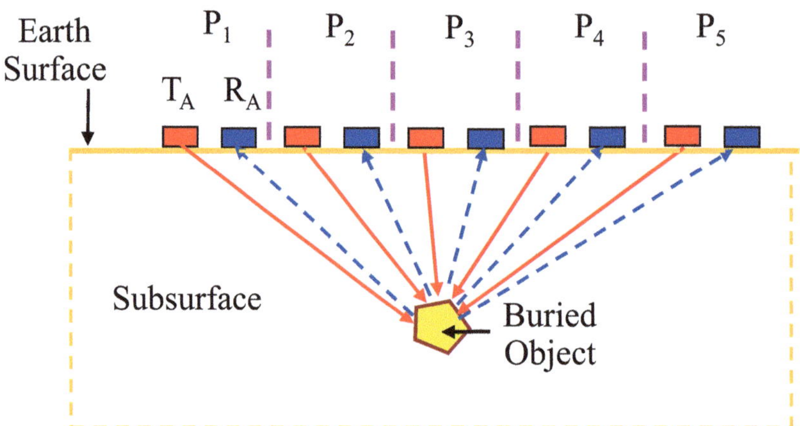

Fig. 8.23: **TA** and **RA** are at a fixed relative distance and are displaced from position P_1 to P_5. At each position a pulse is emitted and its scatter from the buried object is recorded

Drawing the traces for each position, and ignoring the direct pulse through the air, will produce a picture as shown in Figure 8.24a. The amplitudes of the recorded pulses are larger and have a lesser delay when the GPR is closer to the scattering object (position P_3). Pulses may be transformed in a gray scale ranging from white to black. Black is assigned to the maximum amplitude and white to non reflection. A trace to which this gray scale has been applied is called a **scan.** The scans for these traces are drawn in Figure 8.24b.

The GPR scans form an image of the subsurface where the horizontal axis is the surface position of the antennas and the downward axis is the round-trip travel time of the electromagnetic wave. This record is very similar to that of an acoustic fish finder. The buried object is represented as a vertical parabola or a figure with the shape of an inverted V. The closest approach of the GPR to the buried object occurs at the peak of the inverted V.

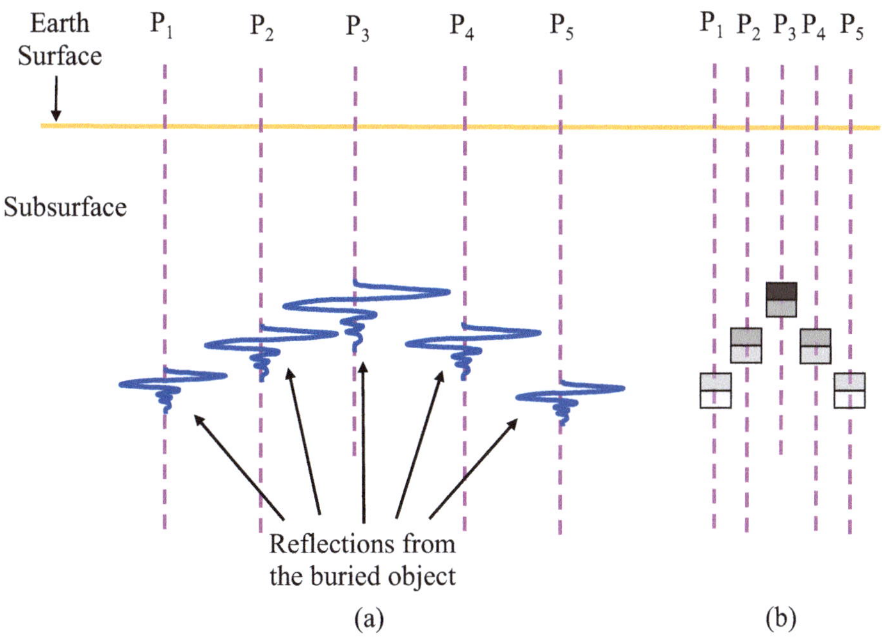

Fig. 8.24: (a) Traces from the buried object. (b) Scans from the buried object.

The electromagnetic wave amplitude decreases exponentially with the distance from the emitting antenna, their scatters becoming undetectable at some point. The attenuation coefficient increases with the electrical conductivity of the material and also with frequency. Then, using radar with lower frequency increases wave penetration into soil, but the price is a loss of spatial resolution.

Radio waves penetrate a few centimeters in sea water, some tens of meters in fresh water and some hundreds of meters in ice (www.sensoft.ca/FAQ.aspx). An approximate way of estimating the exploration depth of a GPR in a given material is by using the approximation of Eq. (8.24), where D is the penetration in meters and σ is the conductivity of the medium expressed in mS/m (www.sensoft.ca/FAQ.aspx).

$$D(m)=\frac{35}{\sigma} \tag{8.24}$$

GPR systems have microprocessor circuits which provide some degree of digital control and allow digital data recording for its post-processing.

So far the GPR has been trolled following only a line, but if it is moved following parallel lines a three dimensional figure of the subsurface can be plotted. The accurate location of each trace is critical for producing reliable 3D pictures. Generally 3D images are constructed from several parallel two dimensional scans as shown on Figure 8.24b.

References

Bahl, M., Gupta, B., John, T., Singh, D., Sharma, O., Garg, S., & Khanna, R. (2011). A phased array acoustic wind profiler for remote atmospheric wind measurements. *IETE Journal of Research, 57* (2), 190-196. doi:10.4103/0377-2063.81738

Bao, X., Webb, D. J., & Jackson, D. A. (1993) 32-km distributed temperature sensor based on Brillouin loss in an optical fiber. *Optics Letters, 18* (18), 1561-1563.

Campbell J.B. (2008). *Introduction to Remote Sensing*. New York: The Guilford Press.

Cariou, J. P. & M. Boquet, M. (2011). *LEOSPHERE Pulsed Lidar Principles, Contribution to UpWind WP6 on Remote Sensing Devices*. LEOSPHERE, Orsay, FR pp. 32. Available at: http://www.upwind.eu/media/576/D6.1.1.pdf

Chandrasekhar Sarma T. V., Narayana Rao, D., Furumoto, J., & Tsuda, T. (2008). Development of radio acoustic sounding system (RASS) with Gadanki MST radar – first results. *Annales Geophysicae, 26*, 2531–2542.

Cist, D. B. (2009). *Non-Destructive Evaluation after Destruction: Using Ground Penetrating Radar for Search and Rescue*. Geophysical Survey Systems Inc., Salem, NH, USA. Non-Destructive Testing in Civil Engineering, Nantes, France.

Coulter, R. (2005). *Radar Wind Profiler and RASS (RWP915) Handbook*. ARM TR-044 U.S. Department of Energy, Office of Science, Office of Biological and Environmental Research.

Daniels, J. J. (2000). *Ground Penetrating Radar Fundamentals*. Department of Geological Sciences, the Ohio State University. Available at: http://www.earthsciences.osu.edu/~jeff/Library/BASICS.PDF

Dean, R. G., & Dalrymple, R. A. (1984). *Water Wave Mechanics for Engineers and Scientists*. Englewood Cliffs, New Jersey: Prentice-Hall, Inc.

DeTect, Inc. (2009). *Radar Wind Profiler Specifications Explained*. Meteorological Systems Group, DeTect Inc Longmont, CO. Document No: 9000190 Availabale at: www.detect-inc.com

Dibbern, J., Engelbart, D., Goersdorf, U., Latham, N., Lehmann, V., Nash, J., Oakley, T., Richner, H., & Steinhagen, H. (2003). *Operational Aspects of Wind Profiler Radars, Instruments and Observing Methods*. Report no 79, World Meteorological Organization. WMO/TD no 1196.

Garfield, N., Hubbard, M., & Pettigrew, J. (2011). Providing SeaSonde High-Resolution Surface Currents for the America's Cup. *Proceedings of IEEE/OES Current, Waves, and Turbulence Measurement Conference, Monterey, CA*. March 20-23, 2011.

Gustafsson, D. (2008). *Remote wind speed sensing for site assessment and normal year correction*. Master of Science Thesis in Energy Technology, KTH School of Industrial Engineering and Management, Department of Energy Technology, Stockholm, Sweden.

Hahn, D. W. (2007). *Raman Scattering Theory*. Department of Mechanical and Aerospace Engineering, University of Florida (USA). Available at: http://plaza.ufl.edu/dwhahn/Raman%20Scattering%20Theory.pdf

Hausner, M. B., Suarez, F., Glander, K. E., et al. (2011). Calibrating Single-Ended Fiber-Optic Raman Spectra Distributed Temperature Sensing Data. *Sensors, 11*, 10859-10879. doi: 10.3390/s111110859

Hennemuth, B., Peters, G., & Kirtzel, H. J. (2012). Temperature profiles with bi-static Doppler-RASS and their correction. *Atmospheric Measurement Techniques and Discussions, 5*, 1399-1408.

Hunter, S. M. (2009). Radar Rainfall Estimation: Capabilities, Limitations and Potential Improvements. WSR-88D. National Weather Service Weather Forecast Office (NOAA) Available at: http://www.srh.noaa.gov/mrx/research/precip/precip.php

Jenkins, F. A., & White, H. E. (1957). *Fundamentals of Optics*. New York: McGraw-Hill Book Company.

Jørgensen, H. E., & I. Antoniou, I. (2002). *Inter comparison of two commercially available SODARS*. Risø National Laboratory, Roskilde, Danmark, Risø-R-1383(EN): Pitney Bowes Management Services Danmark A/S.

Kartashov, V. M., Babkin, S. I., & Volokh, A. V. (2008). Analysis of Current Status of Theory and Practice of Radio-acoustic Sounding Systems. *14th International Symposium for the Advancement of Boundary Layer Remote Sensing, IOP Conf. Series: Earth and Environmental Science* **1** 012052. doi:10.1088/1755-1307/1/1/012052

Kinsman, B. (1965). *Wind Waves*. Englewood Cliffs, New Jersey: Prentice-Hall, Inc.

Kovacevic,V., Gacic, M., Mancero Mosquera, I., Mazzoldi, A., & Martinetti, S. (2004). HF radar observations in the northern Adriatic: surface current field in front of the Venetian Lagoon. *Journal of Marine Systems, 51*, 95-122.

Laws, K. (2001). *Measurements of Near Surface Ocean Currents Using HF Radar*. A dissertation submitted in partial satisfaction of the requirements for the degree of Doctor of Philosophy in Physics, University of California, Santa Cruz.

Lillesand, T., Kiefer, R. W., & Chipman, J. W. (2004). *Remote sensing and image interpretation*. New York: John Wiley & Sons.

Lipa B., Barrick, D., Saitoh, S. I., Ishikawa, Y., Awaji, T., Largier, J., & Garfield, N. (2011). Japan Tsunami Current Flows Observed by HF Radars on Two Continents. *Remote Sensing, 3*, 1663-1679. Available at: www.mdpi.com/journal/remotesensing

Minardo A. (2003). *Fiber-optic distributed strain/temperature sensors based on stimulated Brillouin scattering*. Dottorato di Ricerca in Ingegneria Elettronica, Seconda Universita' Degli Studi Di Napoli, Naples, Italy. Available at: http://www.dii.unina2.it/opto/publications/PhD_Thesis_Minardo.pdf

Mushore, T. D. (2012). *Gauges measurements and ground-radar observation of rainfall over the Water Board district Reggge and Dinkel*. Thesis submitted to the Faculty of Geo-Information Science and Earth Observation of the University of Twente. Enshede, The Netherlands.

Paduan J. D., & Rosenfeld L. K. (1996). Remotely sensed surface currents in Monterey Bay from shore-based HF radar (Coastal Ocean Dynamics Application Radar). *Journal of Geophysical Research, 101* (C9), 20,669-20,686.

Pant, G. B., Joshi, R. R., Damle, S. H., Deshpande, S. M., Singh, N., Vashistha, R. D., Neekhra, P., Chande, J. V., Kulkarni, A. A., & J. S. Pillai, J. S. (2005). Wind profiler and radio acoustic sounding system at IMD, Pune: Some preliminary results. *Current Science, 88* (5), 761-769.

Piccolo, F., & Chirico, G. B. (2005). Sampling errors in rainfall measurements by weather radar. *Advances in Geosciences*, *2*, 151–155.

Rocadembosch, F., Sicard, M., Comerón, A., Rodriguez, A., Muñoz, C., López, M., García, D., & Agishev, R. (2003). Remote sensing mediante lidar en la UPC. *Rama de estudiantes del IEEE de Barcelona, Burán*, *20*, 21-26.

Selker J. S., Thévenaz, L., H. Huwald, H., Mallet, A., Luxemburg, W., van de Giesen, N., Stejskal, M., Zeman, J., Westhoff, M., & Parlange, M. B. (2006). Distributed fiber-optic temperature sensing for hydrologic systems. *Water Resources Research*, *42*, W12202. doi: 10.1029/2006WR005326.

Skilling, H. H. (1974). *Fundamentals of Electric Waves*. Huntington, New York: R.E. Krieger Publishing Company.

Sorensen, R.M. (1993). *Basic Wave Mechanics: for Coastal and Ocean Engineers*. New York: John Wiley & Sons, Inc.

Suárez, F., Aravena, J. E., Hausner, M. B., Childress, A. E., & Tyler, S. W. (2011), Assessment of a vertical high-resolution distributed-temperature sensing system in a shallow thermohaline environment. *Hydrology and Earth System Sciences, 15*, 1081-1093. doi:10.5194/hess-15-1081

US EPA (US Environmental Protection Agency) (2000). *Meteorological Monitoring Guidance for Regulatory Modeling Applications*. EPA-454/R-99-005, United States Environmental Protection Agency, Office of Air Quality Planning and Standards,

van de Giesen, N., Steele-Dunne, S. C., Jansen, J., Hoes, O., Hausner, M. B., Tyler, S., & Selker, J. (2012). Double-Ended Calibration of Fiber-Optic Raman Spectra Distributed Temperature Sensing Data, Sensors, 12, 5471-5485. doi: 10.3390/s120505471

World Meteorological Organization. (2008). Guide to Meteorological Instruments and Methods of Observation., WMO-No. 8 (seventh edition).

http://www.sumitomoelectricusa.com/images/stories/opthermo/ftr3000_e.pdf

http://www.sodar.com/about_sodar.htm, acceded November 2012.

http://www.leosphere.com/products2.php?rubrique=169&cat=wl&item=200sfs

https://docs.google.com/viewer?a=v&q=cache:d9CID27DWkkJ:www.tropmet.res.in/~wprass/Chande-lec.ppt+Radio+Acoustic+Sounding+System&hl=es-419&gl=ar&pid=bl&srcid=ADGEEShoy_yESOfuyqvnOUJ9rTUeTw4SVG7mAzqIrSEs_uZtIrakC4BD_qS1w1dpSgHS0JpWn-tam46zhYMVyaj8HRh_XTbWeXXUmXqZvMESKGFrsBAkA6KC589WGT0_xCOZYJobZhqsh&sig=AHIEtbSBtSSM320I8-sxDKWpFND4L3Iq-Q

http://www.leosphere.com/products2.php?rubrique=157&cat=wl&item=wc200fs

http://www.sensoft.ca/FAQ.aspx

http://www.bom.gov.au/australia/radar/about/what_is_radar.shtml

9 Data Transmission and Storage

9.1 Preliminary Discussion

This subject is so vast that any attempt to address it within the limits of an introductory book has great chances to fail. A way to increase our chances of success is to reduce the subject to a few topics strictly linked to the instrumentation used in environmental sciences, and to the description of some examples of applications. In order to better understand this subject it is recommended that the reader be familiar with the way analog signals are converted into digital ones (Section (3.6)).

All measuring systems produce data as a result, and these data must be evaluated sooner or later, either by an automatic system or a human being; "the final aim of a measurement process is taking a decision" (Ferrero, 2005). The immediacy with which the information should be assessed depends on the purpose for which it was collected. This purpose defines what to do with the information in the steps following the measurement: basically the information generated by an instrument can be stored, transmitted or both.

The storage of information is treated at the end of the chapter, the transmission of data being the backbone of this chapter. It begins with a general introduction and goes on with generic issues on digital data with the purpose of establishing some concepts and the vocabulary frequently used. This generic approach to data communication concepts would hopefully help the reader to understand the transmission methods used at present in scientific instruments. Some characteristics of the transmissions systems, such as transmission delay, local and remote transmissions, network topologies, etc. are described.

A concise mention of analog transmission is done followed by digital data transmission which, due to its importance, is the matter developed most extensively. Several digital transmission concepts, such as signal encoding, transmission modes, serial and parallel transmission, asynchronous and synchronous transmission, error detection and correction, etc., are developed with the aim of providing the minimum needed information to understand how instruments can transmit data.

Finally, some effort is devoted to describe the three most commonly used media to transmit data from a field instrument (or group of instruments) to a central station, namely, private networks, digital telephony and satellite communications.

It has to be stressed that in order to keep the explanations as simple as possible, the descriptions found in this chapter are basic. We do not delve into the details that some communications systems employ to transport data more safely. At present, complex communications systems employ very refined strategies and algorithms to protect the data and correct errors. These approaches, which increase transmitted data immunity and integrity, are merely delineated.

9.1.1 Standardization

Communications in industrial hydraulics have well defined temporal and spatial scales (they need few samples per second, and instruments are commonly disposed in a relatively small area). These characteristics have made it possible to develop communication standards (Modbus, Fieldbus, Profibus, etc.) that are followed by instrument manufacturers. Therefore, to solve the problem of data transmission, the industrial user has only to connect the instrument to the network and setup some parameters in the software. Data transmission in industry is well documented in specific literature and its description is not addressed in this chapter.

As for the transmission of data in environmental sciences, there is a tendency to standardize data management and archiving in specific areas of sciences where data have common characteristics (i.e. marine geology and geophysics). However, due to the very different kinds of data generated by diverse instruments it is difficult to agree on a unique way of data storage and transmission. Moreover, different communication systems have very unlike capabilities of data transportation, which would impede adopting a single transmission way for environmental data transmission. Therefore, in environmental sciences there exist diverse ways of data management that will be addressed below.

9.2 Introductory Concepts

9.2.1 Data Storage

Sometimes the information about a natural phenomenon or an industrial process is collected only for statistical studies or for further analysis. For example, rain precipitation measured for agriculture purposes is averaged over days or weeks and there is no need for the collected data to be processed instantaneously. Also, when directional wave energy is measured for studies on coastal erosion it is necessary to collect information during some big storms before arriving at some conclusion; data is thus not immediately needed. In the above examples it is only necessary to store the information in some kind of storage media for being retrieved and analyzed at a later stage. The process of safekeeping the information for a future use is referred to as **data storage**. The connotation of storage is very broad, even if we reduce its application to the field of instrumentation. Data storage may be performed either in a memory on board an instrument or on a database in an information processing center, which might be located very far from the data collection or measurement point.

9.2.2 Data Transmission

On some occasions real time knowledge of the information is required. For example, when rain is measured in a city with the aim of forecasting floods to evacuate its inhabitants, the data has to be evaluated in real time; when directional waves are measured to decide the transit of ships at the entrance of a harbor, the information has to be available within a few minutes.

In the above examples data has to be known soon, and most of the time measuring systems are not at the same places as the processing and decision maker systems. Therefore, the information has to be conveyed from one place to the other. Distances may be from some meters to thousands of kilometers; the process of sending the information from one point to the other is referred to as **data transmission.**

Also, the significance of transmission is very broad. It may be called the mere action of transferring data stored in an instrument memory to a computer memory adjacent to the instrument by means of a cable. But it can also mean the process of transferring data from a remote field instrument to a processing center through different media; for example via telephone, satellite and Internet networks.

There are different reasons why data generated by instruments has to be transmitted. Very often it is essential for measuring instruments to have the ability of transmitting the information that they generate because they are not accessible and real time knowledge of the information is required, such as in the forecasting of tropical storms. Instruments for measuring snow accumulation in high mountain areas or rain in oceanographic buoys, equipment attached to animals to study their displacements and habits, and obviously instruments onboard satellites, all need to transmit data in order to render the measurements useful.

Besides technical reasons, there may be economic reasons that compel the transmission of the measured data. Simply, it could be cheaper to transmit data than to go to the place where the instruments are installed to collect it. Multiple circumstances may call for data transmission.

9.2.3 Transmission Delay

Transmission of information on a physical medium has a minimum **transmission delay** between the emission and reception instants, which depends on the speed of wave propagation. We stated that measurements are finally used in a decision-making process, and then the transmission delay becomes significant only if it slows down this process. For many purposes the transmission delay due to wave propagation is negligible. In a metallic wire the information is transmitted as an electrical quantity (voltage or current); on a radio link it is transported by an electromagnetic wave that propagates in air or vacuum; and in an optical fiber the information is carried by light.

In all the above examples the propagation speed is close to the speed of light (3×10^8 m/s). Delays are quite short even in long radio links as those through satellites.

It should be noted that even when the wave propagation speed is fast, the data transportation system might introduce additional delays. For example, the time delay between a cellular phone set and a relay station placed 10 km away may be 33 µs, but we all know that a text message between two phone sets may suffer a considerable delay (minutes). It is due to the priority assigned by the phone system to the message. Then in some cases delays introduced by the communication systems have to be avoided. For example, because tide information changes slowly, the transmission system delay would perhaps be unimportant and text messages may be used to transmit it. But if a tsunami warning has to be sent, we would wish no delay introduced by the communication system, and other media should be used.

So far we have focused on communication systems using electromagnetic waves. Acoustic communications are performed by means of a mechanical wave (vibration) that travels generally on a fluid (air or water). The propagation velocity in water is about 1,500 m/s. Thus, underwater acoustic transmissions of information from an instrument to the receiver would introduce a much higher delay than electromagnetic transmission does. Also, the bandwidth of an acoustic link is very restricted, and then the amount of data that can be transferred per unit time is quite limited (about 30 kbit/s). Remember that the bandwidth describes the ability of a system to allow the passage of a signal without significant attenuation and distortion (Section (2.4.4)), and that bits are materialized by pulses of a given frequency. Then, to allow high frequency signals (great amount of data per unit time) to be transmitted, a great bandwidth is needed (http://www.evologics.de). In summary, the capacity of an acoustic communication channel is much less than that of an electromagnetic channel and the delay is much higher. Anyway, in spite of these drawbacks, acoustic links are one of the few ways to communicate underwater instruments, and autonomous underwater vehicles between them or with the surface.

9.2.4 Data Quality and Digital Information

There is a direct relationship between the amount of digital information to be stored or transmitted and the quality requirements for the information being measured and acquired. The first question to answer, even before buying an instrument is: what is the quality of data we need? The answer to this question will define part of the cost of the measuring process. Overestimating data quality could lead to misuse of resources.

In order to know the resources they will need when preparing a measuring campaign, researchers should evaluate the amount of bytes they will store, transmit or process. This amount of digital information is proportional to the wanted resolution and the desired samples per second. Once again, transmitting slow varying phenomena like tide requires less byte than transmitting ocean wave information.

High quality data (high resolution and high sampling rate) requires more bytes per second than a low quality data.

At present, the bottleneck of the field data acquisition flow is the capability of field data processing and transmission. Nowadays memories have great storage capability, and sometimes this leads to the acquisition of more information than needed. Occasionally, it is worth acquiring more information than considered necessary because redundancy prevents data looses due to instrument failures or an incorrect pre evaluation of the temporal scale of the phenomenon being studied. Redundancy in the sense of the field deploying more instruments than are strictly needed is a good practice because a research effort could be incomplete due to the interruption of temporal series if just the minimum amount of required instruments is rigorously used. Also, acquisition of redundant data or high resolution data may be useful for future reprocessing of the information by means of new algorithms.

However, the exaggerated acquisition of information may lead to an unnecessary waste of time and money. When a great amount of data is acquired, a price is paid when the data has to be transmitted. Transmitting large files using satellite channels or cellular networks, which use small-bandwidth radio links, takes long times and generally increases costs. One possibility is to store high quality data but transmitting just that which is necessary. In this case the price is paid in going to the remote site to collect the data. Also, acquiring data with more samples and higher resolution drains more current from batteries and more energy is required. In addition, frequent recharge of batteries may shorten their working lives. Therefore, there is a compromise between quality of the data, storage and transmission, which has to be solved for each particular case.

If it is known that a certain phenomenon has, for example, a maximum frequency of 1 Hz, it would be reasonable to take from 5 to 10 samples per second (sps) to reconstruct the signal easily, but it would be unprofitable to take 100 sps "just in case". Also, if the error of the measuring system is on the order of 5 % it could be useless to store a 10 bits data word for each sample, it would be enough to use only 8 bits (which has a resolution ≈ 0.4 %, Section (3.6.6)). These options (reducing sampling from 100 sps to 10 sps and the resolution from 10 to 8 bits) could reduce the storage, transmitting and processing resources between 12 and 20 times.

The selection of adequate requirements of quality for the data may be crucial in real time processing. The time delay introduced by running models could be decreased by a clever selection of the quality of the information to be acquired.

9.2.5 Local and Remote Transmissions

In the past, local systems were confined to a reduced area such as an industry or a city, where most of the communications were conveyed through copper wires. Because these wires introduce signal attenuation as their length increases, when attenuation becomes significant the signal can be disturbed by noise, thus losing valuable information. Therefore, long distance communications were handled by radio frequency. Then in general, copper wires and radio frequency were synonyms of local and remote transmission systems, respectively.

The arrival of fiber optics (f.o.) with the capacity for transmitting large amounts of data, made it convenient to install very long f.o. cables to replace radio links in remote transmissions. The capacity of optical fibers for transporting huge amounts of data is so great that it became profitable to install transoceanic cables for voice and data communication services. Thus, at present we have communication services over very long distances (remotes) through optical cables. Fiber optics are not affected by electromagnetic noise and do not conduct electricity. Thus, they are very useful even in short distances, when signals have to be transmitted in an electrically noisy environment, or electrical isolation is needed between transmitter and receiver.

Laboratory networks where many instruments simultaneously collect information in a computer are examples of local data transmission. Hundreds of instruments may also be connected in industry to a control process unit through a local network. Networks of computers in universities are also local networks. In these cases short transmission distances are involved (tens to hundreds of meters) and they can be materialized by means of cables, either electrical or optical. Sometimes they are called local area networks (LAN); they have very high data-transfer rates and generally the network is the property of the users, meaning that they do not need to use public service networks. A known representative of the technology employed for these networks is the **Ethernet standard**, which transmits over twisted pair, wireless and optical fiber cable.

9.2.6 Examples of Data Transmission System Selection

The selection of a data transmission system depends on the distance between transmitter and receiver, the amount of data to be transmitted, the noise that could compromise the signal integrity, the maximum time allowed between the data is sent and received, the economic or symbolic value of the data, the transmission service costs, etc. Even when some general rules exist, each transmission problem has it particular solution because the communication's media availability depends on the geographic location of the transmitter and receiver.

There are some remote transmission cases with particular characteristics that constrain the communication solutions. For example, assuming an application case in which distances between transmitting instruments and the receiving unit are of hundreds of kilometers, measuring points may be suppressed or added periodically, and the amount of data to be transmitted is low. This case has a preferential solution based on radio links, either with radio stations spread over the surface of the Earth or through satellite links. A typical example of this case could be a meteorological network, which may be spread all around a country on land and sea. In this case the amount of data to be transmitted may be kept low because instruments can process the information on board, sending only the results.

A quite different case is the remote communication of an intelligent highway that may be hundreds of kilometers in length, and may use many measuring points installed at fixed places along the road. For this application, in general, colored images of the road are needed, which produce great amount of data to be transmitted. Usually, to manage the transit on a highway, the information is required in real time in a control center, and because of the huge amount of data, radiofrequency is not a good option. Since all the instruments are on the trace of the highway, a protection trench along the road may be constructed, and thus this problem is frequently solved by means of an optical cable.

As can be observed from the previous examples, in order to select a data transmission media, many considerations must be taken into account.

9.2.7 Network Topologies

Network topology refers to the structure of a communication network of equipments (Fig. 9.1). It is independent of the physical medium used to materialize the communication, and is related to the way in which each device "sees" the others. Each communication point in the network is called a node and each node has an address. Messages have a header with the address that indicates the node the message is intended for. The most well known communication physical network structures are the topologies called bus, star, ring, mesh and tree.

In the **bus** topology (Fig. 9.1a) all the nodes are connected to the same common transmission medium that can be as simple as a cable. The received header is analyzed by all the nodes, but only the specific recipient, i.e. that whose address agrees with the sent one, will take the message. If the common cable (bus) breaks, the system breaks.

The main characteristic of the **star** topology (Fig. 9.1b) is that there is a central node through which all the traffic passes, and each peripheral node is directly connected to the central node. If the central node fails, all the system fails.

The **ring** topology (Fig. 9.1c) circulates the information around the ring; each node receives the information and retransmits it until it arrives at the corresponding node. As in the bus topology, each node has an address, which permits the recipient node to be identified. If one node fails, the system can change the direction in which the information circulates in the ring, and it can arrive at the corresponding node by a different path.

The **mesh** topology (Fig. 9.1d) refers to a network where all nodes are directly connected to all the nodes of the network. It is a very robust network because if a node fails, the rest of the network continues working due to its great redundancy in communication channels, but the disadvantage is that it is expensive to implement.

The **tree** topology (Fig. 9.1e) has a hierarchical structure with a root node from which the other nodes are connected. There is a direct connection between the root node and the second level nodes. These second level nodes have a direct connection with third level nodes. The root node has a map with the address of the whole tree. Each node has the address of those nodes to which they are connected. If root node fails, all the system fails.

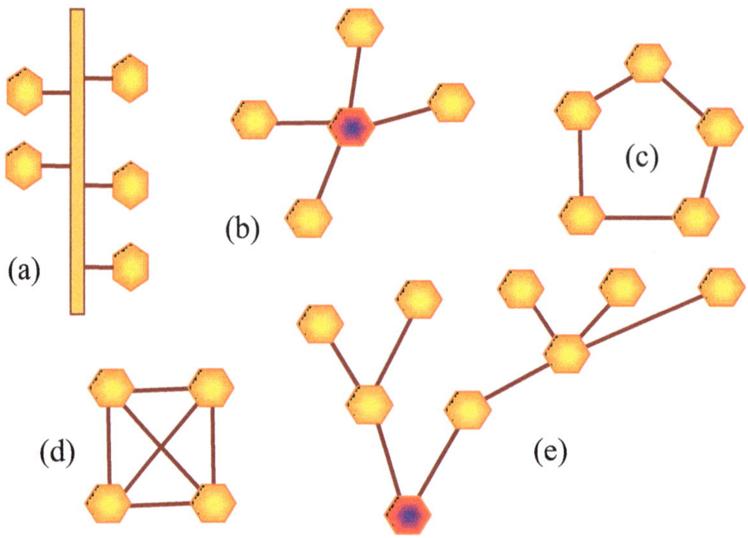

Fig. 9.1: Network topologies. (a) bus, (b) star, (c) ring, (d) mesh, (e) tree. Topologies (b) and (e) have central nodes which organize the network.

9.3 Data Transmission

9.3.1 Analog Data Transmission

Nowadays most of the information collected by measuring instruments is transformed to digital format before stored or transmitted, but some instruments still have analog outputs. Data storage and transmission may be done in an analog way; sending a current through a pair of wires or recording a signal on a paper chart are two analog examples, but today's methods are preponderantly digital.

The voltage output of a measuring instrument may be transmitted by means of an electrical cable just as it is delivered by the instruments' output amplifiers, but if electrical noise exists and the distance between the instrument and the receiver is of several meters, the information at the receiver could be noisy, so transmitted data might be modified. The simplest way to reduce the influence of the electric noise on the transmitted information is to use a cable less prone to noise, such as a twisted pair or a shielded cable. If this is not enough, the instrument's output voltage must be transformed into current before sending it through the cable. The current loop is highly immune to the electrical noise present in the surroundings, such as that generated by large electrical motors deployed in industrial buildings.

A very widespread way of transmitting analog data as an electric current is the 4-20 mA standard, often used in the past. It establishes that when an instrument sends a current of 4 mA it indicates that there is no signal, and a current of 20 mA indicates that the signal is at its full range. For example, a level sensor with a range between 0 and 10 m will send 4 mA to mean that the level is null, 20 mA for a 10 m level and 12 mA for 5 m:

$$4\ \text{mA} + \left(\frac{16\ \text{mA}}{10\ \text{m}}\right) \times 5\ \text{m} = 12\ \text{mA}$$

In general, a 250 Ω resistor is placed in series with the current loop in the receiving unit to convert the standard current to voltage from 1 to 5 V. This voltage is finally converted to digital format through an analog to digital converter at the receiver, thus facilitating control decisions or storage.

Most instruments used in environmental sciences have digital outputs, and some of them also offer additionally the 4-20 mA option as an analog output. The question arises quite logically: why use analog outputs when digital outputs are less sensible to noise? There is a case in which this facility could be advantageous over digital outputs.

When it is needed to synchronize the recording of two fast changing time series, sample to sample, and they are measured by two different digital instruments the existence of analog outputs in both instruments can be useful. In general two instruments with digital outputs acquire their analog inputs independently; samples of two instruments are thus not synchronized at a sample to sample level. For example, when studying rapid changes in pressure and velocity in a hydraulic model, it could be of interest that each pressure sample be taken at the same instant as the velocity sample. In general, it is impossible to collect samples taken exactly at the same instant using different instruments because they control their respective data acquisition and conversion processes with two different internal clocks.

The analog outputs of the instruments offer an opportunity of synchronization. For that to happen, analog outputs from different instruments have to be connected to the inputs of an independent external data acquisition system with a unique clock, thus samples from all the instruments can be simultaneously acquired and stored in the external data logger memory (Fig. 9.2).

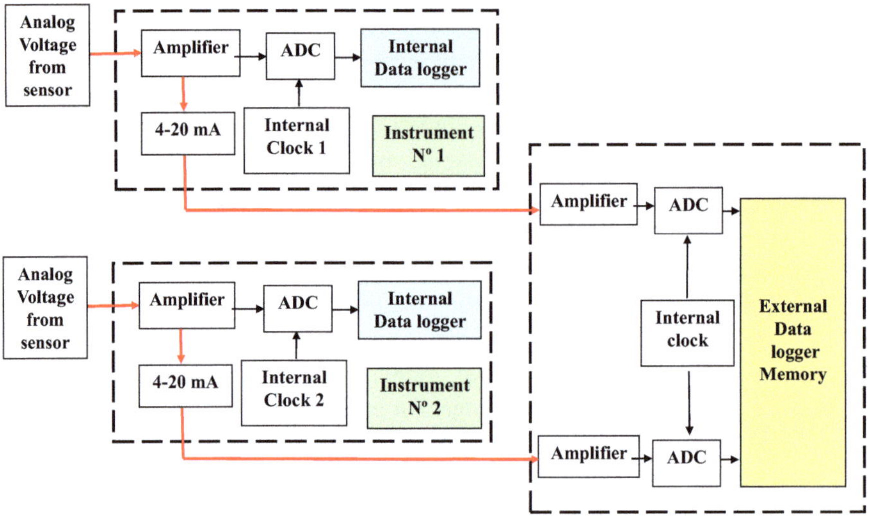

Fig. 9.2: Two instruments with their own internal clock. The data stored in each internal data logger are thus not synchronized. The analog outputs are acquired by an external data logger with a unique internal clock which permits the samples to be synchronized.

9.3.2 Digital Data Transmission

Digital data transmission is the most common way of transmitting data nowadays because digital signals are less prone to noise, and easily amplified and retransmitted. Digital information is coded as 0's and 1's; these bits form strings of bytes, an octet of bits being a byte (Section (3.6.6)). In a given string of bytes, the least significant one is the byte that has the least potential value.

Symbols 0 and 1 must be transmitted as two distinct states over a physical communication medium; for example: two current levels on a wire, the presence or absence of light or sound, etc. The physical medium is also called a **communication line** or a **transmission channel,** which is the pathway for the information and can be materialized for example by a wire, a radiofrequency link, an acoustic link, an optical link, etc.

There is a device that transforms binary data into a physical signal, and vice versa; it is called **Data Communication Equipment** (DCE). Figure 9.3 shows how the DCE interfaces with the physical medium and the digital data at both ends of the communication link. The device that produces the digital data is usually called **Data Terminal Equipment** (DTE). Both devices may be embedded in the measuring instrument. For example, at one end of the physical medium we could think of a field instrument, while at the other end, of a PC.

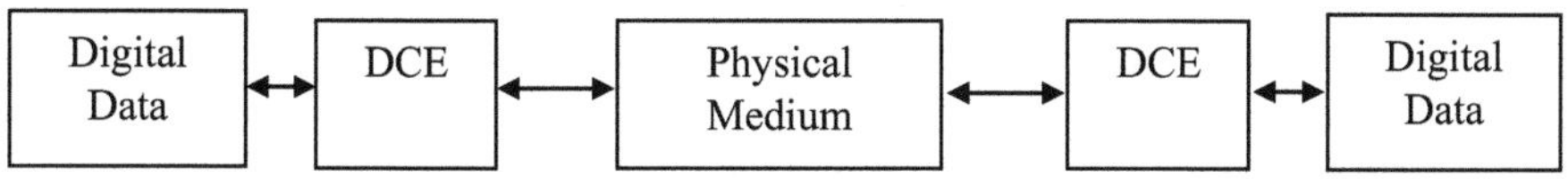

Fig. 9.3: Digital data is converted into a physical signal by the Digital Communication Equipment and vice versa. The physical medium conveys the signal between both DCE.

In order to evaluate the transmission media needed to transport certain information it is important to quantify it on terms of digital units. Storage and transmission of data into computers are done in groups of bits. The most known group being the **byte** (symbol B) which is a word of 8 bits, already defined in Section (3.6.6).

The hardware architecture of old microprocessors used to process 8 bits (one byte) at a time; mathematical and logical operations, as well as memory read and write operations, were based on steps of one byte. More recently, microprocessors extended their internal and external architectures to manage 16, 32 and 64 data bits simultaneously (managing words of 2, 4 or 8 bytes). These processors' capabilities make them more capable of dealing with precision calculations.

In order to optimize data handling by digital systems that are based on microprocessors, each datum should have the amount of bits managed by the digital processing device. For example, handling data with 9 bit resolution in an 8 bit microprocessor requires using two data bytes (16 bits), because the information is handled in modules of 8 bits, thus 7 bits with no information were stored or transmitted, which implies a misuse of resources.

Besides the length of the digital data word, other equally important concepts in data transmission are the **bandwidth of a transmission channel** (B) and the **capacity of a digital channel.** As in the case of the bandwidth of a sensor (Section (2.4)), **the bandwidth of a transmission channel** (B) is the frequency range over which the signal passes through the system suffering low attenuation. For example, a telephone line has approximately a bandwidth from 300 to 3400 Hz for 3 dB attenuation (which corresponds to a signal loss of 50%). This means that the central frequencies of the voice spectrum will pass without attenuation, but low and high frequencies will result somewhat attenuated. As shown in Section (2.4.4) the bandwidth limits the quality of the signal passing through the system.

The **capacity of a digital channel** is the amount of information per second that can be transmitted. In general, for binary symbols, it is the amount of bits per second that the channel can carry, which is also known as **baud rate.** As previously stated, the channel is the media through which the information travels; thus a noisy channel can corrupt the traveling information. Also, as already mentioned, a reduced bandwidth may disturb the quality of the signal. Therefore, the capacity of a digital channel is directly proportional to the channel bandwidth and to the signal-to-noise ratio. That is, the greater the signal and bandwidth, and the lower the noise, the grater the channel capacity. In other words, channels with large bandwidths and low noise are able to transport more information by second, i.e. the channel has a greater capacity.

Part of the energy injected by the transmitter is lost in the medium and then the signal results attenuated as it advances in its way; **attenuation** is proportional to the length of the transmission channel. In addition, the channel will result disturbed by noise; thus the signal degrades in its path through the transmission channel and some information could be lost. Perturbations induced by noise are called **interferences.**

9.3.3 Signal Encoding

In early times, data transmission was analog; signals were varying voltage or current, or modulated radio frequency carriers. Analog signals have the disadvantage that they are prone to noise present in the communication channel. An example will be developed to show the convenience of digital signals.

Let us assume a laboratory water tank in which surface waves with an elevation h varying from -50 to 50 cm are represented by an analog voltage signal that takes continuous values between -5 and 5 V. Suppose that the signal is transmitted on a pair of wires and due to electromagnetic interference a voltage noise of ±0.5 V is superimposed on the signal. When the signal plus the noise arrive at the receiver there will be an uncertainty of ±5% in the wave height due to the noise.

When the analog signal is converted into a digital one before transmitted, it is less disturbed by noise. Using the same example, and keeping the same voltage range, let us assume that the digital signal takes values of -5 V for a digit "0" and 5 V for a digit "1"; suppose also that the signal is sent by the same previously mentioned channel. Due to the channel noise, the digits "0's" will be represented by voltages varying between -5.5 and - 4.5 V at the receiver end, whereas the digits "1's" will be represented by voltages between 4.5 and 5.5 V. Thus, at the receiver we could state that any voltage less than -4.5 V is a digital "0" and that any voltage greater than 4.5 V is a digital "1". Following these criteria, the digital signal is easily rescued from the noise.

This ability to easily distinguish digital signal from noise, makes the transmission of digital signal attractive. In the example we have associated digital values with voltages, but they can be associated with light or any other physical magnitude. In order to transport the digital signal on a real communication channel, the 0's and 1's have to be transformed into a physical magnitude representing them, and this is done by means of the above-mentioned DCE.

Different encoding systems were developed with the aim of better adapting the digital signal to the physical medium. At the beginning of digital communications, the physical medium for transmission was copper wires. When electrical cables are used, one of the goals of encoding is that the average value of the physical signal be null, because it renders the transmission more efficient. For this to happen, the value of the physical magnitude (*S*) for each symbol (0's and 1's) should be of opposite sign (negative and positive). Hopefully, if the average the number of 0's and 1's are the same the average of the physical magnitude will be zero. In the case of electrical cables *S* is a voltage or a current.

One type of signal encoding method utilizes symbols with only two levels (-*S* and +*S*), while there is another that uses three levels (-*S*, 0 and +*S*). The simplest encoding using two levels is known as the NRZ (No Return to Zero) whereas the three-level code is known as the bipolar encoding. In the NRZ encoded 1's are represented by +*S* and 0's with -*S*. In the bipolar encoding, 0's are represented by 0 level, and 1's are alternatively represented by +*S* and -*S*, as depicted in Figure 9.4. The top drawing in Figure 9.4a shows the digital symbols at the input of the DCE while the following drawings represent the DCE output encoded as NRZ (Fig. 9.4b) and bipolar (Fig. 9.4c).

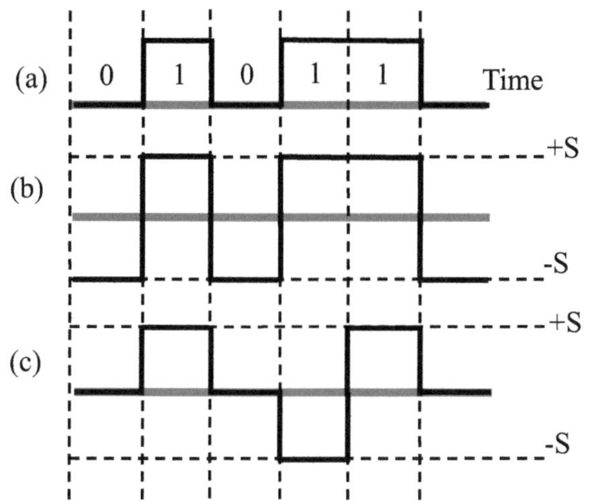

Fig. 9.4: (a) Digital symbols at the DCE input are converted into output levels on the physical medium; (b) NRZ code; (c) bipolar code.

9.3.4 Transmission Modes

Transmission modes describe the way in which the data flow between two points. The DCE that sends the data is defined as a **transmitter**, and that which receives the data is called **receiver**. Data flow can be managed either from the transmitter or the receiver, as will be seen below.

It is our intention to keep the description that follows as simple as possible, and for this reason some sophistication introduced by today's intelligent transmission systems are avoided. Problems inherent to each transmission mode will be mentioned below. However, it should be recognized that modern communication systems have some capabilities that help to decrease the impact of these problems.

Three possible cases are represented in Figure 9.5. In order to fix concepts assume that instruments collecting field data are reporting them to a central station. The Digital Data on the left of the figure is produced by the instrument and sent to the DCE. The right side represents the central station, composed by a DCE and the Digital Data, the latter being materialized by a PC data input.

In this example, the top drawing corresponds to the **simplex** mode in which the information travels in only one direction, from the instrument to the central station in our example (Fig. 9.5a). Thus, the remote station defines when to transmit. The transmission decision could come simply from a clock that periodically activates the instrument and the transmitter (time activated), or in the case of an intelligent instrument, when some event is detected, for example when the measured parameter surpasses certain predefined level (event activated).

The simplex mode requires only one transmitter and one receiver, but if a datum is lost in the transmission process, it cannot be recovered. There is no way to ask the instrument to resend the information lost. Data may be lost due to noise in the communication channel, as could be due to another transmission in the same channel, or to an electric impulse as produced by lightning, electric motor, etc.

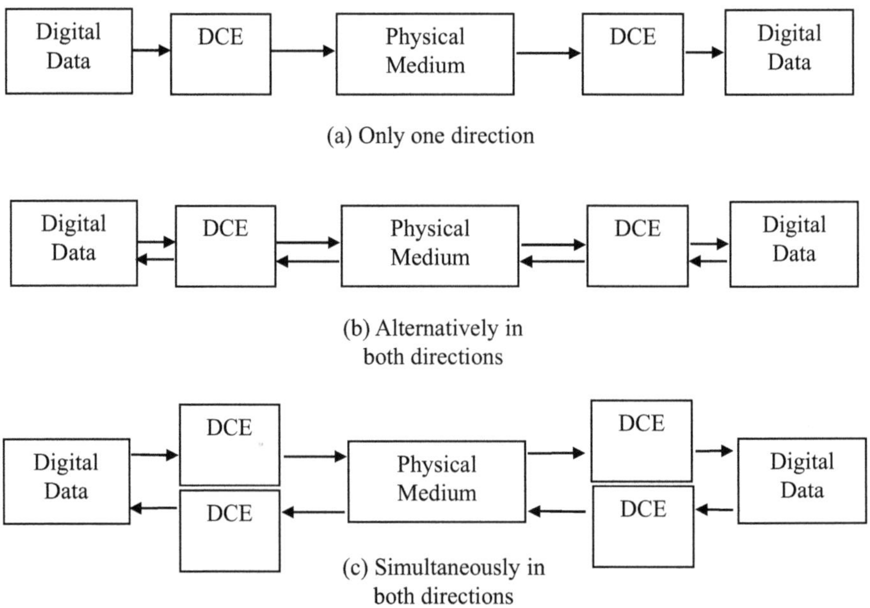

Fig. 9.5: Transmission modes: (a) Simplex; (b) Half-duplex; (c) Full-duplex.

Because each remote station decides the instant in which the transmission is done, if a network of several instruments reporting to a unique central station is implemented using the simplex mode, two remote stations could transmit simultaneously. As the receiver will not be able to manage two concurrent transmissions in the same channel, a collision of the information could appear and the receiver would loose one or both of the two data sets.

In a **half-duplex** connection (Fig. 9.5b) data flows in both directions, but not at the same time. There are one transmitter and one receiver at each end of the communication channel, but there is only one channel available (e.g. in radio frequency both transmitters and receivers share the same frequency).

Following the same example of a remote instrument, this kind of connection makes it possible for the central station to ask the remotes when they have to transmit the data. This kind of controlling action from the central station is usually called **pooling**. The central station sends a message to each remote station, one at a time, asking whether the remote stations want to use the communication line. In this way conflicts between stations transmitting simultaneously can be avoided.

Generally, by means of error detection codes, the central station has the capability of detecting whether some data has been missed in the transmitting process. Therefore, the half-duplex capability allows the central station asking the remote station for a repetition of the missed data.

When a remote station is asked by the central station to transmit a large amount of data, the communication channel will be occupied by the remote station and is useless until the transmission ends; the channel cannot be simultaneously used by the other stations. Thus the central station would loose the control of the communication channel during the transmission from one of the remote stations.

In a **full-duplex** connection (Fig. 9.5c) data can flow in both directions simultaneously because two logical channels are used. In this context, logical channels means that they can be theoretically conceived as two channels, but in practice these channels can be physically implemented, for example, just on a pair of wires (using special modulating techniques). In radio frequency, two frequencies can be employed; the central station transmits in the frequency used by all the remote stations' receivers; and the remote transmitters work at the same frequency that the central station receiver. Each end of the line can thus transmit and receive at the same time.

A full-duplex connection allows sending orders to the remote stations while they are transmitting and can order the remote to stop its transmission. Then the central station always has control over the communication channel and could interrupt the data transmission to attend other more urgent tasks, for example, to interrogate another station which has urgent information to transmit.

9.3.5 Serial Transmission

So far we have described how transmission systems work and how channels are managed. In the following paragraphs it will be analyzed how the information is transmitted inside the channel. A transmission channel as shown in Figures 9.3 and 9.5 sends only a bit at a time; this kind of connection is called serial transmission. A communication standard for wire serial connection, very used in the recent past and still in use in some industrial computers and scientific instruments, is the RS-232 standard. It states the physical characteristics that the hardware should accomplish. For example, it establishes voltage levels, amount of wires, maximum length of cables, connectors' sizes and wiring, pin assignment and signals, data flow control, etc.

RS-232 can send information over a maximum length of cable at a given speed; as a general rule, when the length increases the speed decreases. The maximum length of cable for RS-232 is quite limited; then, when long cables are needed, the standards RS-422 or RS-485 have to be used. An instrument with a serial port could solve communication needs with only three wires (transmit, receive and ground) plus the necessary wires for power supply; in this way handy and low cost cables result.

At present, computers use a serial port called USB (Universal Serial Bus) standard, which has replaced the RS-232. USB can work at higher speeds than RS-232, but over short distances. There are some converters that allow using the USB port of portable computers with instruments that use RS-232 port. Approximated lengths and speeds for the abovementioned standards are shown in Table 9.1.

Table 9.1: Lengths and speeds for some serial connections

Serial connection	Maximum length (m)	Speed (kbit/s)
RS-232	15	115.2
RS-422 and RS-485	1,200	100
USB	From 2 to 5	1,500-12,000

9.3.6 Parallel Transmission

A parallel connection (Fig. 9.6) can transmit a number of N bits simultaneously. That is, there are N different channels that permit a bit to be sent by each channel. Internal connections in a computer are parallel channels called buses that work at very high speeds but over short distances. They are made of tracks on a printed circuit board, each track being a channel.

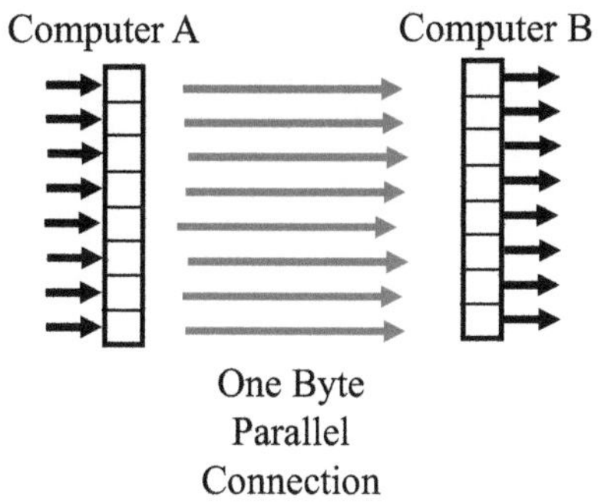

Fig. 9.6: A parallel connection of 8 channels (one byte) allows transmitting 8 bits simultaneously.

In the case of a radio frequency link, each channel needs to use a different frequency. Thus several bits can be sent simultaneously through different frequencies. The same happens with a fiber optic channel.

As is obvious, sending information by means of *N* channels will require less time than sending the same information by means of one channel. Parallel communication is thus faster than serial communication; the higher the number of channels, the more information is sent in the same period of time.

9.3.7 Serial-to-Parallel and Parallel-to-Serial Conversion

In most simple field instruments, their external communications were traditionally solved through serial ports. Perhaps the reasons for this solution are that instruments do not generally require high transmission speeds, computer's serial ports are ease to use, and a few connection wires are needed. If parallel transmissions were used, a considerable amount of wire would be needed. This would result in more heavy and expensive cables and in bigger, and sometimes fragile, connectors.

Modern instruments usually have microprocessors for controlling all the internal activities and to manage the interface with the external world. Instruments handle fast parallel data processing internally, and serial communications externally. When the serial data sent by an instrument arrives at a computer port it has to be transformed again into parallel format, because computers' internal data handling is parallel. Therefore, to perform these transformations some converters from parallel to series and vice-versa are required (Fig. 9.7).

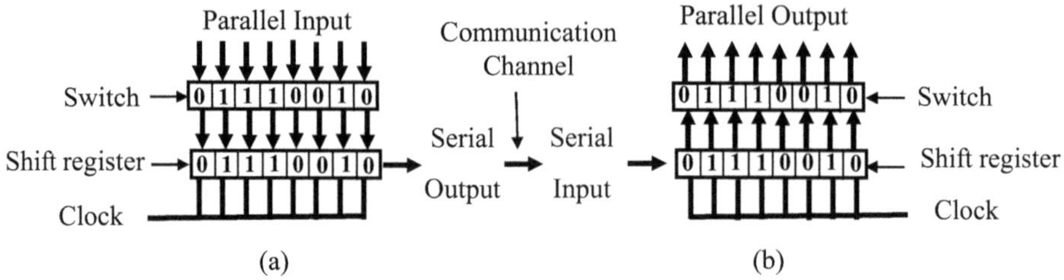

Fig. 9.7: (a) Parallel to serial converter. The parallel input is loaded into the shift register by the switch. The clock shifts the bits to the right, one at a time, into the communication channel. (b) On leaving the channel, the bits enter the serial to parallel converter and the clock accommodates them into the shift register. Once all the bits have arrived they are switched to the parallel output.

9.3.8 Asynchronous Transmission

Usually, when digital information is transported through a channel the speed of transmission (baud rate) is previously agreed between transmitter and receiver. Thus, the receiver knows how much time lasts for each bit. This information is needed by the receiver to recognize the transmitted information, but this is not enough. Suppose that an instrument sends one datum after another, thus it would produce a stream of 0's and 1's on the communication channel, therefore, the receiver needs to know when each datum begins and ends. In order to solve this ambiguity there are basically two transmission strategies: synchronous and asynchronous.

The main characteristic of asynchronous transmission is that each datum is sent in an undefined instant, that is, intervals between data are irregular, as is the case of a person typing on the keyboard of a computer. Each stroke is a datum but the time between strokes is variable. Thus, some extra information has to be sent to the receiver to indicate the start and end of a datum. Asynchronous transmission solves the problem by sending a **start bit** before the datum and one or more **stop bits** at the end. In the case of the keyboard of the computer each datum is called a character; a character is a digital word of 7 bits.

The ASCII Character Set, which stands for the "American Standard Code for Information Interchange", was designed in the 60's, but the character set is still used in modern computers, HTML and Internet. This 7-bit word represents 128 characters. They contain the numbers from 0 to 9, the uppercase and lowercase letters from A to Z, and some special command characters. An example of asynchronous transmission using ASCII characters is presented in Figure 9.8 to clarify ideas. Figure 9.8a shows the start and stop bits.

Assume that when no datum is sent (periods of inactivity, idle) the state of the transmission line is fixed at binary "1" (see right end of Figure 9.8b). Then the start bit will be a binary "0" which warns the receiver that the next bit will be the beginning of a character. Obviously, the receiver knows how many bits compose a character but at the end of the character a binary “1” is sent as a stop bit to reset the state of the transmission line to “1”.

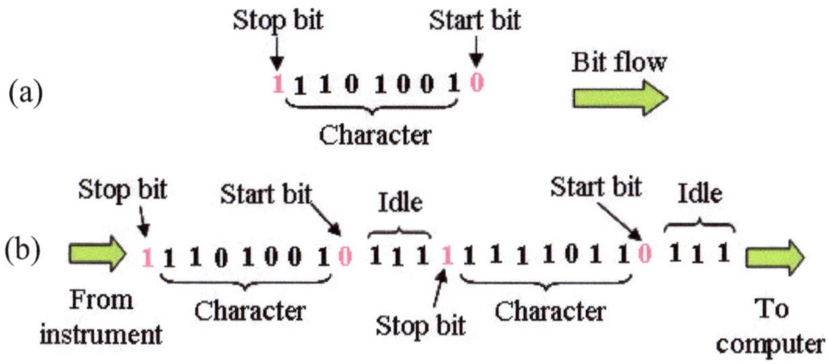

Fig. 9.8: (a) One start bit and one stop bit are added at the beginning and at the end of an ASCII character. (b) Idle periods between characters are at binary “1”. The first “0” after an idle period indicates that the next bit is the start of a new character.

9.3.9 Synchronous Transmission

In synchronous transmission, transmitters and receivers are continuously synchronized at the same speed by the same clock. This transmission mode is used when a large amount of data has to be transmitted; data is arranged in blocks called frames or packets. The data frame is headed by special characters called synchronous characters. There are not any gaps between transmitted characters and, in some way, the same timing is supplied at both ends of the channel. Eventually, both ends could go out of synchronism and errors occur. To solve this problem some means are provided to resynchronize clocks and check if the received information has been corrupted.

9.3.10 Error Detection

Special circumstances on transmissions channels can distort the information that travels over them and then the digital word arrived at the receiver would be different to that sent by the transmitter. The information at the receiver could have some bits changed. Thus, it will process or store a wrong data. To reduce this kind of risk some error detection strategies have been developed. All error detection codes are based on redundancy; this means that some extra information has to be added and sent together with the information we want to protect.

The simplest strategy is to add a **parity bit** to the message, which means that an extra bit that has some logical relation with the information we want to protect is added

to the message. For example, if odd parity is desired a parity bit will be added at the end of the message so that the total number of ones in the word to be transmitted (including the parity bit) is odd. For most applications odd parity is used, so that at least one "1" is transmitted.

Figure 9.9 shows an example of a twelve-bit word protected with a parity bit, using odd parity. Because the word has an even number (8) of "ones" an extra "one" has to be added, in order to have a total odd number.

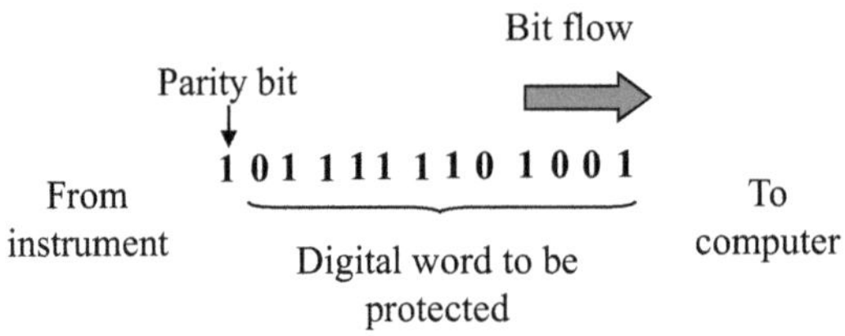

Fig. 9.9: A parity bit "1" is added at the end of the word to be protected to complete an odd number of "ones".

If, due to noise present in the transmission media, an even number of bits is changed in the transmission channel the message will be erroneous but the parity of the message will be unchanged and the error will be undetected. This means that the error detection code is not robust enough. The longer the word to be protected the more the chances that an even number of bits results changed by noise. Furthermore, this method permits an existing error to be known, but nothing is known about which is the erroneous bit.

Another simple method is called the **checksum**; it consists in summing up all the data in a message and, attaching the less significant byte of this sum at the end of the message. Then at the receiver end the same sum is performed and the least significant byte compared with the checksum byte sent.

Let us introduce an example. If a string with 300 bits is to be sent, its sum expressed in binary will have two bytes, namely 00000001-00101100, then, the second of these bytes (00101100, i.e., the least significant byte) is attached to the end of the string. At the receiving end, the checksum byte is separated from the string and the rest of the bits of the string summed. The least significant byte of this sum is compared with the sent checksum. If both agree, there is a great chance that the message be correct.

Also, to detect a corrupted word it could be sent twice. Thus, after a bit to bit comparison of both words at the receiver, a simple algorithm will decide that the word is correct if all bits are identical. This method allows identifying which bit has been changed. In case that a difference appears between the two words, the receiver asks for a retransmission of them. Probabilities that noise corrupts the same bit in both words, keeping words identical are low, but not null.

9.3.11 Error Correction

Previous strategies permit knowing with some degree of probability that a digital word has survived the transmission process unaltered, but do not allow the error to be corrected. In order to have the correct information it is necessary to ask for a retransmission of the information previously identified as corrupted. This requires at least a half-duplex connection.

If a simplex mode connection is used, a simple error correcting code method could consist in sending the same word three times; if the three are identical the chances of being wrong is insignificant; if one of the words is different than the other two, there is a high probability that it is wrong. Both words remaining identical have a great probability of being true and they are selected as the correct word. More efficient methods to correct errors have been developed; the Hamming code (Proakis & Salehi, 2008) is among the most spread correcting codes.

As seen from the previous examples, to increase the probability that the received word be free from error it is necessary to increase the redundancy of the message. In other words, **the price to pay when increasing the certainty of a data is sending more information than the data we really need.**

9.4 Transmission Media

In private networks, users are their owners in contrast with public networks, where users rent a service to the owners (i.e. the owners being telephone or satellite companies). Private radio frequency networks for collecting field measurements from scientific instruments are, in general, of the star or tree topologies.

9.5 Private Networks

It is possible to distinguish two types of private networks: serviced by cables, as in a network of instruments and computers in a laboratory or industry, where most of the time the communication is solved by means of fiber optic and copper cables; and networks where communications are carried by means of radio frequency links (as in field stations).

Private radio frequency networks were frequently used in the past, when satellite communications had a great cost per minute of use, and cell telephony had only services in urban areas. They are still in use in places not covered by the above public services, and also when a big amount of data has to be handled in a medium size area (of a few tens of kilometers in diameter). In these cases the cost of installing the antennas, transmitters and receivers could be less expensive than the cost of a public service.

In some cases where cell telephony and private radio frequency networks could compete, the user should compare both solutions in costs and capability. The advantages of private networks for the users are: (i) they do not pay for the service as in a public network (this could be an important issue if the amount of data to be transported is great); (ii) they do not depend on a carrier company and on the availability or service in the area. The disadvantages are: (i) they have a great initial installation cost; (ii) periodical maintenance service is required; (iii) they need provision of energy (power supply); (iv) in some places it is necessary a license to use the radio frequency channel.

In countries with low population density, digital telephony service in the rural areas does not cover beyond the most important roads and highways; private networks or satellite communications are thus the only alternative solution. There no fixed rule to define when the network has to be public or private. Each case has to be solved on an individual basis, and in general a mixed network using private and public services can result the most efficient solution and has the lowest installation and operative costs.

9.6 Public Networks

9.6.1 Digital Telephony

The evolution of digital telephony was so fast that it would take several pages to explain it, even briefly. Moreover, it would not provide useful knowledge to potential users and any description will be quickly obsolete. There are so many acronyms that only the professional can understand this synthetic language. For these reasons, we will only refer to some features that this technology can provide to transport information from a field instrument to a central station and vice-versa.

In the early times of telephony, voice was transmitted as an analog signal. But because digital signals are less prone to noise and easily regenerated (amplified and retransmitted), in 1962, for special links of commercial telephone systems, voice began to be digitized (Bellamy, 2000). Today almost all telephone services are digital, which facilitates multiplex (placing several transmissions in one physical communication medium) and encryption.

At present, there are basically two telephone networks, one called landline or fixed-line and another called mobile or cellular line. The first uses solid or mechanical media such as copper wires or fiber optic cables as the physical communication medium, while the second uses radio frequency waves propagating in the air. Also, it could be considered that there exists a third telephone network, because many public and private telephone communications are routed through the Internet network.

In fixed telephone networks, the first section from the subscriber (or client) to some point on the network is connected through copper wires, the voice being an analog signal (it is a variable voltage over which the voice is "impressed"). Telephone voice bandwidth is approximately between 0.25 and 3.3 kHz. The copper lines from all the subscribers in a certain area are run to a point in the network called **telephone switch**. It consists of a building where electronic circuits interconnect the lines arriving from the subscribers.

The copper wires used for voice has a transmission bandwidth that much exceeds the required for the voice bandwidth, so the wires have extra capability of transmitting information on them. This additional potential is used to send digital information. The copper lines have a useful capacity as a digital channel (Section (9.3.2)) that can be used for applications such as data transmission.

There are technologies known as ADSL, HDSL, VDSL, etc., that get maximum utility from the copper wires dividing their bandwidth, which enables normal voice telephony **(public switched telephony network – PSTN)** and digital data transfer to be achieved simultaneously. Some telephone companies offer Internet service up to a transfer speed of 54 Mbps, using these technologies.

9.6.2 Global System for Mobile Communications (GSM)

GSM is a cellular mobile communication system adopted worldwide as the standard system for mobile communications (Bellamy, 2000). It is based on radiofrequency transceivers (transmitter and receiver) and the basic unit consists of the central station (or **base station**) and a limited number of remote stations (also called **subscribers** or **mobile units**) working in a restricted geographical area called **cell**. An elementary description of how a cell works to connect two remote stations within the cell will be conceptually described using some concepts explained before.

The example of a **full-duplex** connection shown in Figure 9.5c, refers to a central station (CS) communicated to some remote stations (RS). Let us assume that this network has a star topology (Fig. 9.1b), where the CS is the central node which organizes the network. Two channels (frequencies) are needed to send data simultaneously in both directions between the CS and one RS. With only two channels, the CS can establish a communication with only one remote station at a time. But adding two extra channels to the CS allows it to communicate with another RS simultaneously (a total of four frequencies should be available for this purpose).

If more RS are needed, two new different frequencies are needed for each RS added, because if the same channel (frequency) is used by two transmitters in the same area at the same time, a receiver will take in both transmissions, being unable to distinguish between them. Thus it would not be possible to have a comprehensive message and it is said that there is communication **interference.** This is an unreal oversimplification of the problem, just to give a conceptual idea of how cellular telephony works.

Assume that there exists a set of contacts in the CS (called switch in Figure 9.10), which connects RX1C with TX2C and RX2C with TX1C. Then, both remote stations could communicate between them through the central station. Let us identify the link to send information from remote station 1 to 2, as: TX1, RX1C, TX2C and RX2; and to send information in the opposite direction as: TX2, RX2C, TX1C and RX1.

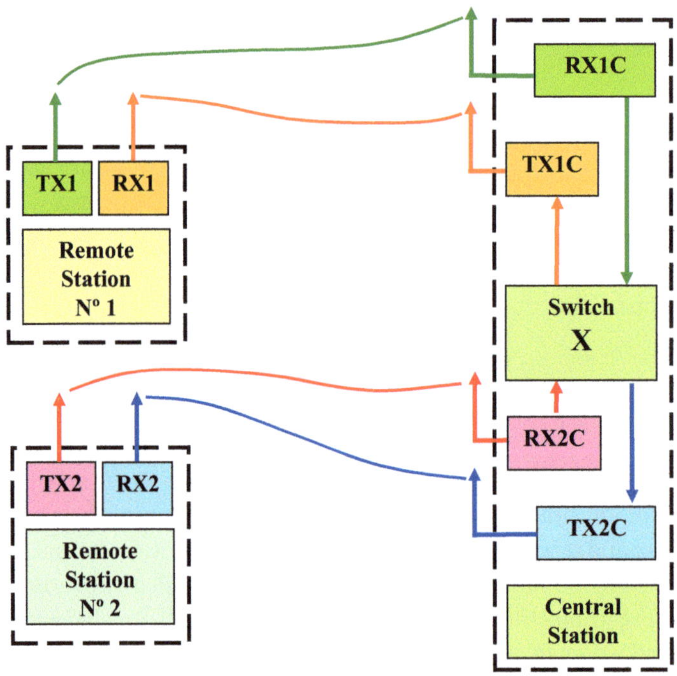

Fig. 9.10: Remote stations 1 and 2 have transmitters TX1 and TX2 and receivers RX1 and RX2. The CS has also transmitters and receivers at the same frequencies as the remote stations in such a way that, for example, TX1 at the RS1 sends data to RX1C at the CS, and TX1C sends data from the CS to RX1 at the RS1. TXs and RXs linked by a solid line indicate that they work in the same frequency. CS switch permits the dialog between RS1 and RS2 to be established.

Now let's think that the RS are mobile phone sets and the data sent is voice. Then two persons could talk through the CS while they are walking in the area covered by the CS radio link.

Each new person (subscriber) added to the network will require two new frequencies, one to send information and another to receive it. Then, because the available bandwidth to allocate the radiofrequency channels is physically limited, the system would have a maximum amount of subscribers in the given area known as **cell**. For the purpose of implementing a realistic system, it is needed to divide the service area in cells reutilizing frequencies used in one cell, into other cells at a certain distance from the first. The size of the cells and the power transmitted by the base station are in accordance with the population density. If the population density increases in a cell, it could be divided into several smaller cells.

A group of cells covering a big geographical area are shown in Figure 9.11. In general, cells have the shape of a hexagon and the serviced area is covered by adjacent hexagons; hexagons having the same letter indicate the same group of frequencies used in it (in the web version of the book hexagons have same colors). The same groups of frequencies are used in areas separated a certain distance to avoid interferences.

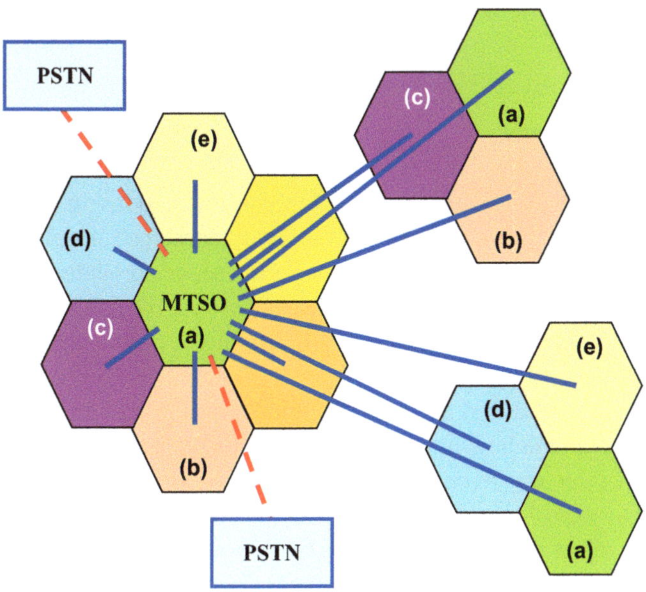

Fig. 9.11: Letters inside the hexagons [(a) to (e)] indicate the group of frequencies used in each cell. Same group should be separated a certain distance to avoid interference. MTSO manages all the functions of the network. It allows communication to be established between mobiles and with the fixed telephone network through the links to the PSTN.

Each cell is connected to the Mobile Telephone Switching Office (MTSO) (solid lines - blue). Cells can be connected to the MTSO by radio frequency, fiber optic or special copper cables (Tomasi, 2001). In this way, the interconnections of cells allow communication between subscribers placed in different cells. Also, the (dashed - red) lines connect the MTSO to the Public Switched Telephone Network (PSTN) to communicate mobile subscribers to subscribers belonging to the fixed or wired network.

The function of the MTSO is to manage and control the calls, to allocate radio frequency channels and to supervise the passage of mobiles from one cell to another. This last function is based in the measurement of the power received by the base stations; when the mobile travels between cells, the power decreases in one base station and increases in another; the MTSO decides which base station will handle the communication.

The mobile station (MS) consists of two components: the terminal, which is a radiofrequency transmitter and receiver, and the Subscriber Identity Module (SIM), which permits the user to access the contracted services at any terminal. The SIM is identified through the International Mobile Subscriber Identity (IMSI) and can be protected by a password to avoid unauthorized use.

GSM has a messaging bidirectional service called Short Message Service (SMS) this is an interesting possibility to send data from field instruments if the amount of data is low and delays are accepted.

9.6.3 Summary on Digital Telephony

Potential users of these technologies, who have scientific applications in mind to be implemented in densely populated areas with a highly developed telephone network, will not face difficulties to find a data transportation service provided by a telephone company. They will have to care only about the way to send the information at the lowest price and to deal with the company taking into account the technical requirements and the service-level agreement offered by it. These users will have to understand how their data are produced and processed to select the most adequate service. They will have to evaluate how much data they will need to transport and the delay they can accept in the transference of the information. The selection of the service to be contracted has to be done with the help of the service provider company because there are services offered by some companies that are not offered by others. Also, services may be charged by time or by the amount of information transported.

9.6.4 Satellites

From the point of view of space aeronautics, an artificial satellite is an object launched by humans into space, describing orbits around the Earth or other celestial objects. Obviously, satellites that will interest us are those that orbit around the Earth. The orbits of the Earth artificial satellites can be circular or elliptical. In the first case, the speed of the satellite is constant; in the second one, the satellite has higher speed when it is close to the Earth and lesser when it is far (Kepler's second law (Feynman et al., 1967)).

Satellites can be passive or active, the first can be used to reflect (or scatter) electromagnetic waves on their surfaces; some tests were performed using the moon as a reflecting satellite. In this way, two points on Earth can communicate impinging a wave on a passive satellite, but to have a detectable signal at the receiver, the impinging wave needs to be of high power.

Active satellites have receivers and transmitters and they work, not merely reflecting, but retransmitting the signal that they receive from the Earth stations. Also, because they have directional antennas, the energy needed by the transmitters and receivers decreases considerably. Commercial satellites are of the active type.

Basically, as regards the orbit, there are two types of satellites: asynchronous and geosynchronous. Asynchronous satellites continuously change their position relative to a fixed point on the Earth, so they can be used only for a short time to transmit and receive data in each pass over that point. For example low-Earth-orbit (LEO) or low-altitude satellites are at around 160 km from the surface of the Earth and it takes them about an hour and a half to turn a full orbit. Then they are visible (available to communicate with a certain ground station) only for 15 minutes in each orbit (Tomasi, 2001). Other more remote satellites employ about 12 hours to complete an orbit and are available for 4 hours at a point.

All the orbits are in a plane passing through the center of gravity of the Earth. If the plane passes through the poles, the orbit is a polar one, if the plane lies on the equatorial plane, the orbit is an equatorial one, and the remaining cases are known as inclined orbits. Equatorial orbiting satellites travel along a circular orbit, and because the distance is about 35,800 km from the surface of the Earth, it takes them 24 hours to turn one orbit; therefore, they stay fixed relative to a point on the Earth surface. Because of this they are called geosynchronous (Tomasi, 2001).

The advantage of geosynchronous satellites is that they remain substantially stationary with respect to a ground station, so they can provide permanent communication between two points that are in the beamwidth of the satellite antenna. The disadvantages are that, by being so far away from the surface of the Earth, they need to use powerful communication equipment. There is a communication delay of about 0.25 s; and propulsion systems are needed to keep satellites in orbit. Moreover, being away from the poles, they are not effective for communication in these areas.

They are ineffective at high latitudes because the Earth's curvature disrupts the transmission path.

The area on the surface of the Earth with which a satellite can communicate is called footprint; it depends on its orbit, communication frequency and antenna. There are maps that allow knowing which zones on the Earth can communicate with the satellite and when.

In the case of a set of instruments for use in environmental sciences that are distributed in a very large area, satellite communication is an interesting, and in some cases, the unique alternative. In environmental studies that include geographic places accessible only for a certain period of the year, or sea places far from the coast (places where there is no cell phone coverage), satellite communication is the only option. As explained in the introduction, the type of satellite to be used depends on the urgency with which the data is needed. In the majority of cases, instantaneous data is not required and asynchronous satellites can be used.

Figure 9.12 depicts an example of a set of instruments distributed in different geographical areas communicating their data by means of an asynchronous satellite. The satellite's receivers and transmitters antennas are focused on a specific area of the Earth's surface from which the satellite can receive or transmit. While the satellite moves along its orbit and the Earth rotates, the antennas' beams sweep a footprint on the Earth's surface.

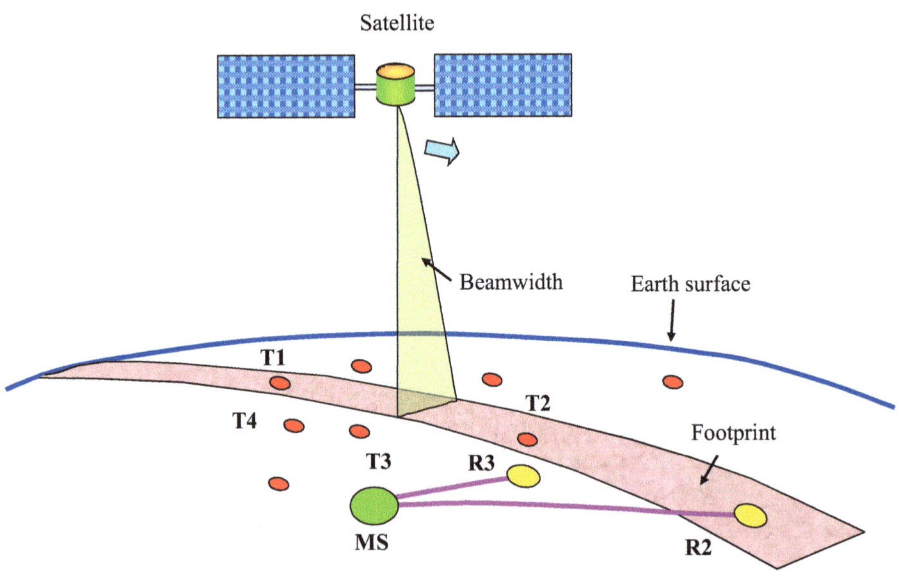

Fig. 9.12: Smallest dots (red) indicate field instruments; R2 and R3 (yellow ones) are Earth stations, and the MS (green one) is the master station. Those instruments on the footprint of the satellite are able to send and receive messages through the satellite when it is passing by. In this example it is assumed that in the next pass the footprint moves to the left passing over T3, T4 and R3.

In Figure 9.12, we assume that the satellite's orbit combined with the Earth rotation will move the footprint from right to left. The instruments from which we want to transmit the data are represented by the smallest dots (red). When the satellite passes over the instrument T1, the instrument has some time available to send data that are stored in the satellite memory. Next, instrument T2 will also send its data. If data are not urgently needed, and the satellite has enough memory to collect the data from all instruments, the information could be downloaded the next time that the satellite will pass over the master station MS.

For a more frequent retransmission of data from the satellite to Earth, there may be other Earth receiving stations R2 and R3 (yellow) such that the data from T1 and T2 are downloaded to the ground station R2. This station could be linked by a public data service (e.g. Internet or telephony) to the master station MS, gathering in this way the information more quickly than if it had to wait the next satellite pass over MS.

In the case of using geosynchronous satellites, the problem is simplified because the communication can be achieved permanently. Although this kind of satellite covers an area on the Earth surface that is well determined, data between satellites can be transferred to cover larger regions. The main problem of these satellites is that they do not cover the areas close to the poles satisfactorily.

9.6.4.1 Some Satellite Constellations

The provision of satellite communication services is such a dynamically changing business that to give names of satellite constellations could make this material become quickly obsolete. Thus, satellite constellations characteristics and services will be described avoiding any commercial names; they will be named as Constellations A, B and C. Because some satellites stop working and new satellites are launched into space, fleets are changing continuously, and thus the amount of satellites composing fleets is offered only with the purpose of giving a rough idea on its size. Communication services also change frequently because new satellites usually have novel technologies that allow offering innovative services.

9.6.4.2 Constellation A

The International Maritime Satellite Organization was originally an international non-profit organization created in 1979 to provide a satellite communication network to enhance safety at sea. Later on it expanded their services to aircraft and land users; in 1999 the organization was converted into a private company (http://www.inmarsat.com).

This constellation is composed of 11 satellites flying in geosynchronous orbit 35,786 km above the Earth. Through this fleet of satellites the constellation provides voice and high speed data services on land, on the sea and in the air.

The constellation has a product that provides access to services such as voice, email and Internet where land networks do not exist. It is based on IP technology and delivers data rates of up to 492 kbps. The service is accessed by means of portable satellite terminals of the size of a laptop.

9.6.4.3 Constellation B

Satellites in this constellation have polar orbits with an average altitude of 780 km; they are distributed in six polar orbital planes, each one having 11 equally spaced satellites. The constellation has 66 active satellites and some spare ones. There is always a satellite visible to all subscribers. Thus, permanent data transmission is possible, unlike what happens with the example shown in Figure 9.12.

This constellation provides a personal communication network for voice and data all over the world. This network provides communication service in both directions between subscribers of this constellation and with those of the public network. Each subscriber has a personal number; service costs are independent of where the communicating points are placed (Tomasi, 2001).

The orbital period of these satellites is 100 minutes and 28 s. Each satellite communicates not only with the Earth's stations but also with satellites in front, behind and in adjacent orbits. This system is of particular interest for the maritime community; ships traveling through any sea may communicate with the coastal stations. The communications follow three steps: the ship communicates with the closest satellite; the message arrives at the satellite nearby the final recipient by means of some bridges between satellites, and this satellite call the recipient to establish the link between both ends. This constellation is becoming popular for transmission of weather and oceanographic data from instruments mounted on marine buoys and ships. Because they use a polar low Earth orbit they can provide truly global satellite communications, including in Polar Regions.

There are manufacturers of instruments which provide modems to connect field instruments to a central station through this constellation (http://www.hko.gov.hk) (http://www.iridium.com). The equipment consists of a transceiver which operates like a standard modem. It has a data input for a RS232 port and they work with 12 VDC; when transmitting they take about 0.6 A from the power supply. In general, a solar panel of about 50 W and 12 VDC is used to keep the battery charged. Obviously, the solar panel has to be selected according with the solar energy available in the place where the remote instrument is located.

9.6.4.4 Constellation C

Another constellation of communication satellites uses about 30 low-Earth-orbit satellites of reduced size and weight (http://www.orbcomm.com). They transmit at a much lower frequency than other commercial satellites (in the 137-150 MHz VHF band), which requires a larger antenna but needs less power to work. Although a satellite may be available for communication during a certain time interval over a given area, subscribers in the area are allowed to use the satellite only 1% of the time the satellite takes to cross the entire area to avoid interferences (about 36 s per hour). Due to this characteristic, this constellation is best suited for small amounts of data such as the case of Global Positioning System reports of vehicles, animal migration or drifting buoys.

There is a delay between the delivery and the reception of messages that for most messages containing less than six bytes may range from 1 to 15 minutes. This delay is usually called latency. Typical data transmission capacity is adequate for sending GPS position data or simple sensor readings. The communication network charges fees are based on the amount of data transmitted. Each field station has an identification number that permits the satellites to be communicated with it.

There are providers of equipment to communicate through this constellation who offer devices for directly connecting instruments to the network (http://www.globalw.com). They consist of modems with RS-232 inputs, power supply and antenna. Some of them include a GPS built in that is used to report ships and trucks positions. Other devices have several analog inputs for instruments with 4-20 mA or voltage outputs. Also, some modems can be used in the reverse sense sending information from the central station to the field station. They have digital outputs which allow activating actuators in the field station, for example, to turn on and off a motor or heater.

In some cases Earth stations receiving satellite communications are connected to Internet storing the received messages in a dedicated web page.

9.6.4.5 Inter Constellation Communications

As Figure 9.12 suggests, low-Earth-orbit (LEO) satellites have a data latency of some minutes to download the information generated by the instruments onboard the satellites because they need to communicate with a ground station network. Until a few years ago, reducing latency meant to increase the number of ground stations, which increased costs. An idea that changed the scene was to transmit the information, not to ground stations but to space stations (McCormick et al., 2010; Lenz et al., 2010).

There is a constellation consisting of three modern geosynchronous satellites spaced at 120° which provides ground, maritime and aeronautical communication services. A terminal type as those used on ground was evolved and adapted to be used onboard a LEO satellite. Thus, this recently developed transceiver allows a LEO satellite to communicate with the ground station in real-time via the existing constellation of geosynchronous satellites.

This new technology, which allows communicating LEO satellites with the space-based platforms, permitted data latencies to be reduced from tens of minutes to milliseconds. It opened new possibilities for the collection and application of data captured by sensors in space.

9.7 GPS (Global Positioning System)

The GPS positioning system has among its objectives the positioning of military platforms in real time with high accuracy and of civilian platforms with moderate accuracy. It provides global coverage 24 hours a day under any climatic conditions and allows synchronizing clocks with accuracy better than 500 ns. It consists of three segments, the space segment, the control segment and the user segment.

9.7.1 The Space Segment

The space segment consists of 24 satellites arranged in six orbital planes (four satellites in each plane); planes are offset approximately 60°. Satellites are in circular inclined orbits (they are not geosynchronous). They are at 20,200 km from the Earth surface and it takes them approximately 12 hours to complete a revolution around the Earth. Satellites are ordered in the orbits so that at any point on the Earth's surface any GPS receiver can always view five to six satellites.

Satellites continuously transmit reference signals. All satellites transmit on the same frequency, but each has a different code; all receivers know each satellite's code. Then they can distinguish which satellite they are receiving. Receivers can determine their positions from the information sent by the satellites. The observation of at least three satellites is needed to calculate the place of the surface on which the observer is.

The user's positioning accuracy depends on the precise knowledge of the position of the satellites' orbits and the measurement of the distance between the satellite and the receiver. To do so, each satellite sends a data packet containing, among other things, the following information: the time at which the packet was sent, measured with an atomic clock on board the satellite; the position of the satellite at the instant of transmission (called ephemeris); summary information on other satellites (called almanac); and a pseudo random code used only by this satellite. A pseudo random code is a digital word (ones and zeros) whose sequence is not repeated for a certain period, so it looks like a random sequence of ones and zeros, but in reality it is a periodic signal (Fig. 9.13). The ephemeris is updated every two hours and is valid for four hours, the almanac is updated every 24 h.

The receiver has stored in its memory the pseudo random sequences of all satellites and seeks what sequence coincides with that of the satellite being received. In the next step the receiver determines the time difference (Δt) from the moment when the sequence was sent from the satellite until the instant it was received. This difference is the time it took for the satellite signal (electromagnetic wave) to travel the distance that separates the satellite from the receiver. As the speed of wave propagation can be assumed to be approximately constant ($c = 3 \times 10^8$ m/s), then the distance (d) between the satellite and the user can be readily known, which is called pseudo range.

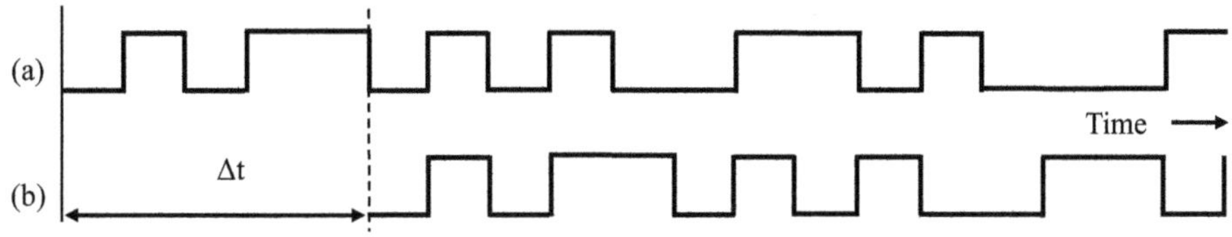

Fig. 9.13: Pseudo random sequences. (a) Sent by the satellite. (b) Receiver generated. Δt is the time needed to travel from satellite to receiver.

$$d = c\,\Delta t \tag{9.1}$$

The solution of Eq. (9.1) allows the receiver to know only the distance d from the satellite, but there are infinite points on a sphere of radius d that meet that condition. To find the three spatial coordinates of the receiver it would be needed to resolve at least three independent equations, each one corresponding to a different satellite. Because the receiver's own clock is not very accurate, one additional equation is required to set the receiver's clock to the satellite clock.

If (x_i, y_i, z_i, t_i) are the ephemeris and the atomic clock time sent from the satellite (where the subscript i denotes the satellite number), we can write the equation of a spherical surface for each satellite as

$$(x - x_i)^2 + (y - y_i)^2 + (z - z_i)^2 = d^2;\ \ d = (\tilde{t} + e - t_i)\, c \tag{9.2}$$

where (x, y, z, t) are the unknowns corresponding to the position, $\tilde{t}$ is the approximate time of the receiver and e is the offset between the satellite actual time and the approximate time of the receiver. Then, simultaneously solving four equations, the real time and the position of the receiver are obtained.

9.7.2 The Control Segment

The control segment of the GPS is composed of fixed Earth monitoring stations called Operational Control System (OCS) scattered throughout different parts of the Earth's surface, and a Master Control Station (MCS). Monitoring stations are receivers that track the satellites as they pass over them and collect telemetry data. As the positions of the OCS and MCS are accurately known and they have precise clocks, it is possible to verify if satellites are displaced from their theoretical orbits. The MCS receives the data from the OCS and determines if satellites have undergone changes in their clocks or their orbits. Then the ephemerides are recalculated based on this information and they are reloaded to the satellites a number of times per day.

9.7.3 The User Segment

The user segment consists of the receivers of the signals sent by the satellites. They receive the signals from the satellites and calculate their position and the local time.

The signals sent by the satellites give two qualities of services for the absolute positioning of the receiver. The Precise Positioning Service (PPS), exclusive for military use and the Standard Positioning Service (SPS), which is available to all users. For GPS receivers that meet with a set of particular assumptions, the SPS can assure that 95% of the time, the horizontal error is less than or equal to 17 m and the vertical error less than or equal to 37 m (Department of Defense (USA), 2008). For specific locations and with special receivers these errors are significantly reduced (Hughes Technical Center, 2011). But if the above particular assumptions are not satisfied the performance is somewhat degraded.

An additional use of the messages sent by the satellites is the synchronization of time bases. Since the satellite sends the atomic-clock on-board hour, solving the four equations allows the clock time of the receiver to be corrected by using the satellite time that is very accurate. For the SPS, the time transfer domain accuracy is less than or equal to 40 ns (Department of Defense (USA), 2008). This information is used to synchronize systems such as cellular telephony, in which several stations must work synchronized.

9.8 Data Storage

The storage of information is a process quite familiar to most of the readers. Therefore, only some general concepts and problems that could happen with instruments' memories will be mentioned.

9.8.1 Storage Systems

In general, present systems that store information are based on electromechanical magnetic disks like hard disk drives (HDD) used in PC's; on optical-mechanical devices as rewritable compact discs (CD-RW) or DVD, which have a mechanical component to rotate the discs and an optical device to "read / write" it; and on solid-state memory, which has no mobile parts. Rotating disk devices consume a considerable amount of power and occupy a greater volume than solid-state memory, so most measuring instruments use solid-state memories as storage media.

Solid state memories can be either volatile or non-volatile; the first loose the information stored in it when the power supply is turned off, the second keeps the information even without power (as it is the case of a pen drive). The second type is used in instruments as storage media. The non-volatile memories may be of two types: read only memory (ROM) and random access memory (RAM). The first type, once written, can only be read; the second type may be rewritten.

A solid-state drive (SSD) is a solid-state memory that has input/output interfaces compatible to HHD, which facilitates its use. There are two types of SSD, one is of the volatile type (backed up by a battery) and the other is non-volatile. In general, there is a compromise between speed and persistency of the data; memories with a fast access to the data are volatile, non-volatile memories have longer access times. A memory can be visualized as an array of boxes where each data is stored in a particular box (Fig. 9.13). Data writing in memory is the process of placing data in a box. The process of reading the memory is copying the data found in the box to an external device. Data writing alters the content of the memory, data reading does not. The procedures for placing or retrieving data from boxes require knowing the addresses of the boxes. The same happens with writing and reading from memories. Therefore, before to transfer data from and to the memory, it is necessary to identify the address of the memory cell.

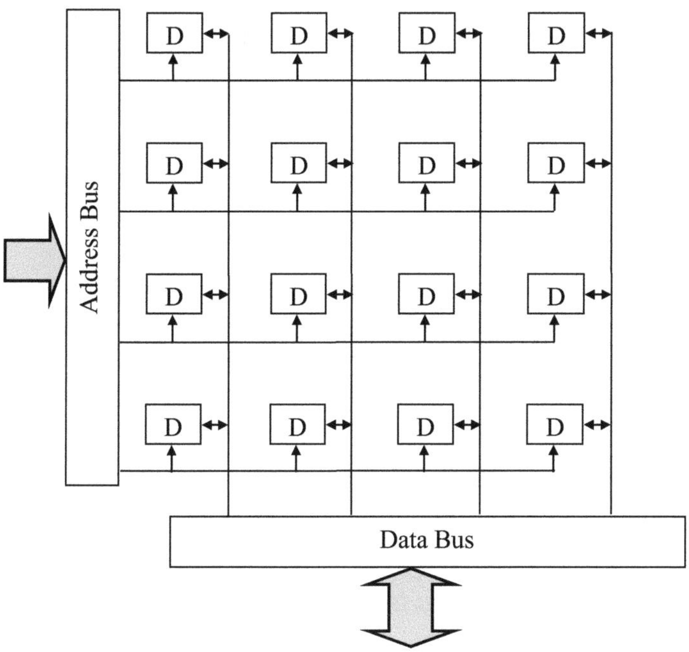

Fig. 9.14: Schematic of a memory

As was explained in Section (9.3.6), internal connections in a computer are parallel buses made of tracks on a printed circuit board, each track carrying a bit. Likewise, solid state memories have internal tracks that form buses with the specific purpose of handling memory information. They are the **address bus** and the **data bus**. The address bus selects a specific cell, and its information is conveyed through the data bus. Put another way, a digital word in the address bus selects which "memory box" will be connected to the data bus. Then if a read process is initiated, information is collected from the bus by other devices, i.e. a register of the processing unit, or if a write process has begun, the information from the bus is introduced into the cell.

An array of 16 cells is depicted in Figure 9.13; each cell has a data (D). The address bus sends the information just in one way, from the bus to the cell; it selects which cell will be connected to the data bus. The data bus is bidirectional, because data can be read (copied from the cell to the data bus) or write (cell data is changed taking the value presented in the data bus).

9.8.2 Comments

Instrument users should be aware that together with the non-volatile memories, where the results of the measurements are stored, instruments might have volatile memories. Because they have short writing and reading access times they are used for some internal applications such as fast signal processing. Sometimes, transference of information from volatile memory to non-volatile is performed in blocks. Then if the battery is disconnected before all the information has been transferred from one memory to the other, some data could be lost. Users should be aware that all instruments' internal processes have to be finished before disconnecting the main battery to avoid data loosing.

In general, an instrument's real-time clock has internal volatile memories and they are supplied by a secondary special long-life battery. Thus, when the main battery is disconnected, the clock keeps working. But when the life of the secondary battery ends and the main battery is disconnected, a reset process of the clock will occurs and the real time is lost. When the reference of time is lost, it is required to set the real time clock before using the instrument again. In general, a manufacturer advises the user about these operational features, but it seems useful to remember them here.

9.9 Tips to Select a Communication Media

These tips could help the user to select a communication media for a field instrument network reporting to a central station. They are based on some questions that the user should answer before selecting a data communication media. Among the parameters to be analyzed are: the period between transmissions, the amount of data to be sent and the accessibility to the remote sites where instruments are installed.

The first question to answer is what real time means for the problem being evaluated. In some control applications, real-time measurements mean to have a set of data inputs for supplying a numerical model at the same speed that the model produces a data output. For example, let us assume that a numerical model is running in a microprocessor on board a model plane to control its trajectory. The numerical model processes the entering information to produce control actions every 0.1 s. In order to obtain meaningful control results, the model should be supplied with a set of data inputs at least ten times per second. If 5 variables, each one with a byte (8 bits) of resolution, are measured onboard the plane, these variables have to be sampled ten times per second to keep the system working in real time; thus, the amount of bits per second resulting from the measurements would be $5 \times 10 \times 8 = 400$ bits/second.

Suppose now that the numerical model is run on a computer placed on land, and that the measured information onboard the plane has to be sent from the plane to the control station and the numerical model results transmitted back to the plane. At least a half-duplex transmission system with a baud rate in excess of 400 bits/s is needed. But if the amount of information to be sent from land to plane is similar to that in the opposite sense, a baud rate about 800 bits/s would be needed or a full-duplex system should be used. This kind of real-time problem is generally solved by means of radiofrequency links.

A real-time system tracking whale migration could be comprised of a GPS receiver attached to a whale and a transmitter to report its position. In this case, perhaps it would be enough to know the whale's position once every 10 minutes. Thus, for this slow speed motion problem, real-time means taking few measurements each hour. Thus, the problem could be solved by transmitting the GPS position through a satellite communication system; a few bits per second are required.

The real time requirement is one of the parameters that most often influence the characteristics of the communication system needed. Then, as already pointed out, it is important to correctly define the time scale of the problem to be solved.

If the maximum delay accepted between the information is sent and received is less than one second, most satellites are not useful as communication media. Perhaps the solution has to be sought in radio frequency links or fiber optic.

The second great question is how much data is really necessary to transmit in real time.

The answer to this question also conditions a lot of the characteristics of the communication system. A field station could have several instruments and all of them store the information on non-volatile memory, and perhaps data from only few of them are needed in real time. Then selecting to send just those really needed could facilitate selection of the communication system, thereby reducing costs.

Another limiting factor is the possibility of reaching the instrument sites at any time, in all seasons and at low cost. If continuous access to the sites is not possible, or if it is expensive to arrive at them, users will be compelled to send the data using any available communication system. Perhaps they should have to modify their measuring expectations, adapting data transmission to the capability of the available communication system.

In selecting a transmission media it should be evaluated whether information has to be sent in both directions. If this is the case, a bidirectional media is required, and simplex mode transmissions are excluded of the selection. This means that a transmitter and a receiver are needed at both ends of the link.

When the information has to be sent in both directions it should also be analyzed whether this has to be done simultaneously. For both ways simultaneous communications, two communication channels are required; in radio frequency, this means that two frequencies are needed. This requirement leaves out some satellite signals that are not bidirectional.

Transmitting large amounts of data precludes the use of some satellite services which are conceived to send small data packets; then, attention should be paid when contracting a satellite service. In these cases, Short Message Service (SMS) of digital telephony cannot be a solution either, because they have been devised for short messages.

For large amounts of data, when possible, a private network should be studied as an alternative, because public networks charge their services by time of use or amount of data transported. Private networks do not have service costs; however they have installation and maintenance costs.

Communication solutions are very different for fixed or mobile instruments. We understand for mobile instrument one which is on board a vehicle or attached to a living moving being. Satellite services are practical solutions for instruments on mobile platforms sending low amounts of data, especially when the platforms are disseminated in a very wide area. In populated areas, perhaps, cellular telephony could be used to solve this kind of communication problem. When mobile platforms are underwater, they require specially designed acoustic networks.

Instruments that simultaneously store and transmit data and can be acceded at any time and low cost, could allow using low quality (generally low cost) communication media, because the data is safe on board the instrument and if the communication fails, the user could recover the data directly acceding to the instrument.

9.10 New Trends in Data Processing and Storage

Since the emergence of Internet the storage and transmission of information have suffered a revolutionary change. This new way of communication gives rise to the availability of a great infrastructure that can be used in measuring systems for the transportation, processing and storing of information in science research activities. An overview on this issue will be addressed below.

9.10.1 Internet

It is a tool well known by all students and it is not necessary to spend time on it. Only a few ideas will be introduces as a first step to the Cloud Computing concept. The Internet is based on a set of protocols which specify how data have to be handled. The most important are known as the Transmission Control Protocol (TCP) and the Internet Protocol (IP). The Internet allows accessing the World Wide Web (WWW) which is a system composed of web pages whose contents are information stored in different formats: images, texts, audio, video, etc. These pages are usually written in the Hypertext Markup Language (HTML) which is a structured language. Such texts contain links to other web pages or other kind of codes inserted inside the texts.

This interlinked system can be navigated by means of web browsers, which are software applications for searching, introducing or retrieving information from the WWW. Each source of information is identified by a Uniform Resource Identifier/ Locator (URI/URL) also known as web address. It consists of a string of characters that identifies where to connect, how to connect and what to ask for.

The Internet allows us to access a wide distributed network where we are able to found numerous information resources and services. Individual nodes of the distributed network can share with other nodes their resources such as process capabilities and storage capacity. These abilities give rise to the concept of cloud computing.

9.10.2 Cloud Computing Description

Cloud computing (CC) is a relatively new concept in Information Technology (IT). Since its introduction computing and data storage is moving from desktop and portable PCs to large data centers. It is the latest step of a long way which begun with the concept of parallel computing, followed by distributed computing and grid computing. The main advantage claimed by vendors of cloud services is that customers do not have to pay for infrastructure installation and maintenance (Jadeja & Modi, 2012).

The National Institute of Standards and Technology (NIST) of the United States defines: "Cloud computing is a model for enabling convenient, on-demand network access to a shared pool of configurable computing resources (e.g., networks, servers, storage, applications and services) that can be rapidly provisioned and released with minimal management effort or service provider interaction" (Jadeja & Modi, 2012).

Cloud computing is also called dynamic computing because it provides computing resources according to the client's needs. It consists of a virtual group of information resources (networks, storage, central processing unit and memory) to accomplish with the user's requirement. CC makes it possible parallel batch processing, which allows users to analyze huge amount of data in small periods of time and at low costs. Many computers can work together to perform computing and data processing services that are impossible to do with personal computers (Mollah et al., 2012).

CC refers to the capabilities provided by real-time communication for sharing information and services among computers. For example, by summing up the hardware abilities of several real servers it is possible to simulate a virtual server that does not physically exist but which provide complex information resources or services to the users.

CC is a set of Information Technology (IT) services that meet certain characteristics: multiple clients share a common technology platform and instead of paying fixed costs for a service, they pay according to its consumption, users can scale services according to their needs and, theoretically, there are no limits to growth. This concept of network-based services could be useful in some applications to the environmental science researcher community.

9.10.3 Cloud Computing Types

There are different types of cloud computing according to the kind of users that can access the cloud. A cloud is called a **public cloud** when the network can be accessed by any customer on Internet. The services provided in these clouds can be accessed by any organization (Mollah et al., 2012). In public clouds all the resources and applications are managed by the service provider (Jadeja & Modi, 2012).

When their requirements and policies agree, a group of organizations can merge to construct and share a cloud infrastructure. Such a cloud models are called **community clouds**. They can be used only by members of the community who have common interests. The cloud infrastructure could be hosted by a third-party provider or within one of the organizations in the community (Jadeja & Modi, 2012).

Either public or community clouds are accessed usually via Internet; for this reason security precautions should be implemented. Because they share resources such as infrastructure and services, the network is more efficient. For example, services that are mostly required during day hours (such as email service) can be used in other parts of the world during night hours http://en.wikipedia.org/wiki/Cloud_computing.

Some disadvantages of these clouds are that the permanent availability of the services is not assured, and that the immediate accessibility to the stored information may fail because the reliability of the complete system depends on the reliability of Internet. Furthermore, users run the risk that the information could be disclosed or that the service provider could access the stored information. The service provider could accidentally lose the hosted information as well.

Sometimes cloud computing services have suffered outages and in such cases users are impeded to access their services and data stored; public clouds are more prone to malicious attacks (Jadeja & Modi, 2012).

In contrast, **private clouds** offer IT services to a predefined group of customers; it is a proprietary architecture which provides hosted services to the users within the organization, with access through Internet or private networks. The private cloud can only be used by the members of the specific organization and could have direct connectivity. This kind of cloud is more secure than both previously described because data and service availability depend only on the own organization. Technologically speaking, private clouds are similar to public clouds, and provide similar benefit such as agility and cost savings. Services are pooled together and made available for the users at the organizational level (Jadeja & Modi, 2012). Cloud is protected by firewall barriers that prevent access to hosted services from those outside the cloud (Mollah et al., 2012).

There are also **hybrid clouds** as a combination of a private cloud and one of the two firstly mentioned clouds. This combination could profit from the benefit of both kinds of networks. For example, some critical information could be managed in the private cloud to avoid information disclosure and to have immediate usability, avoiding relying upon Internet connectivity. At the same time some public cloud services could be used for less sensitive information processes. Cloud users should take a look at the works they should place on a controlled private cloud and what other works they can put in the public cloud, where the risk of service interruption and loss of data is higher.

There is also another classification of clouds as **internal** and **external clouds**, the first being a subgroup of the private clouds. They provide services within the same company or corporate group. The second may be public or private and provide services to other companies http://www.networkworld.com.

Cloud computing could decrease initial investments, facilitating future scalability and increasing availability of services. However there are some risks that must be taken into account. There are no safe regulations for storage and backup of the data handled and hosted in the cloud. When outsourcing their own data to a provider, customers are still responsible for the security and integrity of them, even when they are held by a third party provider (Mollah et al., 2012).

Service providers do not assure that interfaces between the services taken from the cloud and those of the company itself will be sustainable over time http://www.networkworld.com. Furthermore, it is only possible to use applications or services that the provider is willing to offer (Mollah et al., 2012).

9.10.4 Cloud Computing Stack

There is a tendency to classify the kind of services that the CC is offering. There are several classifications but we will mention only the simplest and known.

1) Software-as-a-Service (SaaS)

SaaS applications are designed for end-users, it is software which is deployed over Internet, thus eliminating the need to install and run the application on the users system. This is a "pay-as-you-go" model. An advantage of SaaS is that it does not need specific hardware to run the software (Mollah et al., 2012).

2) Platform-as-a-Service (PaaS)

PaaS provide development environment as a service. The provider's equipment can be used by the client to develop its own program and deliver it to the final users through Internet. In other words, PaaS gives the set of tools and services needed to make coding and deploying the generated applications as a service. The advantage of PaaS is that there is no need to buy special hardware and software to develop and deploy enterprise applications (Mollah et al., 2012)

3) Infrastructure-as-a-Service (IaaS)

IaaS is the hardware and software that support all the information system, such as servers, storage space, networks, operating systems, etc. Also, these tools are offered as services.

There is an analogy that could help to understand the stack (www.rackspaceclouduniversity.com). The infrastructure is compared to a road transportation system, the trucks and cars are analog to the platforms, and people and goods could be considered as software and information. Even when each service is shown as a layer of the stack, their differences are not so clear in reality.

9.10.5 Application Example

Researchers working in environmental sciences may need to manage great amounts of data collected by means of field instruments i.e. radars, satellites, etc. Also, some software may be needed to analyze these data, which requires a large amount of computing resources.

In this example, several research institutions could provide and share data over a community cloud where the valuable information (field data) is safely stored and backed up. Some public cloud could be used as provider for the computing services to run particular private owned software.

Another case could be that some commercial software, available among the services that one public cloud provider offers, be useful to analyze the data. Then the use of the software could be bought as a service from the public cloud. Suppose that the results of the data analysis are a great amount of three dimensional maps which require a large amount of memory to be stored. They could be hosted in a public cloud. Therefore, the selection of the best kind of cloud depends on the problem at hand.

References

Bellamy, J. C. (2000). *Digital Telephony*. New York: John Wiley & Sons, Inc.

Department of Defense, United States of America. (2008). *Global positioning system, Standard positioning service, Performance standard, Navstar Global Positioning System,*

Ferrero, A. (2005). *Fundamentals of Metrology and Measurement Science*. IEEE Educational Courses. IEEE Instrumentation & Measurement Society, eLearning.

Feynman, R. P., Leighton, R. B., & Sands, M. (1967). *The Feynman Lectures on Physics*. Reading, Massachusetts: Addison-Wesley Publishing Company, Inc.

Hughes (William J.) Technical Center. (2011). *Global Positioning System (GPS), Standard Positioning Service (SPS), Performance Analysis Report*. Report #74, NSTB/WAAS T&E Team, Atlantic City International Airport, NJ.

Jadeja, Y., & Modi, K. (2012). Cloud computing - concepts, architecture and challenges. *Computing, Electronics and Electrical Technologies (ICCEET), 2012 International Conference on*.

Lenz Ch., McCormick, C., Goldsmith, R., & Trachtman, E. (2010). Real time, near global, low earth orbit communications using geostationary Inmarsat BGAN system as a relay. *24th Annual AIAA/USU Conference on Small Satellites*. Pp. 11.

McCormick, C. C., Lenz, C., & Yunck, T. (2010). Real-Time Delivery of Global Environmental Observation Data From Space-Based Sensors using the Inmarsat BGAN System. *American Geophysical Union, Fall Meeting 2010*. Abstract #SM51A-1768.

Mollah, M. B., Islam, K. R., & Islam, S. S. (2012). Next generation of computing through cloud computing technology. *Electrical & Computer Engineering (CCECE), 2012 25th IEEE Canadian Conference on*.

Proakis, J. G., & Salehi, M. (2008). *Digital Communications*. New York: McGraw-Hill Higher Education.

Tomasi, W. (2001). *Electronic Communication Systems: Fundamentals through Advanced*. New Jersey: Prentice Hall, Inc.

Understanding the Cloud Computing Stack SaaS, Paas, IaaS, Available at: www.rackspacecloudui-versity.com

http://www.hko.gov.hk/wservice/tsheet/pms/images/Iridium.pdf

http://www.iridium.com

http://www.orbcomm.com

http://www.globalw.com

http://www.inmarsat.com/services/index.htm

http://www.evologics.de/files/DataSheets/EvoLogics_S2CR_4878_Product_Information.pdf

10 Satellite-Based Remote Sensing

10.1 Introduction

Those beginning to read the book at this chapter could find it troublesome with so many references to previous chapters, but it is the only way we found to make this chapter both short and straightforward. We think that these references will also help the readers to refresh the needed concepts that they might have forgotten since they read the first chapters.

We will focus this chapter on the technical characteristics of the instruments on-board satellites, avoiding any reference to the way in which the geometric aspects of the flight determine the composition of the images. Nor are we going to treat how images have to be interpreted. These subjects can be found in most books on remote sensing (Campbell, 2007; Lillesand et al., 2004). As in the rest of the book our interest will be centered on how instruments work in order to understand their applications and limitations.

10.2 Preliminary Discussion

Satellite-based remote sensing is a vast and rapidly changing field. Battery degradation, unexpected failures and exhaustion of propellants, among others, are problems that make satellites' life quite limited. For example, geosynchronous satellites require propellants to keep them in orbit and to maintain their altitude so that the solar panels and antenna can be pointed adequately. Therefore, because they run out of propellant, the useful lifetime of geosynchronous satellites averages about fifteen years (Kurtin, 2013). Also, some missions with polar orbits planned for several years and carrying sophisticated remote sensing equipment lasted less than two years due to failures in the energy system or in the instruments (http://coaps.fsu.edu/scatterometry/about/overview.php).

Similar problems are faced by communication satellites (Section (9.6.4.1)).

The short operating life of satellites, along with the rapid changes in electronics and satellite technologies, makes remote sensing from space a very dynamic field. There are many different instruments working on board spacecrafts, as well as many names, e.g. active microwave instruments, synthetic aperture radar, wind scatterometer, radar altimeter, scanning radiometer, ozone monitor, microwave sounder, precise range and range rate equipment, laser reflector, microwave radiometer, LIDAR, precise orbit Doppler locator, laser tracker, GPS tracker, precipitation radar, microwave imager, visible infrared scanner, clouds and Earth's radiant energy system, lighting imaging sensor, etc.

For inexperienced researchers willing to start in the satellite instrumentation field, the problem looks very confusing because they observe instruments with similar operating principles and different names. But from the point of view of the experts, these instruments are really very different due to the specific characteristics that allow them to gather different information from the same piece of the Earth's surface. For this reason, experts create new names to underline the differences from former instruments, even though the operating principles are quite similar.

It is difficult to describe in a book all the instruments flying in one modern satellite. Besides, due to the fast changes in remote sensing strategies any description runs the risk of becoming quickly obsolete. For these reasons, only lasting concepts on the technologies used in remote sensing and general operating principles will be addressed.

In our efforts to synthesize these issues we will make some oversimplifications. We hope that we will have the comprehension of readers who know about this subject, because they could consider these approximations somewhat rough.

10.3 Introductory Concepts

Some concepts on satellites have already been introduced in Section (9.6.4) when satellite communications were explained. The spacecrafts used for remote sensing have many similarities with those described first. They are space platforms whose main differences are the payloads they carry. In some cases there are also some differences in the orbits they travel. Instead of carrying communication equipment, satellites used in remote sensing applications transport instruments for sensing the electromagnetic energy coming from the Earth's surface.

From the point of view of the measured energy, instruments onboard satellites can be differentiated into two major groups: **passive** and **active** instruments. In the first group are those instruments that measure the sun energy reflected or re irradiated by the Earth's surface. In the second, instruments that measure the energy sent by them and scattered by the Earth's surface.

Some passive remote sensing instruments gather information in the wavelength range from tenths to tens of micrometers. Usually, instruments designed for this range are tailored either for information in the visible spectra or in the infrared range of the spectra. The latter are known as thermal radiometers or scanners.

There are other instruments that work in the microwave band. Whereas thermal radiometers work with wavelengths about 10 μm, microwave radiometers work with wavelengths between 1 mm and 1 m. We find it convenient to clarify this point because the word microwave could be misleading.

Active remote instruments, instead, are similar to radar (Section (8.2)) or lidar (Section (8.8)). A well known selected electromagnetic perturbation is emitted from the instrument towards a target and the backscattered signal is analyzed to obtain information about the target illuminated by the perturbation. Active remote sensing in the radio frequency range uses wavelengths on the order of centimeters. Lidar are active devices in the wavelength range from about 10 μm (near infrared) to about 250 nm (ultraviolet).

The different possible types of satellite orbits were already introduced in Section (9.6.4). Many remote sensing spacecrafts follow north-south orbits which, for an observer fixed to the Earth, appear as inclined with respect to the north-south direction due to the Earth's rotation (Coriolis force, Section (3.10)); they are called near-polar orbits. The usual altitude for meteorological satellites is about 900 km but, as it was explained in Section (9.6.4), geosynchronous satellites need to be at 35,800 km, either if they are used for communication or for metrological purposes. Because the geosynchronous satellites are above the Equator, they do not cover high latitude areas. Satellites in polar orbit provide a frequent coverage of the Polar Regions.

A particular characteristic of some remote sensing satellites not observed in communication satellites is that their near-polar orbit is sun-synchronous. Thus, satellites always cover the same area of the world at the same local time, ensuring similar illumination conditions. Images having the same illumination are easier to compare and changes in one specific area are readily appreciated.

In Section (9.6.4) it was called as footprint the area of the Earth's surface with which a satellite can communicate while following its orbit. In remote sensing field a similar concept, known as **swath**, is used to define the portion of the Earth's surface that the sensor "sees"; Figure 10.1 depicts a satellite and its swath. The Earth's rotation determines that in each orbit the instrument on board the satellite cover a new swath, thus the complete Earth's surface is imaged after some regular time.

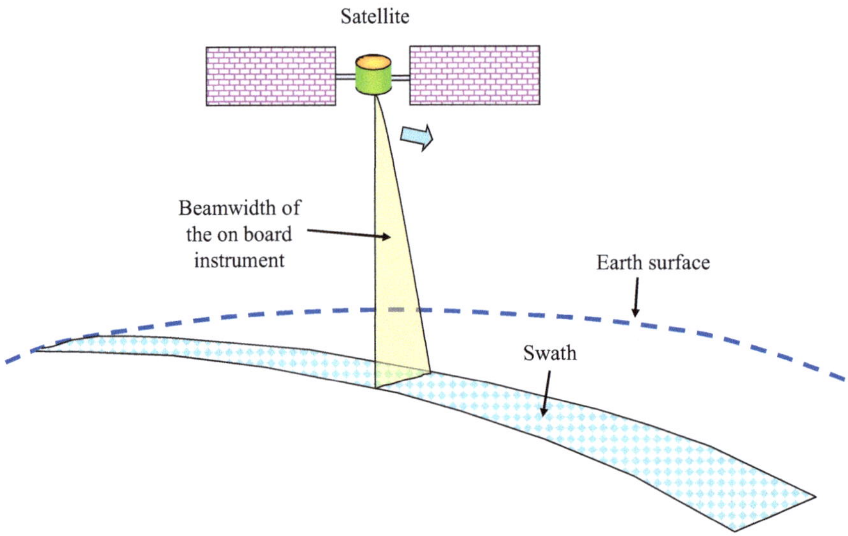

Fig. 10.1: The beamwidth of the instrument and the orbit of the satellite define a swath on the Earth's surface.

Figure 10.2 shows the nomenclature used in the geometry of satellite's radar. The angle between the radar beam direction and a line directed towards the nadir is called the **look angle**. Recall that the nadir is the point where a descendent line oriented perpendicular to the Earth's surface and passing through the satellite and the center of the Earth seems to intercept the celestial sphere (the opposite, upwards point, is the zenith). The complement of the look angle is the **depression angle**. The **incident** angle is the angle between the radar beam direction and the normal to the Earth's surface at the point where the radar beam hits the surface (Lillesand et al., 2004).

Satellites devoted to acquiring images of the Earth can be classified into two large groups according to their uses; some are designed to study the Earth's resources, while others are devoted to study environmental phenomena.

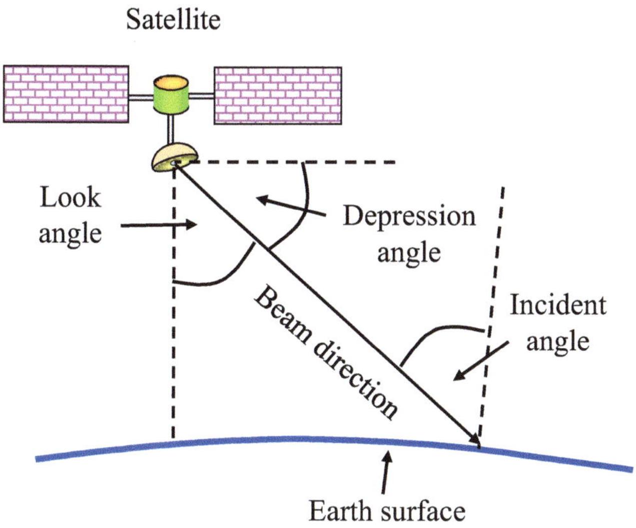

Fig. 10.2: Nomenclature used in satellite's geometry

Earth's resources such as crops do not change very fast, but to estimate the amount and degree of growth it is necessary to have good spatial resolution. Therefore, the swath width of these images is less than 200 km, and the spatial resolution is less than 100 m (Sabins, 1999). Due to the relative small width of the swaths, these satellites require several days to obtain a complete image of the Earth.

Environmental satellites, in contrast, need to collect information on the complete Earth's surface daily or hourly. Thus, swaths are of hundreds to thousands of kilometers and the spatial resolution of several hundreds of meters (Sabins, 1999). Once again, the concept of spatial and time resolutions of the instruments dedicated to measure environmental parameters is essential in the quality of the data gathered; thus, the concept of image resolution is introduced below.

10.4 Image Resolution

The image resolution gives an indication of the ability of the instruments to capture the Earth's surface information with a certain degree of detail. When speaking of satellite images, in addition to the spatial resolution already mentioned, it is usual to define the terms: spectral resolution, radiometric resolution and temporal resolution. All of them will be described below (http://concurso.cnice.mec.es/cnice2006/material121/unidad1/i_resolucion.htm).

10.4.1 Spatial Resolution

In Section (4.7.7), where infrared thermography was described, we mentioned that the cameras detect the energy emitted from individual points of an object and convert them into a visual image. An image consists of a matrix of pixels; the number of pixels depends on the number of detectors (optical sensors) that the camera has. For example, some cameras for domestic use have 1600 × 1200 pixels (1.92×10^6 pixels). The amount of pixels defines the maximum spatial resolution of the camera (the minimum point of the image that can be distinguished). For instance, if the camera flying on board a spacecraft has 4000 × 3000 pixels and the swath is 40 km × 30 km, the spatial resolution would be one hundred square meters.

In active remote sensing instruments (radar, lidar, etc) it is more complex to quantify the spatial resolution because it depends on several factors: the beamwidth (linked to the transmitted frequency and antenna), the pulse length and the data processing.

10.4.2 Spectral Resolution

Spectral resolution is the amount of spectral channels through which the instrument on board the spacecraft detects the electromagnetic energy coming from the Earth's surface. Different kinds of objects on the Earth's surface can present different spectral responses when illuminated by the sun. Therefore, studying the surface response over distinct wavelength ranges gives information about what objects there are, e.g. to differentiate soil surfaces it is needed to measure on several narrow wavelength ranges.

Some radiometers on board satellites are called multi-spectral sensors because they record energy over a few separate wavelength ranges. For example, the SPOT 6 satellite has a spectral resolution of five, because it records the following wavelengths: panchromatic (450 – 745 nm), blue (450 – 520 nm), green (530 – 590 nm), red (625 – 695 nm) and near-infrared (760 – 890 nm) (Astrium GeoInformation Services, 2013). More recent radiometers can record hundreds or narrow spectral bands called hyperspectral sensors. Obviously, this high spectral discrimination allows acquiring more information about the targets.

Active instruments use few frequencies to "sense" the targets, and then we could say that they have a reduced spectral resolution. The electronics and the transmitting and receiving antennas define the spectral resolution.

10.4.3 Radiometric Resolution

The radiometric resolution is the ability to discriminate very slight differences in the amount of the received energy. Each optical sensor, which defines a pixel in the image, receives the electromagnetic energy from the Earth's surface and produces a certain analog output, which has to be converted into a digital one. The brightness of each pixel is discretized according to the analog-to-digital converter (Section (3.6.6)) used by the on-board instrument. For example, some satellites had 8 bit resolution, whereas others had 10 bit; i.e. the brightness was converted to 256 or 1024 levels of gray, respectively. In hyperspectral radiometers this discretization is done for each spectral band, thus increasing considerably the amount of resources needed to store or transmit the data.

10.4.4 Temporal Resolution

Temporal resolution in satellite vocabulary refers to the period of time needed for a satellite to sense again the same area. The time that it takes for a satellite to complete one entire orbit cycle is called the revisit period; it is the period for the satellite to pass again over exactly the same swath. Because some satellites have the ability to change the pointing angle of their optical systems and antennas, the same area could be "viewed" from another orbit when the satellite travels in a near, but different, orbit. Thus, some satellite can define a temporal resolution smaller than the revisit period. It has to be emphasized that they sense the same region but with a different viewing angle. High resolution satellites have small swaths and a long revisit period, e.g. satellites with a swath 60 km wide could pass over the same point every 26 days.

10.5 Instrument's Scanning Geometry

10.5.1 Introduction

It is worth recalling that instruments devoted to study the Earth's surface from on board satellites only measure electromagnetic waves coming from it. Therefore, the purpose of the scanning systems is to point the instrument towards some specific spot on the surface (ground resolution cell). They are merely pointing systems (opto-mechanical devices) conceived to select pieces of the surface over which the measure will be made. They are designed having in mind how the composition of these pieces (spots) will result in an image of a region of the Earth.

In order to simplify our explanation we will describe these scanning geometries associated to a passive or active instrument, but the way the Earth's surface is scanned is independent of the instrument itself. They can be used either to point a passive or an active sensor.

10.5.2 Across-Track Scanning

In passive instruments there are different ways in which the electromagnetic signal coming from the Earth's surface is conveyed to the sensor through the optical arrangement. One of them is called across-track scanning and consists of a rotating mirror that scans the terrain on the swath in a direction perpendicular to the satellite track. Thus, it measures the energy from one side of the swath to the other traveling an arc, e.g. 90°. As the satellite advances in its trajectory, a new scan is performed. Successive scans form a bi-dimensional image (Fig. 10.3). It has to be emphasized that in this geometry there is only one point on the instrument where the spots from the Earth's surface are sensed. Namely, the conversion of the spots from electromagnetic waves into electrical signals able to be recorded on-board the satellite or transmitted to the receiving stations is done by only one sensor. Furthermore, in this system the response of the sensor has to be fast (short time response) because its output has to fluctuate as the brightness of the spot changes (Lillesand et al., 2004).

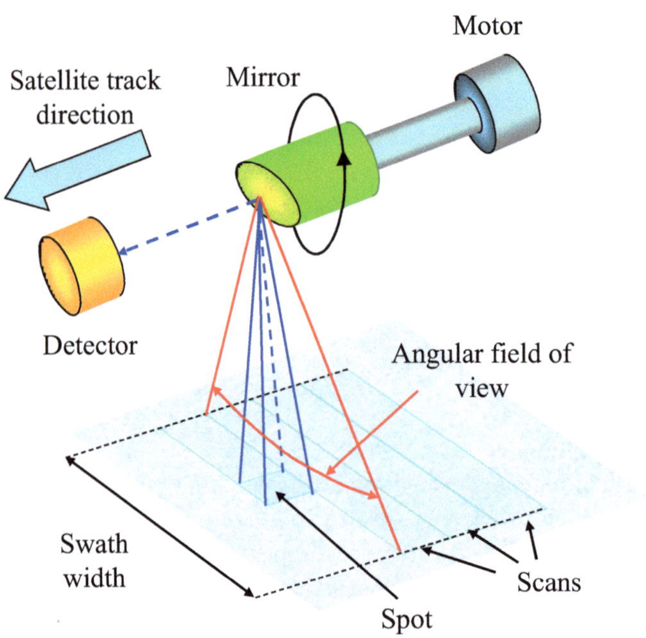

Fig. 10.3: Across-track scanning

Although there is only one point (called detector in Fig. (10.3)) where the successive spots are converted from optical into electrical signals, multiple sensors or spectral filters can be placed at this point to obtain the same surface's spot information with different spectral resolutions.

The outputs of the sensors are analog functions of time, which are sampled at fixed intervals and converted into digital signals (Section 3.6.4). Digitalization has to be synchronized with the scanning to assign a digital value corresponding to a specific spot on the Earth's surface. Due to variations in the scan mirror velocity some geometric errors are introduced in the sensing process and some pixels could be incorrectly positioned on the image. The main advantage of this arrangement is that all the spots composing the images are weighted by only one calibration constant because the sensor is unique.

10.5.3 Along-Track Scanning

In this case the optical system is simpler than in the previous case because many points in a complete line across the track are sensed simultaneously and no rotating mirror is needed. The sensing system consists of a linear array of sensors positioned side by side; the array can include thousands of sensors (Fig. 10.4).

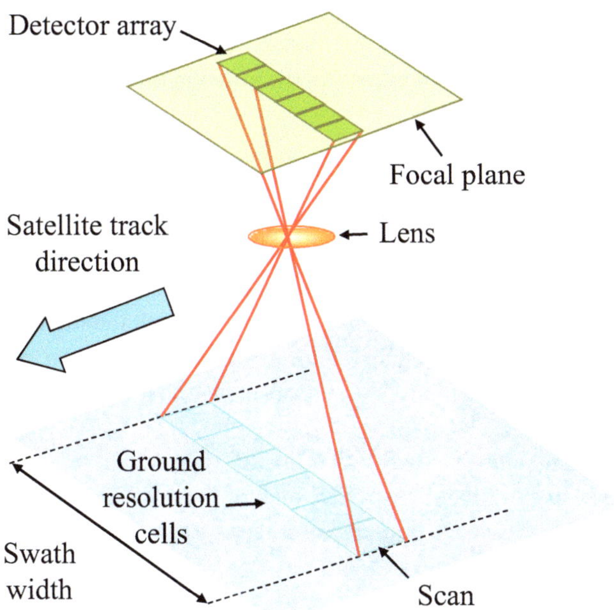

Fig. 10.4: Along-track scanning

The forward motion of the satellite generates a bi-dimensional image by integrating successive across lines. In this array, the dwell time of an optical input over a sensor can be increased with respect to the across-track scanning because the sensing process is divided in multiple sensors. This enables stronger signals to be recorded, thus higher signal to noise ratios are achieved. Due to the fixed relationship between the sensors of the array, geometric errors are minimized. Also, this array has longer lifetime expectancy because they have no moving parts (Lillesand et al., 2004). A disadvantage of the along-tracking is that to provide reliable images, sensors should be individually calibrated to have their calibration matched. Differences in calibration lead to a lack of homogeneity in the pixels of the same line.

10.5.4 Rotating Antenna

Some active instruments use a rotating parabolic antenna as was shown in Figure 8.4. In remote sensing from satellites the antenna are installed on the spacecraft looking down. The antenna is pointed in such a way that the beam describes a cone of vertical axis. It is an inverted version of the system shown in Figure 8.15 where the LIDAR to profile the atmosphere was described.

The antenna is mounted on a spin mechanism that makes the antenna rotate at several revolutions per minute. The spin mechanism provides very accurate spin control, which, in turn, assures an accurate position or the pointing direction, allowing a correct positioning of the spot on the image. Optical encoders such as those described in Section (4.14.2) are used to resolve the antenna direction within a small fraction of degree.

10.5.5 Digital Video Camera

The digital camera is conceptually similar to those used for home photography; it senses a certain number of complete lines across the track simultaneously. It is similar to the concept described in Section (4.7.7) for infrared thermography. An image is generated by a matrix of sensors, the number of sensors being the number of pixels in the image.

Multiple-spectral sensors can also be used to acquire information not only in the visible spectra but in other spectral bands as well. Some of these cameras provide images similar to the infrared film (used in aircraft remote sensing) by acquiring three spectral channels, namely in the green, red and infrared.

10.6 Passive Instruments

These measure the Earth's surface radiance, from which meteorological quantities are then derived; other information such as amount of vegetation and volcanic activity can also be acquired. Images are two-dimensional distributions of brightness composed of individual points (pixels), each one with a particular intensity (World Meteorological Organization, 2008). Some parameters that influence the sun reflected radiation are: sun elevation, satellite-sun azimuth angle, satellite viewing angle, transparency of the object, reflectivity of the underlying surface and overlying layers such as thin clouds or aerosols (World Meteorological Organization, 2008).

Water, carbon dioxide and ozone in the atmosphere produce absorption in the infrared bands. Thus, studying absorption in these bands permits the atmospheric temperature and amount of absorbers to be known. Fortunately, there are two regions of the thermal spectrum where atmospheric absorption is low; they are in the wavelength regions from 3 to 5 μm and from 8 to 14 μm (Lillesand et al., 2004). These atmospheric "windows" allow measuring the energy coming from the surface of the Earth or from the top of the clouds.

In the past, remote images of passive instruments taken from aircrafts were recorded on films. At present, opto-electronic devices are used for recording Earth' surface images. The optical part of these devices is similar to that used in aerial photographic cameras. The electronic components consist of sensors that convert electromagnetic waves to a voltage or a current signal (as it was shown in Sections (4.7.6), (4.7.7) and (7.4)). In the next step of the on-board process the analog signal is converted to a digital one and transmitted or stored (Chapter 9).

10.6.1 Calibration

Calibration of the **visible** channels is performed before a satellite's launch, neglecting the atmospheric effects. This procedure has the disadvantage that the satellite orbital circumstances can be quite different from the calibration conditions. When using data from the satellite it has to be taken into account the radiometer's degradation with time, furthermore, images from different satellites can have different calibration coefficients. For these reasons it would be interesting to develop on-board calibration techniques for the visible channels (World Meteorological Organization, 2008).

Conversely, thermal scanners in the **infra red** range, which have operating principles similar to that shown in Sections (4.7.6) and (4.7.7), are continuously calibrated on board the satellite. They have two internal temperature references that are acquired together with each scan; thus, a calibration coefficient may be derived for each image's line. These two calibration points are set at both extremes of the expected temperature range to be measured. Additionally, an absolute cold reference can be obtained by pointing the instrument to the space, which can be considered as a black body at about 3 K (World Meteorological Organization, 2008).

10.6.2 Accuracy

A common method for evaluating the accuracy of satellite measurements is to analyze the differences between satellite derived parameters and radiosonde profiles. Also, satellite estimations are compared to field data collected on meteorological stations on the land, or on the ocean by means of buoys or ships. Due to the great differences between the temporal and spatial resolutions of the satellite acquired data and those utilized to contrast them, it is quite difficult to assess the accuracy of the satellite measurements. For example, a work aimed to evaluate the quality of the Tropical Rainfall Measuring Mission (TRMM) found differences of concurrent satellite and rain gauge rain rates at various time scales. Correlations for 5, 30, 60 and 1440 minutes were 0.6, 0.8, 0.85 and 0.93 respectively. The authors said that the major obstacle in evaluating data quality is the lack of "ground truth" reference at the radar pixel scale. A tipping-bucket rain gauge measures rainwater impinging on a circular surface of few decimeters in diameter. In contrast, radar measures reflectivity factors over an atmosphere column with a base on the order of 1 to 10 km^2, and then it estimates rainfall using reflectivity – rain rate relationships. The authors conclude that gauge measurements cannot be directly treated as the ground truth reference due to the lack of areal representativeness (Wang & Wolff, 2009).

10.6.3 Passive Microwave Sensing

Instruments used for passive microwave sensing are called microwave radiometers because they essentially work with the same operating principle as thermal radiometers, but in the **microwave** range. Because they are passive instruments they sense naturally available energy in the microwave range (wavelength between 1 mm and 1 m) using antennas as sensors (these antennas are similar to those used by active radar).

The information gathered by these instruments is more difficult to interpret than with other instruments because the amount of energy received by the antenna is low and may proceed from several different sources. For example, it may have components due to surface temperature and material characteristics of the object, emission from the atmosphere, reflection of the sunlight and transmission from the subsurface (Lillesand et al., 2004).

Instruments for microwave sensing are classified as radiometers and scanners. The first have an antenna that generates a profile of the energy received but not an image. Instead, the second have the ability to move the antenna or use several antennas scanning different swaths; from this information an image is derived.

The interpretation of passive microwave information is not yet well developed, but some interesting characteristics of this technology have emerged. It can be used during day and night under most weather conditions, it is sensitive to water vapor and oxygen, it can be used for identifying sea ice, it can detect soil temperature and oil spills. It has been found that the analysis of passive multi spectral signals in the microwave range gives information from objects beneath the soil; hence it has obvious applications in geology (Lillesand et al., 2004).

10.6.4 Cloud Drift Winds

An indirect way for estimating winds from satellites is by evaluating cloud tops' evolution in time. This is carried out by means of **infrared** images taken from geostationary satellites every half hour. The accuracy of the method relies on the assumption that the cloud motion is due to the local wind, which is not always true because clouds may move due to mesoscale atmospheric disturbances. Another factor that limits the accuracy is the time interval between images.

The process of estimating the wind begins with the selection of target clouds of a size about 20 × 20 pixels. This selection is based in the brightness of the clouds. Once the targets are chosen, their motion is evaluated from one image to the next. Results are presented as a wind vector field of about 1200 vectors covering both hemispheres (World Meteorological Organization, 2008).

When the results of this method are compared with wind radar measurements there appear differences of 3, 5 and 7 m/s on the mean vector for low, middle and high clouds respectively. This method is not usable in areas without clouds, nor can it be applied for high latitudes because geostationary satellites do not reach high latitudes. In high latitude areas, sequential images from polar-orbiting satellites can be used (World Meteorological Organization, 2008).

10.6.5 Precipitation

Some relationships between the brightness of clouds in the **visible / infrared** spectra and precipitation have been developed such that from images taken every half hour, the precipitation can be inferred. However, these techniques are valid only for certain regions and seasons and they cannot yet be extended in time, or applied to other areas.

One of the methods is called the cloud indexing method. It is based on the definition of some relations between precipitation and the amount and structure of clouds. It consists in evaluating the amount of pixels over the cloud surface, which has brightness above a certain threshold. This procedure can be performed with information from the infrared and the visible spectrums. The result of this evaluation is a number called precipitation index.

Some empirical coefficients are adjusted to match the results from satellites to the observations from rain gauges or radar data. Comparisons with rain gauges are difficult because they integrate the precipitation over time in a particular place, while satellite images are instantaneous observations integrated in space.

The cloud indexing method can produce erroneous results due to the presence of non-precipitation clouds, such as cirrus. It has been proposed that satellite results should be used as a complementary tool to ground-based radar observations. Some authors declare that it is easy to adjust coefficients for a particular case, but is difficult to obtain a method that works for most of the cases (World Meteorological Organization, 2008).

10.6.6 Sea Surface Temperature

Estimates of ocean surface temperature may be derived from the measurement of radiation emitted by the top surface layer of the ocean (about 1 mm thick). These estimates are mostly made in the **infrared** spectrum, and to a lesser extent, in the **microwave** spectrum. A great advantage is that they are available in real time for meteorological applications.

The measuring process requires removing any pixel covered by clouds and correcting brightness for water vapor attenuation. When the infrared band is used the spatial resolution of the method ranges from 1 to 5 km (World Meteorological Organization, 2008). Measurements in the microwave band suffer much smaller cloud attenuation and they are an alternative when infrared measurements are not available. The disadvantage of microwaves is that the spatial resolution of these radiometers is of several kilometers.

When the infrared band is used (from 10.5 to 12.5 μm) one way for correcting water vapor attenuation is by taking measurements of the same places from two different look angles because atmospheric attenuation is proportional to the path length. Another way is measuring the same area in two different infrared channels. There are some linear relations between attenuation and wavelength that allow discounting the atmospheric attenuation.

The temperature of the ocean surface measured by radiometers can differ from temperature at a depth of few meters. Some of the factors with direct impact on this difference are: the wind that mixes the surface layers, the sunlight heating the ocean and the rainfall. To reduce these discrepancies daytime and nighttime algorithms are used. In general, nighttime measurements are less disturbed by unwanted conditions.

Measurements from buoys and ships are made at a given point, the depth being about 1 m to 10 m, respectively. Instead, radiometric measurements are made averaging over a large surface layer. In spite of these differences, satellite derived sea-surface temperature compares very well with drifting buoys and ship observations; they show differences lesser than 1 K on average (World Meteorological Organization, 2008).

10.6.7 Other Parameters Estimated From Satellites

There are many parameters estimated from instruments on board satellites, and fast advances are going on to refine some measurements and to obtain new ones. For example, the radiance measured at the 6.7 μm wavelength is used to estimate the amount of tropospheric humidity in a deep column (not very accurately defined); for this application the segments of the troposphere analyzed should not contain medium and high clouds.

Also, the total ozone density on a column may be measured with radiometers in the ultraviolet band. Some methods combine the information gathered in six wavelengths between 0.28 and 0.3125 μm giving ozone measurements within about 2 % of ground based instruments. With a lesser accuracy total ozone can be derived from its absorption in the thermal infrared (9.7 μm). Volcanic ash clouds, fires, and vegetation indices are also obtained from the satellite images.

10.7 Active Instruments

10.7.1 General Approach to Active Sensing

The main advantage of this kind of instrument is that the electromagnetic sources that illuminate the Earth's surface are selected by the instrument designer. The designer can choose the frequency and time characteristics of the wave that will be used to sense the Earth's surface. The benefit of this degree of freedom is that since the instrument knows the characteristics of the signal it is expecting back, it can reject all other signals that do not match the expected pattern. This is done by correlating the received signal with a replica of the sent signal. This selection improves dramatically the signal-to-noise relation.

Also, since the "sensing" signal is produced by the instrument on-board the satellite, the precise instant at which it was emitted is known. Therefore, by computing the time elapsed between the instant in which the signal was emitted and the instant in which the correlation is a maximum, the flying time of the signal between the emitter and the target is calculated. The flying time contains information about the distance between the spacecraft and the target. Furthermore, because the satellite is moving, the signal received back will be of a different frequency than the signal sent. The change in frequency will depend on the speed of the satellite and on the position of the satellite and the target (remember the example of the train horn and the concepts introduced in Section (3.3.1)). This information is also used by the measuring systems as will be seen below.

Since sun illumination is not required to get an image, active instruments can be used during night hours. This means that radar images can be acquired on both passes of the satellite, from south to north and from north to south. The radiant sources used by active instruments are generally designed to have frequencies between 0.3 and 300 GHz, or wavelengths between 1 m and 1 mm. Most of these frequencies break through clouds and rain and some targets are "seen" better with them than when illuminated with visible light (GlobeSAR Program, Canada Centre for Remote Sensing).

The microwave spectrum is classified in bands identified by letters; Table 10.1 shows those most often used.

Table 10.1: Microwave bands

Band name	Frequency range (GHz)	Wavelength (cm)
Ka	26.5 - 40	1.13 - 0.75
Ku	12 - 18	2.4 – 1.66
X	8 – 12.5	3.75 – 2.4
C	4 – 8	7.5 – 3.75
S	2 – 4	15 – 7.5
L	1 – 2	30 – 15
P	0.3 – 1	100 – 30

It is convenient that radar wavelengths are similar to the sizes of the targets from which information is desired (remember the Bragg law (Section (3.5.1)). The C band is a good choice for most targets. The L band is the best for geology. Some instruments on board satellites use more than one frequency band (GlobeSAR Program, Canada Centre for Remote Sensing).

Microwaves are electromagnetic waves formed by the oscillations of electric (**E**) and magnetic (**B**) fields that are perpendicular to each other and to the direction of wave propagation (transverse waves) (Section 3.4). The oscillations of the electric (**E**) and magnetic (**B**) fields lie thus entirely in the plane of the wave front. The most common mode of oscillation of the vector **E** (or **B**) is the elliptic mode, in which the linear and circular modes are but particular cases. To see this, suppose that two mutually orthogonal sinusoidal electromagnetic waves of the same frequency and different amplitudes are composed. In general, if the waves are not in phase, the tip of the resultant electric **E** (or magnetic **B**) vector travels around an ellipse, and the resultant wave is *elliptically polarized*. If the phase difference is zero or any integral multiple of π the ellipse becomes a straight line, and the resultant wave is *linearly polarized*. Finally, if the amplitudes are equal and the phase difference is any odd integral multiple of $\pi/2$, the ellipse becomes a circle, and the resultant wave is *circularly polarized*. By the way, it is worth noting that in the case of light, which is also an electromagnetic wave, it has been shown, both experimentally and theoretically, that the electric vector **E** should be identified as the luminous or optic vector of the electromagnetic wave (Jenkins & White, 1957; Rossi, 1957).

The designers of the satellites' instruments can select the polarization of the electromagnetic wave transmitted, which is another tool to extract information about the target. The radar signal transmitted from the satellite can present different modes of polarization depending mainly on the antenna system. In general, they are linearly polarized and have polarization planes made horizontal (H) or vertical (V). When the transmitted and received waves are in the same polarization plane they are like-polarized, for example HH means horizontal transmitted and received. When the transmitted and received signals are in orthogonal planes they are cross-polarized, for example HV for horizontal transmission and vertical reception.

It has been observed that some targets produce different backscatters depending on the wave polarization. The way in which radar waves interact with surface features depends on their slope orientation, soil moisture, vegetation, roughness, etc. Therefore, there may be conditions where the HV images have more information than the HH ones. For this reason more information is obtained using waves with both polarizations to illuminate the target, thus resulting more detailed images.

Most of the previous chapters deal with instruments such as current meters, wave meters, anemometers, rain gauges, etc that are designed to measure specific parameters. In these cases the names of the instruments have implicit the expected data to be acquired. In active satellite-based remote sensing some instruments are better suited for estimating some particular parameter, but, in general, because they acquire a certain broad spectral band they can estimate several parameters simultaneously. Hence it is not yet simple to name such instruments. Furthermore, the designers of new instruments used to add to the instruments' names some flamboyant words describing the data processing aimed to extract the information from the targets. These characteristics lead to instruments' names that at first sight do not clearly reveal what they measure.

In our opinion, it would be convenient for didactical purposes to separate the concepts on data processing from the instruments' applications. The data-processing concepts are generally used by most of the instruments regardless of the target's parameter to be estimated, so they deserve a separate treatment.

As part of the efforts to synthesize these issues, we will roughly group the on-board data acquisition systems according to three functionalities. One of the functionalities concerns the instruments ability to accurately know the position of the spacecraft at every instant of its trajectory. The accurate knowledge of the satellite's orbit to within a few centimeters requires several locating systems on board satellites working cooperatively. Doppler shift techniques are used to determine the satellite's velocity; GPS or similar systems are also used together with laser measurements to know satellite position (Rosmorduc et al., 2011). They are very special instruments giving important information, but this information is generally indirectly used by the environmental science researcher, and for this reason these instruments will not be addressed here.

Another functionality of the acquisition systems is the on-board data processing, which is performed in both analog and digital manners and is implemented by means of electronic circuits or numerical algorithms. They are used to extract information about parameters of interest to the environmental science researcher. These aspects of the problem will be addressed here because they have a direct impact on the final data used by the researchers. We consider it fundamental that data users know how the information has been processed on board the satellite before it arrives at their computers, because this previous process defines the representativeness of the data they will use.

The last functionality to consider is the application of the instruments to estimate some parameters. This subject deals with how the components of the acquisition systems (including the data processing) are disposed to be most adequate to acquire certain specific characteristics of the targets; i.e. it is attempted to describe how instruments are applied and why some of them are more suitable to acquire certain parameters more accurately than others.

Therefore, the first part of the remainder of this chapter deals with data-processing general concepts used in almost all systems, and the second part with a few of the most renamed applications called scatterometers, altimeters and lidars.

10.7.2 Data Processing On-Board Satellites

We will describe some operating principles associated with systems such as Synthetic Aperture Radar (SAR), deramp technique, Doppler frequency shift, frequency modulation and Chirp radar. As stated before, this nomenclature is a little confusing because one could infer that SAR and Chirp radar are specific instruments developed to measure a particular parameter, but they are instruments of general applications. In reality one might think that they are general **concepts on data processing**.

All active instruments are devoted to extract information from the backscatters received at the satellite antenna. Thus, they are simply emitters of electromagnetic waves towards the Earth and collectors and processors of the backscatters coming from the Earth. The differences between these instruments are in the frequency they transmit, the kind and number of antennas, the on-board electronics and the way they process the data. But, in general, even when an instrument is designed to be better suited to perform a specific measuring task, it will provide much more information than that for which it was tuned.

10.7.2.1 Synthetic Aperture Radar (SAR)

SAR does not name a specific instrument but a group of radar systems that use a particular way of acquiring and processing the received backscatters. One feature that differentiates this measuring method is that it is used in moving active instruments (which emit a well-known wave) and whose spatial positions are well known at any instant. The main characteristics of the operating principle that distinguish SAR technology will be described using some concepts already introduced in previous chapters; therefore it would be convenient to have them present now.

Section (3.5.2) illustrates an array of sensors located at different spatial positions with the aim of measuring a propagating wave field. It was stated that such an array permits the wave field at different places to be known as a function of time. These data may be combined to enhance the signal-to-noise ratio beyond that of a single sensor's output.

Also, the beamforming concept was introduced in Section (3.5.3). It was also stated that processing signals from several antennas arranged in a certain spatial allocation permits the signals to be processed to reinforce some signal coming from a particular direction, rejecting others.

Figure 3.8 shows a linear array of transducers receiving the signal from a scattering source. A similar case is now presented in Figure 10.5, where the transducers are replaced by receiving antennas and the scattering source is a particle that was irradiated by an electromagnetic pulse coming from a transmitting antenna. Since the distances from the source to each receiving antenna (d1, d2, d3 and d4) are not the same, the scattered signals will follow different paths, suffering different delays to arrive at the receiving antennas and showing a different degree of attenuation.

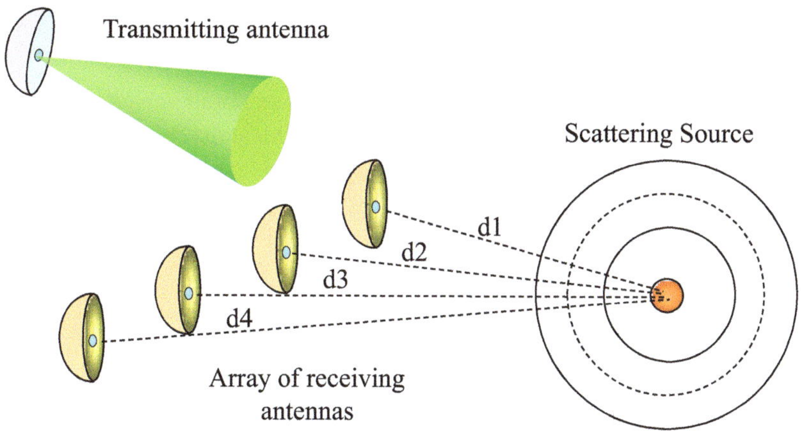

Fig. 10.5: An array of antennas receiving electromagnetic energy scattered by a particle (source). The distances between the particle and the antennas are d1, d2, d3 and d4.

Recall now Section (8.10.1) where it was shown an array of receiving antennas fixed on a beach and a couple of transmitting antennas that irradiate the sea with electromagnetic pulses. In this case all the receiving antennas simultaneously acquire the signals coming from the sea (scattered by the waves). The signal received at each antenna can be processed (basically phase shifted) and the resulting signals added. Thus, only signals coming from a given direction will add in phase (beamforming). This array of antennas and the data processing permit focusing on some small area of the scene. The signals received over the antenna array can be recorded and reprocessed differently several times (De Chiara et al., 2007). Thus the results for each processing step can be combined and displayed together.

There is a relation between the spatial resolution, the length of the antennas' array (A) and the transmitted wavelength (λ) (Campbell, 2007):

$$\beta = \frac{\lambda}{A} \tag{10.1}$$

where β is the beamwidth of the array of antennas. A narrow beamwidth means a higher spatial resolution and according to Eq. (10.1) the longer the antenna array (A) the better the spatial resolution. For this reason several antennas were equally spaced on the beach. As explained in Sections (2.3.2) and (2.3.3), the beamwidth defines, in part, the spatial resolution. The other factors defining the spatial resolution of a radar system are its frequency and the pulse width (Section (8.2)).

Consider again the antennas fixed on a beach. If the scattering of the target under study would change very slowly with time, successively placing the same antenna at the position of each one of the antennas of the array would sequentially record "similar" information as all of them would also do. In other words, by displacing only one antenna, "similar" information as that simultaneously recorded by the whole array could be obtained. Later on all this information can be processed (delayed and summed up) to enhance the signal-to-noise ratio, which improves the information quality, since it is usually done for the fix complete array.

Suppose now that we have one transmitting (Tx) antenna and one receiving (Rx) antenna (Fig. 10.6) and that an orbiting satellite carries this pair. The position of the antennas as a function of time is known due to the on-board instruments to accurately position the spacecraft. The target scene is repeatedly irradiated by electromagnetic pulses sent from the transmitting antenna. The scattered signals are received back by the receiving antenna at its successive positions and stored.

In the example shown in Figure 10.6, signals from four different positions are stored in a similar way as the four fixed antennas in Figure 10.5. Later, the stored signals are read out and combined using specific phase shifts. The result is equivalent as if the signals were recorded simultaneously by an array of equally spaced antennas fixed at four points of the satellite orbit.

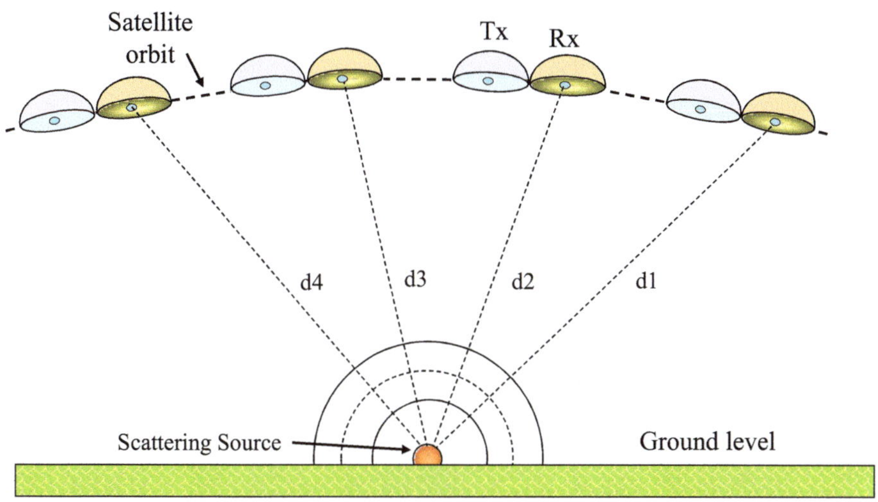

Fig. 10.6: The scattering source is irradiated by the transmitting antenna from four different positions. The scattered signals are received back by the receiving antenna at its successive positions.

In the example, two antennas, one for transmitting and other for receiving were used only with didactic purposes. As it was mentioned in Section (8.2), only one antenna can be used to transmit and receive; a device called *duplexer* can switch the antenna between the transmitter and receiver. Only one antenna is used by the SAR. This is the working principle of a Synthetic Aperture Radar (SAR); **a moving antenna can gather "similar" information as an array of fixed antennas can do**.

Recapitulating, radar on board a satellite moving at constant speed and acquiring information from different positions performs like an array of antennas. Remember that the array provided with an intelligent data processing (beamforming) performs as a narrow beamwidth, which allows focusing on some small area of the scene giving a higher spatial resolution. Therefore, SAR is synonymous to high spatial resolution due to a specific data processing.

Going a little deeper with the data processing and keeping in mind the Doppler effect, the reader can infer from the concepts developed in Section (3.3.1) that the backscatter's frequency is different than the sent frequency. This is so because, albeit the targets are fixed, the radar mounted on the satellite is moving. At a given instant, a target illuminated by the leading edge of the radar beam will reflect a frequency higher than the transmitted one (see Section (3.3.1)). Instead, when the target is in the trailing edge of the beam the reflected signal experiences a decrease in frequency. Because the radar system is able to calculate these frequency shifts, it is possible to use them to accurately know the relative position of the backscatters at each instant.

Thus, due to the frequency shifts, when composing and image using the information provided by the backscatters it is possible to assign each scatter to its correct positions on the image.

We want to stress once more that the acronym SAR does not describe an instrument for measuring a particular parameter, but several instruments with different names can be said to be SAR systems (e.g. scatterometers, altimeters, etc).

10.7.2.2 Chirp Radar

This is another data processing method to improve signal-to-noise ratio. It is also known as linear Frequency Modulation (FM) pulse compression. The name chirp refers to 'chirping', i.e. to make short, sharp notes or sounds (as small birds or insects), and it is also the acronym for "Compressed High Intensity Radar Pulse".

In contrast to old radars, which relied on the transmission of short-duration continuous wave pulses with a peak power of several kilowatts, most modern instruments transmit linear frequency-modulated pulses of longer duration and a relatively lower peak power. These frequency-modulated pulses are also known as "chirps". This concept was initially used in military applications and the simplest way to explain how chirp radars work is by following one of the first public documents on this subject (Klauder et al., 1960).

Suppose that the transmitting signal of a radar, instead of being one short pulse of a single frequency, is comprised of a sequence of adjacent pulses each one of a particular frequency (f_N); frequencies are arranged in increasing order as in Figure 10.7a, and this resulting signal is transmitted to illuminate a target.

Assume that in the receiver, there is a network possessing delay versus frequency characteristic as in Figure 10.7b, and that it is introduced in series in the backscatter signal path so that the signal will be delayed in steps. Then the signal with frequency f1, which was the first emitted, will be the first received, but will be the most delayed. Each successive pulse of a different frequency is delayed in Td until the last signal arrives to the receiver. Then, all pulses are made to emerge together from the delay network. It may be thought as though the signals are summed up in phase. In this way, the original subsequent pulses of Figure 10.7a are compressed in time and increased in amplitude.

In the previous example, the frequency is changed in steps. In real chirp radar the signal transmitted has also constant amplitude but it is linearly modulated in frequency (Fig. 10.8a). It is centered at frequency f0 and has a bandwidth Δf. In this case the delay network is a linear function of frequency (Fig. 10.8b). Thus, the backscattered signal at the network output will also produce a compression in time and an increase in amplitude.

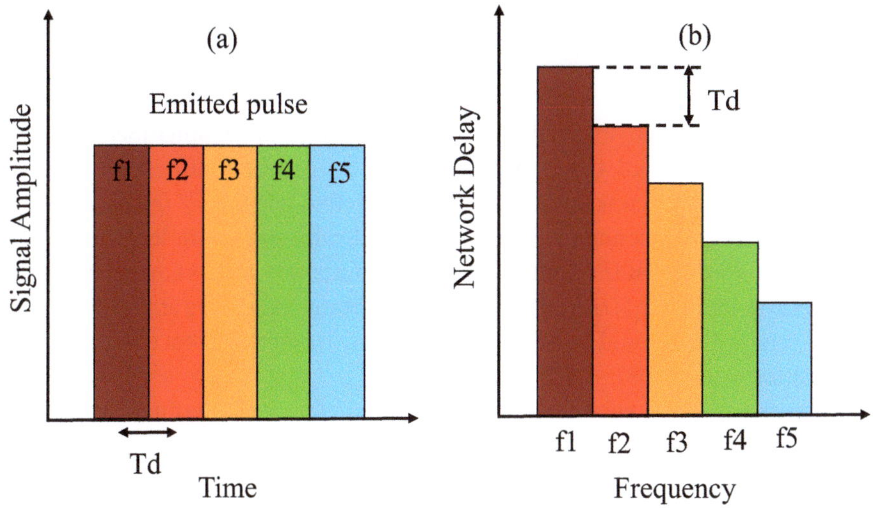

Fig. 10.7: The emitted pulse is an array of consecutive pulses of time length Td and different frequencies. The received signal enters the Network Delay, which retards each frequency a time Td.

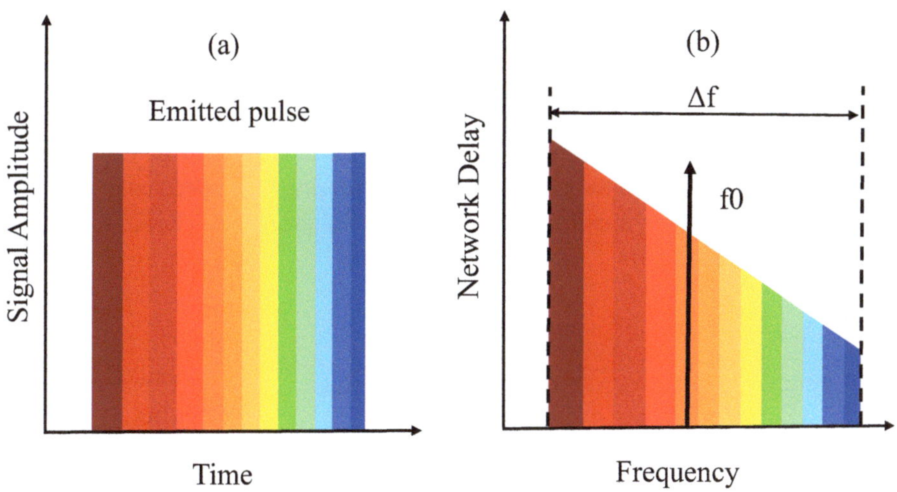

Fig. 10.8: The emitted pulse is a signal linearly modulated in frequency whose central frequency is f0 and the bandwidth is Δf. The Network Delay produces a linear retard as a function of frequency.

This process may be imagined as a number of echo signals of different frequencies arriving in a certain period of time that are superimposed, signal overlaps and the random noise averages, raising the signal-to-noise ratio.

Hitherto, the signal processing was aimed at rescuing the signal from the noise. Remember that the received signal is an echo from a target whose characteristics we want to know. Hence a new step has to be performed to rescue the flying time of the pulse in both directions that contain the information of the target range (Section (8.2)).

The output pulse of the delaying network is retarded with respect to the transmitted chirp for an unknown time due to the round-trip time of the pulse. Figure 10.9 shows *Tx* as the transmitted chirp, *B* as its bandwidth and *T* as its length. A replica of the transmitted signal (*Reference*) is retarded purposely for a known amount (*D*) called *deramp* time delay. *D* is intended to match approximately the arrival time of the received echo which should be equal to the two way travel time from the radar to the target (Cotton et al., 2008).

When the received echo is compared to the *Reference*, the difference in time between both signals (*td*) will be the difference between the real flight time and *D*. As *D* is known, measuring *td* will give the flight time.

Since the emitted pulse was linearly modulated in frequency as a function of time, a difference in time between these two signals will appear as a difference in frequency. Therefore, the time difference can be extracted from the difference in frequency which is related to the original chirp slope by

$$td = \frac{\Delta f T}{B} \tag{10.2}$$

By mixing the received echo and the reference, differences in frequencies will be made evident (as explained in Section (8.2) for synchronous demodulation). Low pass filtering and spectral analyzing the mixer output allows finding these frequency differences and then *td*. Hence, the time of flight can be calculated as $D + td$, and consequently, the target's range (considering light speed constant). This data processing method is called a full-deramp technique (Chan & Koo, 2008).

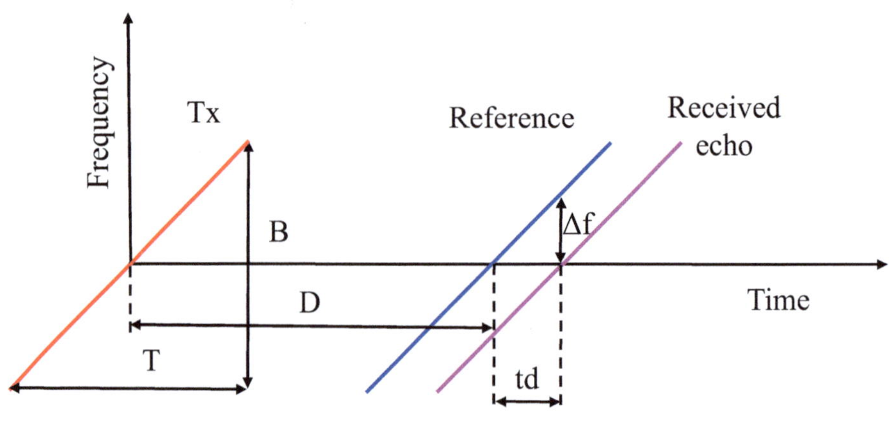

Fig. 10.9: Tx is the transmitted chirp, B is the chirp bandwidth and T the chirp length. Reference is a replica of the transmitted signal delayed a deramp time D. The received echo is delayed td with respect to Reference. The time td can be obtained from Δf.

10.7.3 Examples of Applications of the On-Board Instruments

On board instruments make use of most of the data processing techniques described above. SAR, chirp radar, Doppler shift, frequency modulation and deramp techniques are used in scatterometers, lidars and altimeters to extract Earth's surface information.

10.7.3.1 Scatterometers

Scatterometers used to measure near-surface ocean winds are basically microwave radar that transmits high frequency electromagnetic waves towards the ocean surface. This energy is scattered by the ocean waves and, as explained in several sections of Chapter 8, the backscattered signal carries information on the target. The data processing operation allows drawing the information of interest from the backscatters received at the satellite antenna.

In addition to the data processing there are other design features that define the ability of the radar to extract particular information from the target. For example, it has been shown in Section (8.9) that the maximum of the scattered energy depends on the relation between the surface roughness of the sea and the illuminating wavelength of the radar (Bragg law). It should also be considered that some high electromagnetic frequencies are strongly attenuated by the clouds, thus being useless to measure surface ocean winds during cloudy days. Therefore, the selection of the scatterometer frequency determines the kind of information that can be gathered from the Earth's surface.

The number, direction and polarization of the antennas which receive the backscattered signals define the quality and the redundancy of the information collected. More "points of view" or "viewing angles" means more signals to be processed and more possibilities of getting detailed pictures of the target. The beamwidth of the antennas and the satellite's orbit altitude play an important role in defining the spatial resolution of the radar "images".

10.7.3.1.1 Wind Measurements

The concept behind the measuring method of sea surface winds from satellites with a scatterometer is that the roughness of the sea surface depends on the wind speed. It is a daily experience that a flat lake surface exists when there is no wind; small ripples appear with moderate winds, and a rough surface is observed under high-wind conditions. With no border conditions limiting wave propagation, wind direction and wave front direction have a defined relation.

Electromagnetic waves from radar are used to evaluate the sea surface roughness; although they do not penetrate sea water, the backscatters on the sea surface are completely influenced by surface roughness. Backscatter presents a wide level range of energy which is directly related to wind speed and direction (for a constant radar incident angle). Therefore, the fraction of energy returned to the satellite (backscattered by the waves) is a function of wind speed and wind direction (Center for Ocean-Atmospheric Prediction Studies, Florida State University).

Radar pulses in the C-band have wavelengths of about 5 cm while ocean wind-generated capillary waves and gravity waves are in the order of the radar wavelength; hence Bragg resonant scattering is highly probable. These short-length wind-induced waves (mounted on longer waves) modify their amplitudes according to the wind speed, thus changing the roughness of the sea surface. Therefore, the local backscattered energy increases with wind speed. But if the wind produces high sea states, the radar ability to detect sea features decreases due to the higher level of turbulence. When the direction of the wind and the direction of the radar beam agree, the backscatter is the highest. Also, some difference is found between upwind and downwind directions (GlobeSAR Program, Canada Centre for Remote Sensing).

The theoretical function describing the sea surface backscatter coefficient as a function of wind direction, and for a given incidence angle of the scatterometer, is shown in Figure 10.10 (Gelsthorpe et al., 2000). For each wind speed it has two minima, one relative maximum and one absolute maximum. Each wind speeds have a different letter color from A (red - maximum speed) to E (blue - minimum speed). Theoretically the absolute maximum corresponds to the true direction of the wind, the relative maximum to the opposite direction, and the two minima to directions approximately perpendicular to the wind direction.

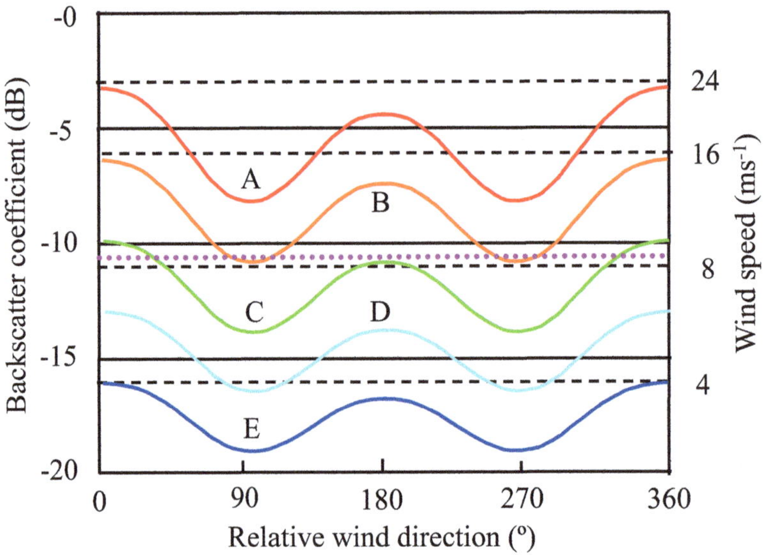

Fig. 10.10: Schematic of the sea surface backscatter coefficient as a function of wind direction for several wind speeds (different letters A-E) and for a fixed incidence angle.

As is obvious from Figure 10.10 there exist many possible wind speeds generating the same backscatter; hence, estimating wind speed from only one measured backscatter generates multiple ambiguities. For the dotted line (– 10.5 dB coefficient), there are several winds that can generate this backscatter. The process of selecting the direction from the graphic is called ambiguity selection. Problems with ambiguity removal exist for low wind speeds because the signal is weak and easily corrupted by noise. For wind speeds greater than 8 ms^{-1} successful ambiguity removal is highly probable (Center for Ocean-Atmospheric Prediction Studies, Florida State University).

The method of removing ambiguities relies on "seeing" the same backscatering ocean area from several different relative directions between the beamwidth and the wind. For this purpose modern scatterometers have several antennas with different looking angles, and in some cases they are mounted in a spinning mechanism as described early in Section (10.5.4).

Since 1996 three successive scatterometer missions have been operated and some periods of data overlapping exist; two instruments used C-band frequency and one Ku-band frequency. Bentamy et. al. (2013) were interested in constructing a consistent scatterometer time series despite changes in scatterometer technology. The study focuses on scatterometers that use transmitting frequencies 5.3 GHz (5 cm) and 13.4 GHz (2 cm). Because the ocean surface radar signal backscatter occurs from capillary/gravity waves of a few centimeters in amplitude, waves with about 5 cm and 2 cm scatter most efficiently at the above frequencies (Bragg scattering).

Due to the shorter wavelength, Ku-band scatterometers are more sensitive to impacts of rain than the longer wavelength C-band; undetected rainfall preferentially affects the higher frequency. Satellite estimations of winds are perturbed due to several different causes: attenuation by raindrops in the atmosphere, amplification due to volume scattering and changes in the sea surface roughness by impinging drops. Some authors have found systematic biases in the 10 m equivalent satellite wind speed that are a function of radar azimuth angle and wind speed ranges and rainfall, but wind direction does not result affected (Bentamy et al., 2013).

10.7.3.1.2 Rain Measurements

In Section (10.6.5) it was anticipated that precipitation can be estimated from passive instruments through relationships between brightness of clouds in the **visible / infrared** spectra and precipitation, but if it were required to extend rain estimation to night hours, the previous technique is not useful.

The incorporation of data collected by means of active instruments improves precipitation estimation. Some satellite missions devoted to measure rainfall in tropical regions use active sensors called Precipitation Radar (PR). The PR can derive the three-dimensional structure of precipitation. Its operating frequency is 13.8 GHz. The PR has a minimum detectable rain rate of 0.5 mm/h with range resolution of

250 m (Oki et al., 2010). Combining data from the PR and passive instruments greatly improves rainfall estimation from satellites.

Development of another instrument called dual PR (DPR) is underway; it is composed of two radars, one in Ku-band (13.6-GHz) and other in Ka-band (35.5-GHz). Simultaneous observation of KaPR and KuPR would enable accurate measurement of precipitation from heavy rainfall in the tropics to weak snowfall at high latitudes. KuPR is suitable for heavy rainfall while KaPR can estimate weaker rainfall and snowfall (Oki et al., 2010).

Cohen Liechti et al. (2012) analyzed rain data gathered from three different satellites (that will be identified here as A, B and C) during four years (2003 – 2007) with the objective of using them in modeling the hydrological behavior of a basin. These investigators found that global correlation coefficients were 0.54 between Satellite A and B data sets, 0.76 between Satellite A and C and 0.60 between Satellite B and Satellite C. They concluded that although in terms of yearly rainfall the main characteristics are preserved, the rainfall spatial patterns produced by the three satellites show considerable differences, which illustrated the very strong dependency of the satellite product quality on the region studied (Cohen Liechti et al., 2012).

10.7.3.1.3 Moisture

In Section (4.9), when describing humidity sensors, we saw that the conductivity of some materials changes with their moisture content. Also, the complex dielectric constant of a material depends on the moisture presence. The dielectric constant for water is about ten times that of dry soil. Because the dielectric constant defines the ability of a material to reflect, transmit or absorb electromagnetic energy, the scattered energy from a given surface changes according to its water content. Therefore, the same area of soil in a radar image will have a different appearance if it is dry or wet. This is the reason why radar images, when calibrated, can be used to infer the moisture content of a certain soil. In wet areas the radar waves do not penetrate the subsurface, but dry areas such deserts are easily penetrated by radar waves. Thus, the scatters have information not only on the surface properties but on the subsurface too.

Radar from satellites can quantify moisture on the surface of soil from 0 to 15 cm in depth, but the strongest correlation is for the 0-3 cm depth range. A stepped incident angle gives better results because it minimizes roughness contributions. The moisture content of frozen soil cannot be measured with these methods. Also, rain and dew degrade the acquisition of moisture content in soil.

10.7.3.1.4 Ice

Not only can satellite's radar (SR) identify ice, they can also estimate ice type and surface roughness as well. If the ice surface is smooth there is a good reflection and it appears dark. Instead, when roughness increases the area appears bright. This simple relation between surface roughness and brightness is spoiled when melting snow covers the ice surface.

First year ice has a large dielectric constant due to its salt content; hence the radar wave does not penetrate it. Multi-year ice that has several melt seasons has lower levels of salinity and hence lower dielectric constant. In this case the radar wave has a greater probability of penetrating the ice and the scatters come from the surface and subsurface layers. This makes it to appear brighter when compared with first year ice. SR cannot measure ice thickness directly, but discriminating ice age permits inferring ice thickness (GlobeSAR Program, Canada Centre for Remote Sensing).

10.7.3.1.5 Other Applications

SR can be used for ship detection and oil spills monitoring because surfactants produce a localized suppression of the small waves that give rise to Bragg scattering, but SR cannot estimate oil layer thickness. Specular reflections from water produce dark images. Thus, flooded areas can be distinguished, but wet snow can be confused with flooded areas (GlobeSAR Program, Canada Centre for Remote Sensing).

Since the scattering of a volume of vegetation is stronger when the radar wavelength is similar to the mean size of the plants components, some radar frequencies are more suited for specific crops. For example, wavelength between 2 and 6 cm are sensible for detecting corn, soybeans, wheat and tree leaves. Instead, for sensing tree trunks and limbs wavelengths between 10 and 30 cm are more adequate (Lillesand, et al., 2004).

10.7.3.2 Altimeters

This explanation will begin with the operating principle of a simple altimeter, and in a second step the characteristics of more realistic ones will be described.

The altimeter antenna emits pulses toward the nadir hitting the Earth's surface (water or land). Frequencies are in a range from about 3 to 13 GHz. The beamwidth is narrow but the wave front widens as it travels downward. When the wave front hits the Earth's surface it has the shape of a spherical cap. If the hit area is plane, the footprint of the leading edge of the pulse is a disc linearly spreading with time until the arrival of the rear edge of the pulse. From then on, the footprint becomes a ring with increasing radius and constant area (Fig. 10.11) (Dinardo, 2013).

The strength of the backscatter (Fig. 10.11d) is proportional to the footprint area, thus the signal retuned to the altimeter antenna will rise linearly up to a maximum corresponding to the passage of the rear edge of the pulse over the Earth's surface. From this moment on, the returned signal should be theoretically constant, but due to practical reasons such as the antenna beamwidth it decays until it arrives at the noise level.

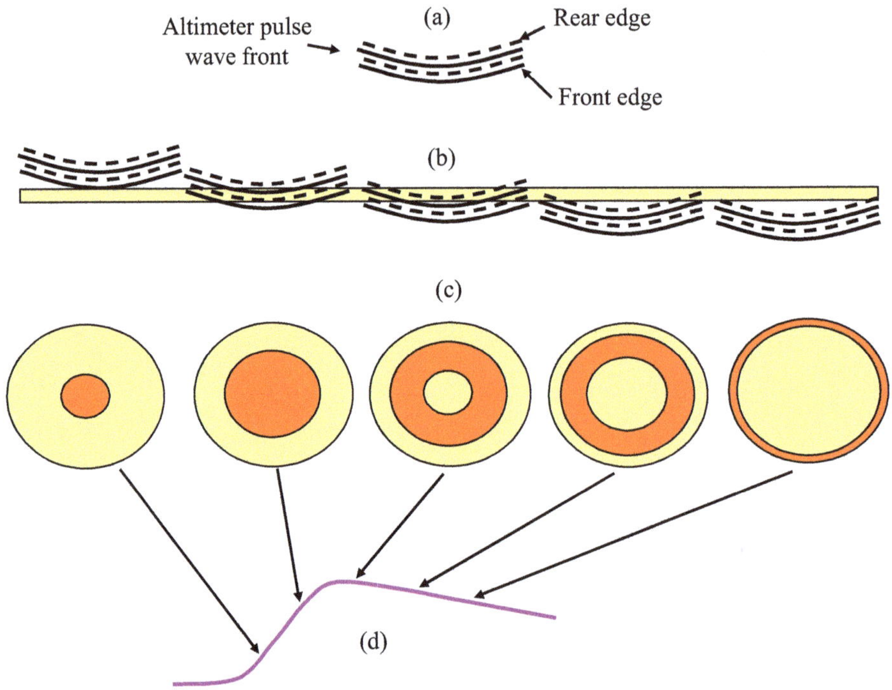

Fig. 10.11: (a) A curved wave front arriving at the Earth's surface. As the pulse hits a plane surface (b) the footprint changes as a function of time (c). The shape of the echo is, up to some point, proportional to the footprint area (d).

The previous explanation was for a plane surface, but if the wave front hits the ocean surface, due to the wave heights, the returned echoes will travel different distances to arrive back at the satellite. This effect produces a change in the footprint, which is a function of the sea state. For this reason, the echo rise time is longer and it is possible to relate the echo slope to the wave height. Therefore, the shape of the signal returned to the altimeter antenna contains information about the wave heights (Rosmorduc et al., 2011).

As could be intuitively perceived, the data are affected by statistical fluctuations because the signals received at the altimeter antenna are composed of many scatters that are not in phase. Therefore, to reduce the noise the instrument averages the echoes.

A more sophisticated altimeter is the delay-Doppler altimeter which, as a first approach, is similar to the instrument just described but with an increased complexity in the data processing method. Since the position of the altimeter as a function of time is known because the position of the spacecraft is known, the same processing methods explained for the Synthetic Aperture Radar (SAR) can be applied. The backscatters are recorded at the successive positions of the altimeter and combined using specific phase.

Once again the beamforming concept is used to obtain higher spatial resolution in the along-track direction.

For example, in the CryoSat-2 altimeter 64 beams per burst are synthesized in the along-track direction. As a result, the resolution is sharpened whereas the across-track remains with the previous diameter of the pulse-limited circle. Figure 10.12 shows how the along-track beamforming increase the resolution on the footprint for three of the cases previously shown in Figure 10.11. For the CryoSat-2 case the resolution is 300 m after the data processing (Dinardo, 2013).

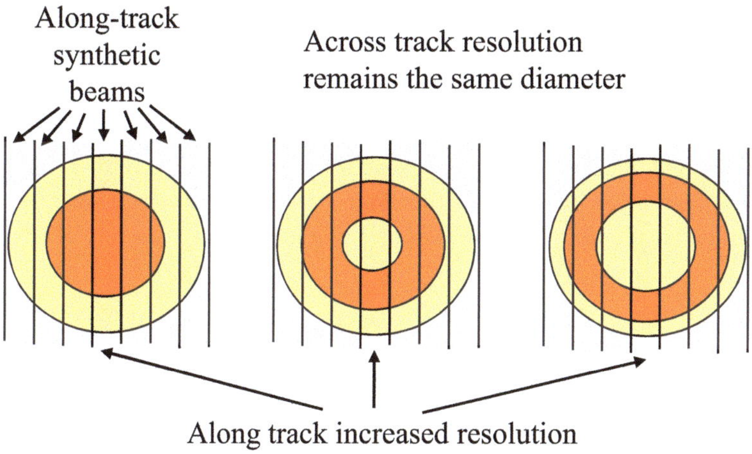

Fig. 10.12: The along-track synthetic beams increase resolution in the satellite track direction, whereas in the cross-track the resolution remains the same.

It was described for the SAR that the Doppler effect due to the satellite movement with respect to the target causes differences between the sent and received signal frequencies. Returned waveforms are frequency analyzed and used to separate the radar footprint into "strips" of increased resolution. Coherent processing (beamforming) of echoes allows for multiple "looking directions" for a given strip, increasing signal to noise relation.

10.7.3.3 Lidar

Lidar operating principle used in satellites is quite similar to that described in Section (8.8). In that case, the laser was pointed towards the atmosphere; in the spacecraft, the instrument is mounted such that the laser pulses are sent towards the ground, thus illuminating a fraction of the surface. The earliest systems were used to profile ground's elevations under the spacecraft by measuring the time elapsed when a pulse is sent and when it is received back.

In those cases where the laser beam can pass through the water it can be used to measure the water depth (bathymetry). Each pulse from the laser illuminating the water body produces two echoes, one at the surface and one at the bottom, the second being the weaker. Measuring the time difference between these two echoes permits the water depth to be estimated.

In an urban area or a forest, each lidar pulse can return several echoes due to the landform. The time differences allow estimating intermediate distances, for example, the height of the top of a building, the top of the trees, etc.

In order to know height with accuracy it is necessary to know very accurately the satellite's orbit. For this purpose ground stations with well-determined positions are used to establish the distance to the satellite. Range resolutions in ground elevation from 10 to 30 cm are common.

The first NASA lidar flying on board a satellite was lunched in early 2003 and worked at 1.064 and 0.532 μm wavelengths. It transmitted 40 pulses per second and measured the echo to within 1 ns. The laser footprint was about 70 m in diameter and they were spaced 170 m along-track. This lidar can measure small changes in ice sheets and was aimed to study the balance of polar ice sheets and sea level (Lillesand et al., 2004).

References

Astrium GeoInformation Services (2013) Available at: info@astrium-geo.com/es/ www.spotimage.com

Bentamy A., Grodsky, S. A., Chapron, B., & Carton, J. A. (2013). Compatibility of C- and Ku-band scatterometer winds: ERS-2 and QuikSCAT. *Journal of Marine Systems*, *117–118*, 72–80.

Campbell, J. B. (2007). *Introduction to Remote Sensing*. New York: The Guilford Press.

Center for Ocean-Atmospheric Prediction Studies (COAPS), Florida State University Available at: http://coaps.fsu.edu/scatterometry/about/overview.php

Chan, Y. K., & Koo, V. C. (2008). An Introduction to Synthetic Aperture Radar (SAR). *Progress in Electromagnetic Research B*, *2*, 27–60. Available at: http://www.jpier.org/PIERB/pierb02/03.07110101.pdf

Cohen Liechti, T., Matos, J. P., Boillat, J. L., & Schleiss, A. J. (2012). Comparison and evaluation of satellite derived precipitation products for hydrological modeling of the Zambezi River Basin. *Hydrology and Earth System Sciences*, *16*, 489–500. Available at: www.hydrol-earth-syst-sci.net/16/489/2012/

Cotton, D., Andersen, O., Cipollini, P., Gommenginger, C., Quartly, G., Martin, C., Marquez, J., & Moreno, L. (2008). *Development of SAR Altimetry Mode Studies and Applications over Ocean, Coastal Zones and Inland Water*. SAMOSA. ESA AO/1-5254/06/I-LG, Satellite Observing Systems.

De Chiara G., Crapolicchio, R., & Lecomte, P. (2007). ERS-1/2 Scatterometer new products: mission reprocessing and data quality improvement. ESA. Earth Observation. Product Control Service. *Second Space for Hydrology Workshop*, Geneva, Italy. Available at: http://earth.eo.esa.int/pcs/ers/scatt/articles/Extended_abstract_DeChiara.pdf

Dinardo, S. (2013). *Guidelines for the SAR (Delay-Doppler) L1b Processing*. Esrin, ESA, Italy. Available at: www.esa.int

Gelsthorpe R.V., Schied E., & Wilson, J. J. W. (2000). ASCAT – MetOp's Advanced Scatterometer. *ESA Bulletin*, *102*, 19-27.

GlobeSAR Program, Canada Centre for Remote Sensing, Natural Resources Canada. Available at: http://www.ccrs.nrcan.gc.ca/ccrs/rd/programs/globsar/gsarmain_e.html

Jenkins, F. A., & White, H. E. (1957). *Fundamentals of Optics*. New York: McGraw-Hill.

Klauder J. R., Price, A. C., Darlington, S, & Alversheim, W. J. (1960). The theory and design of Chirp Radars. *The Bell System Technical Journal*, *XXXIX* (4). American Telephone and Telegraph Company.

Kurtin O. D. (2013). *Satellite Life Extension: Reaching for the Holy Grail.* Publications, Via Satellite. New York: Kurtin PLLC.

Lillesand, T. M., Kiefer, R. W., & Chipman, J. W. (2004). *Remote Sensing and Image Interpretation*. New York: John Wiley.

Oki R., Kubota, T., Katagiri, S., Kachi, M., Shimizu, S., Kojima, M., & Kimura, K. (2010). Cloud and Precipitation Observation by Spaceborne Radar in Japan: Current and Future Missions. *International Archives of the Photogrammetry, Remote Sensing and Spatial Information Science*, *Volume XXXVIII*, Part 8, Kyoto, Japan.

Rosmorduc, V., Benveniste, J., Bronner, E., Dinardo, S., Lauret, O., Maheu, C., Milagro, M., & Picot, N. (2011). In J. Benveniste & N. Picot (Eds.), *Radar Altimetry Tutorial*. Available at: Available at: http://www.altimetry.info/documents/Radar_Altimetry_Tutorial_20110216.pdf

Rossi, B. B. (1957). *Optics*. Reading, Massachusetts: Addison-Wesley.

Sabins, F. F. (1999). *Remote Sensing: Principles and Interpretation*. New York: W. H. Freeman & Co.

Wang J., & Wolff, D. B. (2009). Evaluation of TRMM Ground-Validation Radar-Rain Errors Using Rain Gauge Measurements. *Journal of Applied Meteorology and Climatology*, *49*, 310-324.

World Meteorological Organization (WMO). (2008). *Guide to Meteorological Instruments and Methods of Observation*. WMO-No. 8 (seventh edition).

11 General Measurement Tips

11.1 Field Measurement Tips

With the aim of obtaining correct field measurements some common sense rules have to be followed. A set of suggestions will be offered below with the intention of preventing some common mistakes. Even when some of the ideas herein presented are known, having them collected in an ordered form could help in preparing field works. These rules are divided according to different phases in the data acquisition process, such as deciding the use of an instrument, installing it in the field, assuring data quality, etc.

Instrument's users have to be familiar with the potentialities and limitations of their instrument before installing them in the field. For this reason, at the end of the chapter, a case study with instruments and sensors available in most laboratories is presented. It is intended to show why researchers should perform their own tests on instruments before using them.

11.1.1 Before Deciding the Use of an Instrument

Let us assume that a given instrument already exists at the laboratory, but the future users do not have any experience with it. In this case, it is convenient that users become familiar with the instrument's capabilities before deciding to use it in field work. With the purpose of getting acquainted with the instrument's performance some testing should be done. Perhaps, in these preliminary steps, users will discover that the instrument is not adequate for the work they have in mind.

In the process of recognizing the instrument's capabilities for the desired measuring task we find the following recommendations useful:

1. Read the **original** manual of the instrument.
2. Understand exactly the working principle of the instrument.
3. Know what the **assumptions** that underlie the measuring method are. All methods are based on some conditions that must be met for the data to be reliable. For example, in Section (6.2.3) it was assumed that buoys follow the sea surface water particle orbit, and in Section (5.4.6) that particles are moving at flow velocity.
4. Study how the instrument processes data internally before recording a final data output. For example, Radio Acoustic Sounding System (Section (8.7)) average samples in time and space, so their data are quite different from that acquired by radiosondes or traditional anemometers.
5. Carefully read the manufacturer's specifications to find the instrument's limitations.

6. Test the instrument to make sure that it is performing according to the specifications.
7. Mimic laboratory environmental conditions to those in which the instrument will be used. Tests should be performed under similar temporal and spatial scales and similar environmental conditions (temperature, humidity, electrical noise, etc.)
8. Process data recorded during these tests to be sure that their format is compatible with your software and data processor unit.

11.1.2 Simple Examples to Materialize Previous Ideas

In order to make it clear what we refer to in step 6, we present some simple examples with sensors that have been described in previous chapters.

In the case of non-contact level meters using acoustic transducers (Section (6.2.3)) or radio frequency antennas, a simple test is to place the equipment on the floor pointing toward the ceiling and let it run for as long a time interval as possible. The instrument should measure a constant value. Making changes in the instrument's level, for example by placing the sensor on a table, and then back on the floor, the instrument repeatability could be verified, and some information on time response obtained. It is useful to simultaneously measure air temperature and humidity to verify whether they could have some influence on level measurements.

Also, this kind of instrument could be suspended from the roof with a rope, pointing down and balanced as a pendulum. By measuring the length of the pendulum and its oscillation period with a watch, the real pendulum movement can be calculated and compared with the data as measured by the instrument. This test could give information about the dynamic response of the instrument.

These are examples with one kind of instrument, but each singular instrument requires its own test and readers should find the most adequate for their particular instruments.

11.1.3 Before Installing an Instruments in the Field

Initial functioning tests of the equipments at the factory (or after a repair) are usually short and they are not normally representative of field-use conditions. It therefore happens that a certain percentage of new or recently repaired instruments fail within a short time of use. This early failure behavior is known as "instrument infant mortality".

If the installation of the equipment is expensive, as happens with instruments placed on board oceanic buoys moored at hundreds of kilometers from the coast, it is necessary to exhaust all efforts to ensure that instruments will work for a long time.

For this to happen, it is advisable to follow a strategy of "use prior to installation" of equipment.

For new or recently repaired instruments, it is convenient to have them running for a while, under known and controlled conditions, prior to field installation. Whether a weak component or hidden construction defect exists, hopefully a failure would occur before site installation. Obviously, the operating conditions of the tests should be as similar as possible to those find in the field.

11.1.4 Suggested Steps for a Field Measuring Campaign

The steps we recommend in preparing instruments for the acquisition of field data are summarized in Table 11.1. Instruments' users sometimes skip some of these steps, whereas other steps are not viewed as part of the measuring process. However, all of them have some influence on the evaluation of the data quality, and users should find it convenient to meet them all.

One rule for increasing the chances of assuring reliable data is to calibrate the equipment before and after measuring campaigns.

Table 11.1: Steps for a field measuring campaign

Order	Step
1	Tests to verify proper operation of instrument.
2	Calibration
3	Transportation and installation
4	Measuring
5	Instrument recovery and back transportation
6	Re-calibration
7	Comparison of calibrations
8	Evaluation (to see if the instrument retains its ability to measure; repeat some of the tests made in step 1).

The first step mentioned in Table 11.1 has already been described in item 6 of Section (11.1) The next step is instrument calibration; it consists of comparing its measurements with others taken simultaneously by a standard instrument of better quality (Section (2.2.3)). The following stages are transportation and installation.

The transport and installation must be done carefully enough so that the instrument does not change its calibration. After the measuring step, it is necessary to recover the instrument and transport it with care, keeping the measuring capabilities unchanged. Once the instrument is back at the laboratory, it should be recalibrated and both pre and post calibrations compared. Finally, it is appropriate to make some measurements to verify that the equipment works as before the field campaign.

Whether there were any abnormal or unwanted actions, notes should be taken in a log book for each of the above steps. For example, any hit during transportation or during recovery, or any sign of vandalism observed should be reported. Also, it is important to record what happens during step 8, and also whether some events outside the measuring range of the instrument appear during the data processing. This information could help in validating field data.

11.1.5 Assuring Data Quality

We will use the term "reliable data" for data that has a very high probability of being within the errors specified by the manufacturer, and the term "unreliable data" for data with high probability of being wrong.

When, after following all the steps of Table 11.1, it is found that both calibrations coincide (steps 2 and 7), and that in step 8 it is possible to repeat tests as in step 1, then the probability that the collected data is reliable is very high. Instead, in the event that in step 7 calibrations do not match within a certain range, it is not possible to assure the data quality because it is not usually possible to know when the calibration changed.

If the instrument's transference change is small, some kind of estimate about how the calibration changed with time can be made. For example, changes could be attributed to a linear aging process. But this kind of estimate must be made very carefully while knowing very well the characteristics of the instrument being used. Sometimes, the log book could help to identify the causes of calibration mismatch; for example a hit during transport from field to laboratory. This, in turn, could help to determine to what extent data could be used or should be disregarded.

An undesirable situation that occurs with some frequency is that the instrument fails during the measurement campaign (step 4), and then the suggested strategy to ensure the quality of the data cannot be applied.

11.1.6 Some Kind of Failures Likely Found in Data Records

Sometimes, when analyzing the data acquired during field work we can find a point at which the data suffer an obvious alteration. This "abnormal behavior" could indicate that the instrument has probably stopped working properly and that the data are unreliable. Examples of failures are shown in Figure 11.1.

The first of the failure modes (a) shows that at a certain moment, the instrument continuously records a constant value. We will refer to the fail detection instant

(FDI) as the point in the time series at which the acquired information begins to be "suspicious". In (b), alternately nulls and out of scale values appear from FDI on. In (c) there is noise beginning at FDI. In (d) the instrument alternates periods where it seems to measure well with periods of apparent malfunctioning. In (e) the instrument shows a failure, but then it continues to measure apparently well.

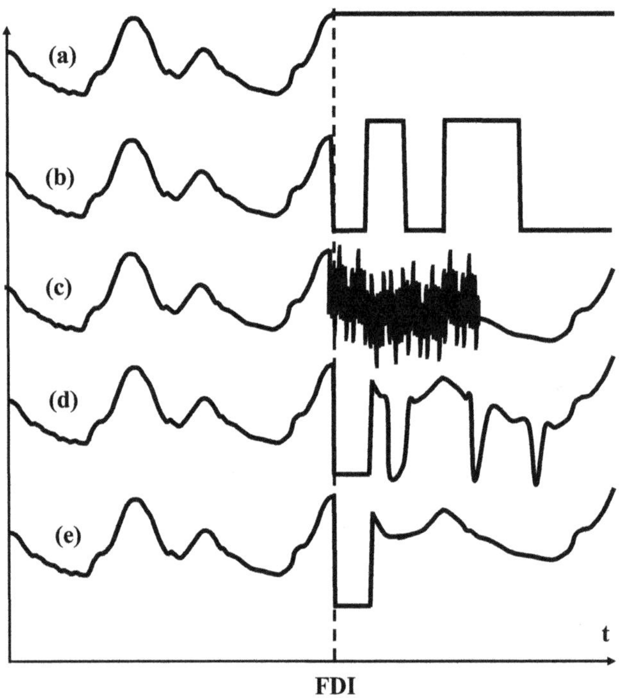

Fig. 11.1: Some failures beginning at FDI are observed in the data recorded: (a) the instrument begins to record a constant value; (b) data alternates between zero and out of scale; (c) noise appears preventing the signal from being rescued; (d) the instrument measures intermittently because of random failures; (e) there is a circumstantial cause producing a partial loss of data.

Following the FDI there could be periods with useful data, but doubts appear about their quality. Sometimes, it is possible to recover data from an instrument that has some of the faults shown in Figure 11.1. For this to be possible a good knowledge of the operating principle of the instrument is required, as well as having followed the steps suggested in Table 11.1. So to use those data it would be necessary to check whether the instrument has regained its ability to generate correct data after the FDI.

Items (a) and (b) in Figure 11.1 suggest that the instrument stopped working and it is not possible to rescue data beyond FDI. In cases (c), (d) and (e), on the other hand, there could be data with errors due to temporary interferences induced by sources external to the instrument. That is to say, the instrument can be measuring well, but temporarily suffering from data losses due to an external factor. For example, failures (c) and (d) could be due to external electromagnetic noise introduced into some stage of the instrument (usually in the sensor). In case (e), the behavior is similar to what happens when an instrument is circumstantially without power. It could be, for example, the case of a battery-powered instrument that is recharged from a solar panel. If solar radiation is not enough to recharge the battery, it may cause a loss of data, but after some time of sunshine, the battery recovers and the instrument continues to record data normally.

If for cases (c) and (e) the user could check that after field work (steps 7 and 8) the instrument works as before the campaign (step 1 and 2), some data after FDI could be rescued and used as reliable data. In case (d) it is a bit more difficult to distinguish reliable data from faults because noise has similar frequency components as the signal.

In the examples (a) and (b) we will assume that the instrument effectively ceased operation and therefore it is not possible to carry out the tests described in steps 7 and 8. In this case we should wonder whether data before FDI could be considered reliable. Researchers sometimes assume that all data collected up to FDI are correct. This is a somewhat risky position, since it is possible that measurement problems, which are displayed from FDI on, have begun much earlier, and what is observed beyond FDI is the end of a process of gradual deterioration of the instrument.

Obviously, in cases where instruments stop working during field work, there is no way to define what data is reliable, but a methodological analysis based on the knowledge of the instrument's working principle could help in defining the degree of reliability of the data. This is another reason why it is important to know how instruments work and the way in which they acquire and process the data.

11.1.7 How Instruments Pass from Correct Operation to Failure

It is important to know when the instrument stopped working properly. In Figure 11.1a and b, the data could be incorrect from some time before FDI. For example, an instrument with mechanical parts could be suffering from some wearing that ends in the failure that appears at FDI. The data collected may have begun to be wrong long before the failure becomes evident.

In searching which are useful data in records from instruments that failed during field work, the way instruments usually fail must be analyzed. There are some instruments that pass from a correct operating condition to a fault condition abruptly (Fig. 11.2a). Others (Fig. 11.2b) alternate periods with correct and incorrect data until they fail completely and others pass gradually from correct operation to malfunctioning (Fig. 11.2c).

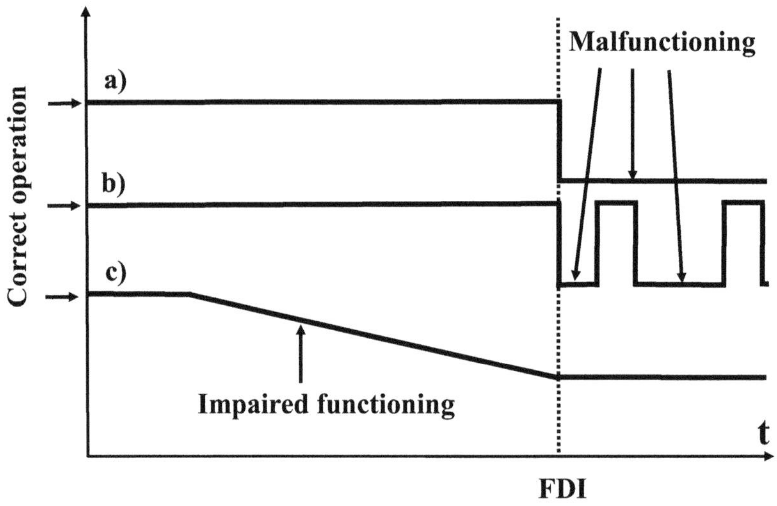

Fig. 11.2: Different ways in which instruments pass from correct operating condition to fault condition. The top horizontal line on the left indicates proper operating condition, and the low line on the right indicates fault condition. (a) The instrument passes abruptly from one condition to the other at FDI. (b) Some failure begins at FDI but the instrument alternates correct operation and fault condition. (c) The data quality suffers from a gradual deterioration until the instrument stops working at FDI.

Unfortunately, the same instrument can have different ways for passing from the correct operating condition to the condition of malfunctioning. But sometimes, instruments have a typical way of changing from one mode to the other, and this characteristic could be useful to decide which data can be taken as correct. Therefore, to decide which data could be valid in cases where the instrument stopped working during the field data collection, it should be checked whether the manufacturer describes some typical form of instrument failure. This feature could be found in the manuals, usually under titles such as analyzing data or system troubleshooting.

Just to illustrate the above ideas, we could say that an inductive conductivity meter (Section (4.10.2)) is more likely to change from a correct working condition to a fault condition as shown in Figure 11.2a and b, while one using electrodes in contact with the fluid (Section (4.10.1)) might behave as in Figure 11.2c. This could happen because the fluid influence on the inductive sensor is low, whereas in the conductivity cell, the electrodes can suffer from gradual changes in their physical or electrical characteristics due to the liquid, beginning with an impaired functioning and ending in a failure.

Even in a correct working condition some instruments can produce data that must be discarded (Fig. 11.3). This can happen with instruments whose measuring capability depends on some conditions of the medium where they are working, as in the case of remote measurement techniques (Sections (8.4) and (8.5)) or in the case of ADCP and ADV (Sections (5.4.4) and (5.4.6)). This kind of instrument performs a statistical analysis of the signal and determines the signal / noise ratio (SNR), which defines the quality of the data being recorded. When the SNR is below a certain value, the manufacturer recommends considering the data as unreliable. Therefore, users should disregard these data, resulting in time series records having periods with useless information.

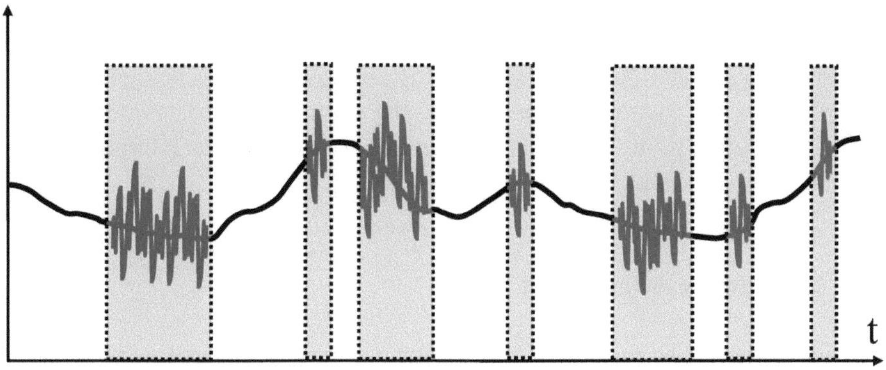

Fig. 11.3: Record of an instrument in a correct working condition that produces a time series with areas in which the signal to noise ratio is very low. In those sections, indicated by a gray rectangle, it is not possible to recover the signal and these pieces of the record must be removed.

11.1.8 Some Ideas for Collecting Continuous Data Series

There are studies in which events to be recorded have very low recurrence. In this case, in order to acquire the data of interest for the study, which could last a few minutes, it is necessary to measure continuously for a long time (could be several years). Measuring a tsunami or an extraordinary rainfall are two simple examples. In these cases where data time series has to be continuously recorded and instruments should not fail it is necessary to know the working life of the instrument producing correct data. Instrument replacement must be done before the possibility of a failure.

The new instrument should record simultaneously with the old one during a certain period in order to ensure that both instruments measure in a similar manner. This procedure would allow numerical adjustments on data collected by the new instrument to coincide with the instrument that is being replaced (Section (7.2.4.1)). Sometimes, this is a quite difficult task because the new instrument may be of a different technology (different operating principle) than the instrument already installed. If it is possible to adjust both instruments to give similar measurements, long continuous data series can be achieved.

Sometimes, the life of the instrument under the conditions in which it will be used is unknown even by the manufacturer, so it cannot be estimated when to perform the replacement of the instrument in use by a new instrument. In this case, the method known as "canary test" can be employed to detect when the time of possible failure is approaching.

The method consists in using an additional measuring instrument made purposely weaker than those used for the real desired continuous measurements. This weaker instrument is called "canary instrument". It is expected that the canary instrument will fail before the instruments performing the continuous measurements. Thus, when the canary fails, the instruments whose series should not be interrupted must be replaced preventively.

11.1.9 Before Buying an Instrument

Some suggestions to be considered before buying an instrument are listed below.

1. Select the instrument according to the problem being solved at the present time. Avoid buying any instrument thinking about potential measurement problems that may (or may not) occur in the future.

A common mistake is trying to buy measuring instruments thinking about problems that do not exist at the time of purchase, or about applications with little chance of occurrence. Typically, the greater the measuring flexibility offered by instruments, the higher the prices. Therefore, overrating measuring requirements may not be the most appropriate solution for the actual study for which the equipment is needed.

For example, if it is required to measure pressures up to 0.1 kgr/cm^2 with an error of 0.001 kgr/cm^2, it is desirable to avoid the frequent temptation to buy a sensor that can measure in a range of 1 kgr/cm^2 for potential future work. For the same quality of instrument (similar price range), the instrument of higher range will measure with greater error (probably around 0.01 kgr/cm^2), thus it would be no longer useful for the initial application.

Instead, if it is decided to buy a sensor for the range of 1 kgr/cm^2 having errors less than 0.001 kgr/cm^2, it must be an instrument of higher quality, so you would have to pay a higher price. Therefore, unless it is certain that the instrument will be used in other work, and that it is therefore convenient to make a major investment to cover all future measurement requirements, you should focus on the characteristics of the instrument required to solve the present case.

1. Clearly define the temporal and spatial characteristics of the phenomenon to be measured. This information is essential to select the right instrument.
2. Clearly specify the environment in which the instrument will be measuring (range of temperature, pressure, humidity, environmental aggressiveness, noise presence, energy availability, probability of vandalism, media to transmit data, etc.). It is essential to verify that the instrument can measure within the expected error in the specified environment.
3. Define the maximum unattended period required, as this will determine the memory needed to store the data and the kind of electrical source required. It should be considered whether instruments can be powered with batteries or if they require power sources such as solar panels or wind generators.
4. If there is any doubt on the specifications provided by the manufacturer, it is highly advisable to contact the manufacturer directly and ask specifically if the instrument is able to measure what the researcher wants.

11.2 Case Study (I): Static Pressure Sensor Test

11.2.1 Introduction

As standard practice, before using under field conditions any instrument that has not been in use for a time, it is convenient to make some measurements of well-known quantities in a controlled environment to verify the correct functioning of the instrument. As an example, a real case will be presented in which some tests were performed on water level data loggers with pressure sensors.

It was desired to perform research work where it was very important to measure the difference in an aquifer level between several boreholes. The instruments at hand had been purchased five years before, but they had been scarcely used.

To verify the quality of the data that could be obtained with these level loggers, some simple tests were performed. Some conclusions regarding the results that could be expected from these instruments in future field work will be presented.

There were five water level data loggers of the same model with the following manufacturer's specifications:

Range: 9 m
Resolution: 0.2 cm
Accuracy: 0.1 % of full scale
Accuracy of the internal clock: ± 1 minute per year
Working temperature range: -20 to 80 °C
We called these five sensors L1, L2, L3, L4 and L5.

11.2.2 Test Description and Preliminary Results

The data loggers were made ready and installed in a swimming pool where it was possible to change the depth of the instruments in known steps. At the beginning the data loggers were placed at a depth of 0.93 m for a week, then they were put at a depth of 0.72 m for a day, and finally at 0.48 m for another day. Figure 11.4 shows the data recorded for the last four days approximately. The changes in depth can be seen at about 8,500 and 10,000 minutes.

Even though the sensors were placed at a constant depth, the records show a wave-like pattern due to changes in the atmospheric pressure. It can be seen that most of the records follow those changes in a similar way. The difference between the maximum and minimum of the undulations during the nine-day test was 0.2 m. This indicates that if the aquifer's level in a well is to be measured, it is **necessary** to subtract the atmospheric pressure because it introduces changes that could be easily confused with real water level changes.

In the research in which the five data loggers are intended to be used, all of them will be placed at short distances from each other, so that the atmospheric pressure they will bear will be the same. Besides, the only variable of interest in this investigation is the difference in water level. *Therefore, if all the sensors would have the same response to the atmospheric pressure, it would be enough to measure the pressure difference between the instruments because the contribution of the atmospheric pressure to the hydrostatic pressure would be the same for all the sensors.*

It can be seen from Figure 11.4 that the atmospheric pressure variations are similar for the five tested sensors. However, sensor L4 behaves rather differently from the rest. Not only does it measure smaller values than the other four, it also exhibits rapid oscillations with amplitudes up to 0.08 m. Sensor L4 must hence be rejected for the field work planned because its oscillations could mask the phenomenon to be measured.

11.2.3 Subtracting Atmospheric Pressure

The mean value of the measured atmospheric pressure (**map**) for each sample was calculated from the data recorded by the four instruments that present reliable pressure variations. This mean value was then subtracted from each measurement of the four sensors. The result is illustrated in Figure 11.5. It can be seen that the variations in the atmospheric pressure no longer appear when the mean value is subtracted. This indicates that all the sensors can measure the atmospheric pressure variations in a fairly similar way. *Therefore, they are able to carry out differential measurements with negligible influence of the atmospheric pressure.*

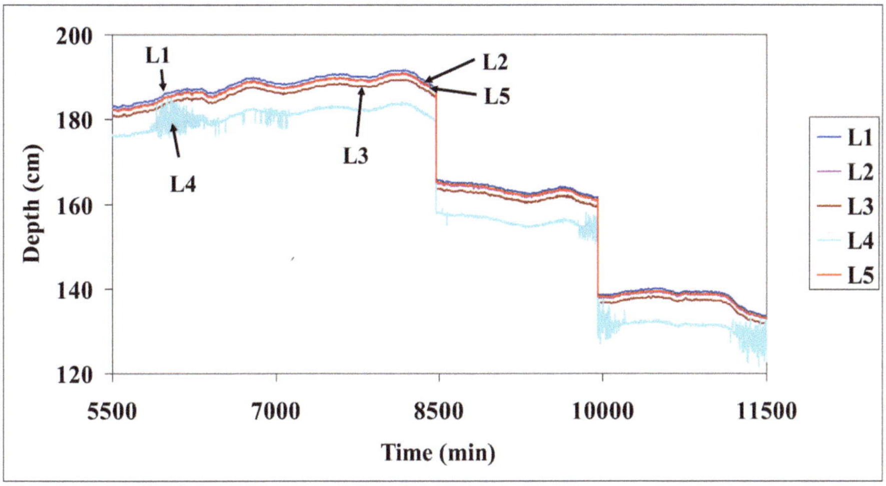

Fig. 11.4: Level outputs as recorded by the five data loggers with pressure sensors (Data courtesy of Leandro Rodrigues Capitulo and Eduardo E. Kruse, Faculty of Natural Sciences, National University of La Plata).

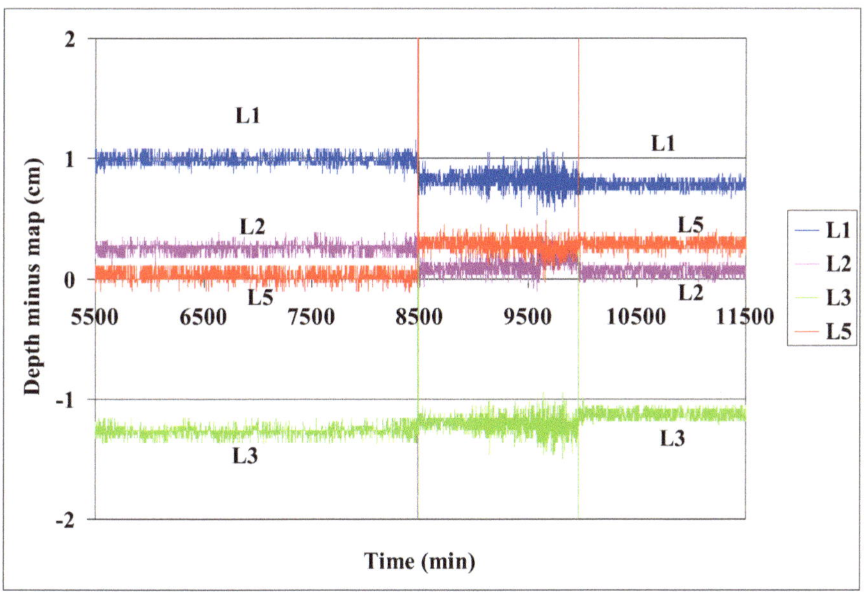

Fig. 11.5: Outputs of the four data loggers selected for field work when the measured average level is subtracted. All instruments have quite stable outputs. Manufacturer calibration was used. L2 and L5 measured closer to the average than the others.

11.2.4 Evaluating Potential Errors for the Particular Field Work

All the sensors have some background noise of about 0.002 m, which corresponds to the resolution specified by the manufacturer. In some cases the background noise reaches 0.004 m. Sensors L2 and L5 exhibit (at different times) a difference of about 0.004 m with respect to the mean value, represented by the horizontal line corresponding to the zero of ordinates. Sensors L1 and L3 show differences up to +0.010 and -0.012 m, respectively. It can thus be concluded that interchanging any one of these sensors, differential measurements could be made with a maximum expected error of 0.022 m (0.010 + 0.012) m. These comparisons were done using the calibration provided by the manufacturer. By calibrating the four data logger simultaneously at the users' laboratory, and calculating a new transference for each one, all of them (L1, L2, L3 and L5) could probably work within differences under 0.004 m.

It must be emphasized that in this example the maximum depth at which the sensors were tested was 0.93 m; therefore the abovementioned maximum expected error of 0.022 m **is valid only for this range**. It must be a standard practice to test the functioning of instruments over the whole range of the variable they are intended to be used. It is not possible to extrapolate this result to other ranges.

If these instruments are used to measure in the full range, the calibration error could produce a measurement error greater than the expected maximum of 0.022 m. For example, if it were considered that the differences found are attributable to different gains (slope of each of the transferences), and the instruments were used to measure a 9 m water column (about ten times greater than the tested depth), errors would probably increase proportionally (0.022 × 10) and would reach a value of about 0.2 m.

11.2.5 Evaluating the Gain of Each Instrument

It can also be noted in Figure 11.5 that the pressure differences change with respect to the average pressure at 8,500 and 10,000 minutes, just when the sensors were moved to different depths. This indicates that each sensor responds in a different way to a same change in depth, a typical behavior of instruments having differences in their calibration. If the instruments are supposed to have linear transferences (Section (2.2)) they would have offsets or slopes, or both, slightly different.

Since a change in depth produces a level difference with respect to the previous value, it allows the gain of each instrument to be assessed independently of the zero or offset error. Table 11.2 shows the real change in depth of the instruments, the measurements made by each one of them, and the gains (G_1 and G_2) understood as the ratio between the change measured by the instrument and the real change.

If the calibration of the instruments were stable, the gain for each instrument, calculated with both depth changes should be the same ($G_1 = G_2$); both right columns should be equal, which is not the case. This could indicate that there are some changes in the calibration with time or that the “real changes in water depth” were measured with some error. Because students without experience collected the data of our example during a practical work, probably the second is true. This assumption is supported by results shown in Table 11.3, where the real changes in depth were modified to 0.218 and 0.231 m, resulting values of G_1 and G_2 much closer to 1 in both columns.

Table 11.2: Real change in depth versus measured change

Time (min)	**8,500**	**10,000**	G_1	G_2
Real change in depth (m)	0.210	0.240		
Instrument L1 (m)	0.220	0.231	1.047	0.963
Instrument L2 (m)	0.220	0.232	1.047	0.967
Instrument L3 (m)	0.217	0.230	1.033	0.958
Instrument L5 (m)	0.215	0.231	1.024	0.960

Table 11.3: Real change in depth versus measured change

Time (min)	**8,500**	**10,000**	G_1	G_2
Real change in depth (m)	0.218	0.231		
Instrument L1 (m)	0.220	0.231	1.009	1
Instrument L2 (m)	0.220	0.232	1.009	1.004
Instrument L3 (m)	0.217	0.230	0.995	0.995
Instrument L5 (m)	0.215	0.231	0.986	1

Since the method used to verify the functioning of the sensors is very basic, it is not possible to quantify accurately which is the zero error and which is the slope error for each sensor. In order to better quantify these errors it is necessary to carry out another type of test, as it will be described below.

11.2.6 Full Range Calibration

To find a new water level sensor transference (i.e. to get a new input/output curve (Section (2.2)) that allows level measurement errors to be reduced), each instrument should be calibrated by applying known pressures and recording the values measured by each one of them.

An efficient and precise way of detecting differences between the transferences is to calibrate all the instruments simultaneously. For this purpose the pressure inputs of all the sensors should be connected in parallel to a common pressure input, and then put at the same level (e.g. over a smooth horizontal floor). Then the common pressure input should be connected to a water-filled hose to apply different pressures to the instruments by changing the water level inside the hose in steps. Special care should be taken to avoid air bubbles in the water, so it recommended using a transparent hose.

The real value of the water level in the hose and the value measured by the sensors must be written down at each step of the calibration process. A table will result with a series of points within the operating range of the sensors (a reasonable number for a quasi linear sensor is between 5 and 10 points). Finally, correction coefficients can be calculated from the values of the table so that all the sensors have a similar transference.

It must be taken into account that the test for determining the transference of the sensors must be performed in a time interval short enough so that the variations of the atmospheric pressure and temperature are negligible.

If the measurements must be very accurate it is advisable that the calibrations be made a few days before the field work, because they could vary with time. If it is desired a greater certainty about the quality of the recorded data, the calibrations should be repeated after the field tests have been carried out.

11.2.7 Temperature Influence

Figure 11.6 shows water levels as recorded by the instruments at the moment they were submerged; it appears as a step with an initial peak and an exponential decay. It can be seen that it is required about ten minutes for the instruments to stabilize their outputs at the new values.

The time required for outputs to become constant is the same time needed by the data logger to change from air temperature (25 °C) to the water temperature (11 °C); this was found through the temperature records given by the same instruments.

It is seen that for a change of level of about 1 m, the additional increase in the mean level registered during the first ten minutes is of the order of 0.05 m, i.e., 5 %. This transient in depth records is not due to a change in depth, but to the change in temperature. In other words, the instruments were affected by a change in temperature that modified the transferences. It is most likely that the thermal shock has modified the behavior of the electronic circuits. It can then be concluded that in order to obtain reliable measurements, the temperature of the sensors should remain as constant as possible. Also, the first minutes of any records should be discarded when the initial instrument temperature is different from water temperature. Furthermore, to reduce errors, it is convenient to calibrate the instrument at a temperature similar to that of the aquifer.

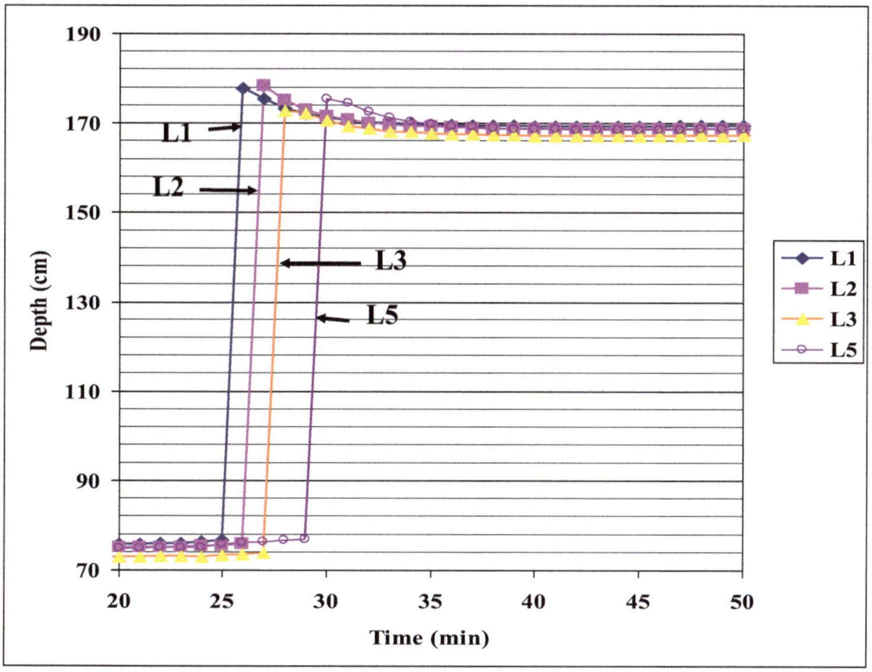

Fig. 11.6: Water level recorded by the sensors at the moment they were submerged in water.

Fortunately, the temperature of an aquifer remains fairly constant and does not change significantly during short periods, so for the application at hand, there would be no need to take into account the gain changes due to temperature variations. But if it were desired to measure water levels in a medium with significant temperature variations, it would be necessary to add the errors due to temperature changes.

It was noted that during the 9 days in which the test was carried out the sensors measured pressure differences fairly constant (Fig. 11.5). This shows that their calibration remained stable with time.

11.2.8 Some Conclusions

A summary of the information obtained from this simple test is given below.

1. It was determined that one of the instruments was not suitable for field measurements.
2. It was verified that if the water level variations to be measured were less than 1 m, the maximum error in level difference with two sensors chosen at random would reach about 0.02 m. If instead, the sensors L2 and L5 were chosen, the error would be less than 0.005 m
3. It is not possible to measure water level variations less than 0.004 m because of the background noise.
4. In order to measure absolute aquifer levels it is necessary to subtract the atmospheric pressure variations. During the 9 days period of the tests the variations in the recorded level data due to atmospheric pressure variations were 0.2 m.
5. Four instruments out of the five tested measure the atmospheric pressure variation in a similar way. Hence it is possible to take differential measurements if the sensors are separated by a distance such that they are subjected to the same atmospheric pressure.
6. The calibration of each of the four sensors chosen is slightly different, and if it were desired to reduce the measurement errors, the transferences of the instruments would have to be found.
7. The calibration of the instruments remained stable during the entire testing period (9 days).
8. A change in temperature of 14 °C in the instrument produced a transient change of approximately 5 cm variation in level. It corresponds to a drift due to temperature (Section (2.2.6)) of 3.6 mm/°C.

11.3 Case Study (II): Dynamic Pressure Sensor Test

In some applications it is important to know the high frequency response of the sensor, but sometimes the manufacturer specifications of the time response are not easy to understand. That was the case of a commercial pressure sensor that triggers our curiosity. The sensor has a range from 0 to 10 m of water column (0 to 1 kgr/cm^2) and the manufacturer specifies the response time in this way: **Passive output ≤ 1 ms; Amplified devices ≤ 10 ms.** These specifications on their own are not clear, even for an experienced engineer.

The test described in Section (2.4.11 – P3) was conducted to learn more about the dynamic response of the sensor. The test consisted in applying a step of pressure to the sensor and measuring its output. The sensor was connected to one of the ends of an elastic cylinder (deflated balloon - orange in the internet version of the book)

(Fig. 11.7). The other end of the cylinder was connected to a manual pump, and the cylinder pressurized to about 0.5 kgr/cm^2 (Fig. 11.7 (bottom)). Figure 11.7 (top) shows details of the valve to inflate the balloon (left) and the sensor plugged to the balloon (right), both fastened with rubber bands.

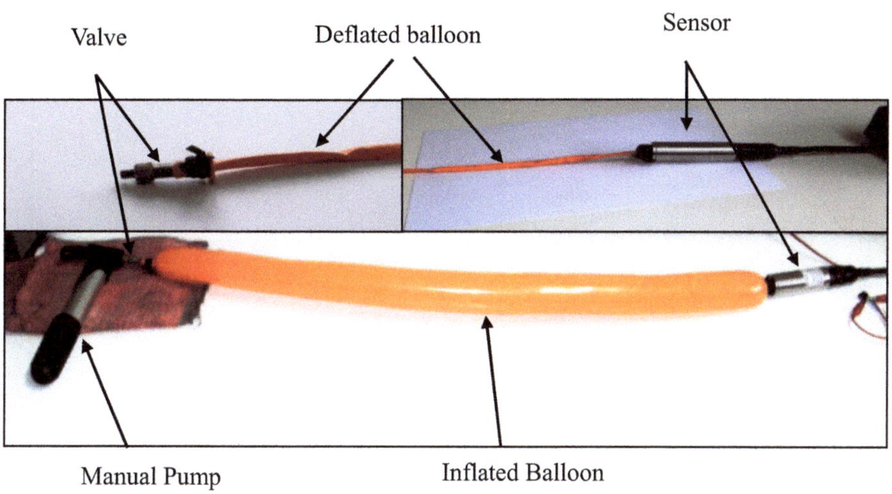

Fig. 11.7: Setup for testing a pressure sensor dynamically. An elastic cylinder (orange balloon) was connected to the sensor (top right) and to a valve (top left). The balloon is inflated through the valve by means of a manual pump.

The pressure sensor electrical output was connected to a digital oscilloscope with internal memory that works as a data logger. The balloon was inflated up to the point when it blows up; the oscilloscope pre-trigger function was used to record data during 12 ms before and after the balloon is blown up. The sudden change in pressure represented a negative pressure step input to the sensor. This step allows us to study the sensor response time.

The oscilloscope record of the sensor response is shown in Figure 11.8. The sensor's output samples before the balloon explosion have almost constant amplitudes of about 3 V. It is estimated that explosion occurred at 2.9 ms where amplitudes begin to fall up to 3.2 ms, reaching a minimum of about 0.5 V, where an attenuated oscillation of about 2 kHz begins. This oscillation is almost completely attenuated 7 ms after the explosion (not shown). This oscillation may be of mechanical or electrical origin. In the first case, it could be related to the resonance of the sensing membrane (Section (4.8.1)); in the second it could be due to the electronic circuit.

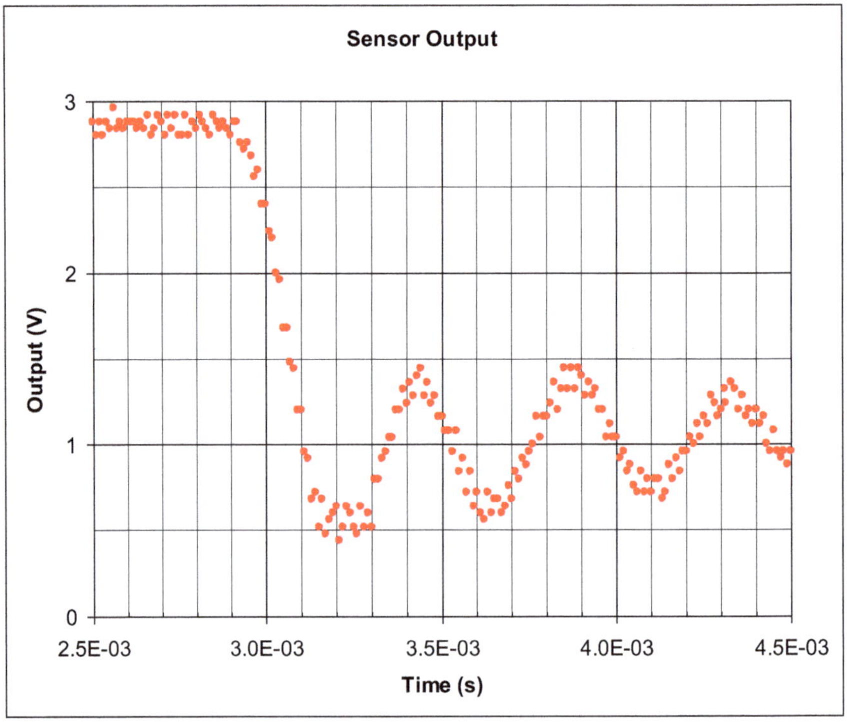

Fig. 11.8: Oscilloscope record of the sensor output when the balloon blows up.

The lesson learned from this test, which is not possible to figure out from the limited manufacturer's specification, is that the sensor has a preferential gain for a certain frequency about 2 kHz. Therefore, if dynamic air pressures should be measure with this sensor, the manufacturer's specifications do not help. The sensor will magnify signals present in the phenomenon with a frequency close to 2 kHz. In order that all frequencies have the same gain, pressure input signals should be limited to frequencies significantly below the oscillation frequency. A possible solution is to place a filter at about one decade below the oscillation (200 Hz) to prevent amplitude distortion due to sensor resonance. It should be noted that the test was carried out using air as the environment in which the sensor membrane is immersed, and this result should be considered valid only for a sensor working in air.

If the sensor will be working in water, the same test should be repeated using water to inflate the balloon and the balloon should also be immersed in water, so that the membrane will be in contact with water after the balloon blows up. If the oscillation were of mechanical origin, we estimate that being in water the oscillation would be more rapidly attenuated than in air because of a faster dissipation of the mechanical energy accumulated in the elastic membrane deformation.

11.4 Final Words

Almost all instrument manufacturers are serious, honest and publish reliable technical information about their products. But sometimes, in their eagerness to promote the benefit of their instruments when compared with those of the competition, manufacturers' brochures tend to confuse customers.

Frequently, customers do not know which characteristics of instruments are important for the desired measurements, how specifications have to be understood and what are the operating principles of instruments. Therefore, the unfamiliarity of customers with instruments contributes to the misinterpretation of manufacturers' brochures.

For example, some manufacturers specify the precision of their instruments, but say nothing about their accuracy, and then some customers think that they can measure with accuracy similar to the specified precision. As is made clear in this book there is a considerable difference between these concepts (Section (1.4)); the instrument can be very precise, but can be measuring with great errors.

Also, when manufacturers specify the accuracy, sometimes they do not make clear under what conditions it has been evaluated. Thus when users measure under other conditions or under variable conditions errors become unacceptable. Manufactures cannot be blamed for this omission because when the specifications are not clear, users should ask for details.

A common mistake of instruments users is to ignore the spatial and temporal characteristics of the instrument being used. In Table 6.1, the volume sampled by the different instruments presented varies 10^{13} times. Then the spatial resolutions of these methods are very different and it is the responsibility of the researcher to select which one is the most adequate scale for the problem being studied. The same happens with the time scale, as seen in the previous example in Section (11.3). The manufacturer specifies the time response in a way difficult to understand, and says nothing about oscillation. Perhaps the manufacturer has never tested the sensor in air, because most customers use the sensor in a medium other than air. Therefore, users should make clear to the manufacturers the specific conditions in which they will use the product to be acquired, asking for instruments performances in that particular condition.

Remote sensing instruments deserve a special consideration. They depend on the conditions of the media where a wave propagates (i.e. atmospheric conditions for wind radars and RASS (Sections (8.6) and (8.7)). Therefore, data is not always accessible. Also, individual measurements suffer a complex data processing and results are highly influenced by the algorithms utilized. Users should know how instruments calculate the final result. For example, if data are passed through a process of consensus averaging it could be of interest to know what the rules are for this averaging because the temporal scale could depend on it.

A good practice for instruments users is to consult manufactures about the specific needs they have before buying a new instrument. Instead, if the instrument has already been acquired an excellent attitude is to test it in known and controlled conditions before using it.

Understanding how a given instrument works and how the parameter recorded by the instrument is related to the measurand we want is essential in understanding the phenomenon we want to research.

List of Figures

List of Tables

Bei Fragen zur Produktsicherheit wenden Sie sich bitte an:
If you have any questions regarding product safety,
please contact:

Walter de Gruyter GmbH
Genthiner Straße 13
10785 Berlin
productsafety@degruyterbrill.com